AF572271

ASTRONOMY EDUCATION:
CURRENT DEVELOPMENTS, FUTURE COORDINATION

Cover photograph: John Ginder,
Royal Astronomical Society of Canada,
Toronto Centre

A SERIES OF BOOKS ON RECENT DEVELOPMENTS IN ASTRONOMY AND ASTROPHYSICS

Printed by BookCrafters, Inc.

First published 1996

Library of Congress Catalog Card Number: 95-83917
ISBN 1-886733-10-4

D. Harold McNamara, Managing Editor of Conference Series
Box 63 KMB
Brigham Young University
Provo, UT 84602-4463
801-378-2298

pasp@astro.byu.edu
Fax 801-378-2665

A SERIES OF BOOKS ON RECENT DEVELOPMENTS IN ASTRONOMY AND ASTROPHYSICS

Vol. 1-Progress and Opportunities in Southern Hemisphere Optical Astronomy: The CTIO 25th Anniversary Symposium
ed. V. M. Blanco and M. M. Phillips ISBN 0-937707-18-X

Vol. 2-Proceedings of a Workshop on Optical Surveys for Quasars
ed. P. S. Osmer, A. C. Porter, R. F. Green, and C. B. Foltz ISBN 0-937707-19-8

Vol. 3-Fiber Optics in Astronomy
ed. S. C. Barden ISBN 0-937707-20-1

Vol. 4-The Extragalactic Distance Scale: Proceedings of the ASP 100th Anniversary Symposium
ed. S. van den Bergh and C. J. Pritchet ISBN 0-937707-21-X

Vol. 5-The Minnesota Lectures on Clusters of Galaxies and Large-Scale Structure
ed. J. M. Dickey ISBN 0-937707-22-8

Vol. 6-Synthesis Imaging in Radio Astronomy: A Collection of Lectures from the Third NRAO Synthesis Imaging Summer School
ed. R. A. Perley, F. R. Schwab, and A. H. Bridle ISBN 0-937707-23-6

Vol. 7-Properties of Hot Luminous Stars: Boulder-Munich Workshop
ed. C. D. Garmany ISBN 0-937707-24-4

Vol. 8-CCDs in Astronomy
ed. G. H. Jacoby ISBN 0-937707-25-2

Vol. 9-Cool Stars, Stellar Systems, and the Sun. Sixth Cambridge Workshop
ed. G. Wallerstein ISBN 0-937707-27-9

Vol. 10-The Evolution of the Universe of Galaxies. The Edwin Hubble Centennial Symposium
ed. R. G. Kron ISBN 0-937707-28-7

Vol. 11-Confrontation Between Stellar Pulsation and Evolution
ed. C. Cacciari and G. Clementini ISBN 0-937707-30-9

Vol. 12-The Evolution of the Interstellar Medium
ed. L. Blitz ISBN 0-937707-31-7

Vol. 13-The Formation and Evolution of Star Clusters
ed. K. Janes ISBN 0-937707-32-5

Vol. 14-Astrophysics with Infrared Arrays
ed. R. Elston ISBN 0-937707-33-3

Vol. 15-Large-Scale Structures and Peculiar Motions in the Universe
ed. D. W. Latham and L. A. N. da Costa ISBN 0-937707-34-1

Vol. 16-Atoms, Ions and Molecules: New Results in Spectral Line Astrophysics
ed. A. D. Haschick and P. T. P. Ho ISBN 0-937707-35-X

Vol. 17-Light Pollution, Radio Interference, and Space Debris
ed. D. L. Crawford ISBN 0-937707-36-8

Vol. 18-The Interpretation of Modern Synthesis Observations of Spiral Galaxies
ed. N. Duric and P. C. Crane ISBN 0-937707-37-6

Vol. 19-Radio Interferometry: Theory, Techniques, and Application, IAU Colloquium 131
ed. T. J. Cornwell and R. A. Perley ISBN 0-937707-38-4

Vol. 20-Frontiers of Stellar Evolution, celebrating the 50th Anniversary of McDonald Observatory
ed. D. L. Lambert ISBN 0-937707-39-2

Vol. 21-The Space Distribution of Quasars
ed. D. Crampton ISBN 0-937707-40-6

Vol. 22-Nonisotropic and Variable Outflows from Stars
ed. L. Drissen, C. Leitherer, and A. Nota ISBN 0-937707-41-4

Vol. 23-Astronomical CCD Observing and Reduction Techniques
ed. S. B. Howell ISBN 0-937707-42-4

Vol. 24-Cosmology and Large-Scale Structure in the Universe
ed. R. R. de Carvalho ISBN 0-937707-43-0

Vol. 25-Astronomical Data Analysis Software and Systems I
ed. D. M. Worrall, C. Biemesderfer, and J. Barnes ISBN 0-937707-44-9

Vol. 26-Cool Stars, Stellar Systems, and the Sun, Seventh Cambridge Workshop
ed. M. S. Giampapa and J. A. Bookbinder ISBN 0-937707-45-7

Vol. 27-The Solar Cycle
ed. K. L. Harvey ISBN 0-937707-46-5

Vol. 28-Automated Telescopes for Photometry and Imaging
ed. S. J. Adelman, R. J. Dukes, Jr., and C. J. Adelman ISBN 0-937707-47-3

Vol. 29-Workshop on Cataclysmic Variable Stars
ed. N. Vogt ISBN 0-937707-48-1

Vol. 30-Variable Stars and Galaxies, in honor of M. S. Feast on his retirement
ed. B. Warner ISBN 0-937707-49-X

Vol. 31-Relationships Between Active Galactic Nuclei and Starburst Galaxies
ed. A. V. Filippenko ISBN 0-937707-50-3

Vol. 32-Complementary Approaches to Double and Multiple Star Research, IAU Collouquium 135
ed. H. A. McAlister and W. I. Hartkopf ISBN 0-937707-51-1

Vol. 33-Research Amateur Astronomy
ed. S. J. Edberg ISBN 0-937707-52-X

Vol. 34-Robotic Telescopes in the 1990s
ed. A. V. Filippenko ISBN 0-937707-53-8

Vol. 35-Massive Stars: Their Lives in the Interstellar Medium
ed. J. P. Cassinelli and E. B. Churchwell ISBN 0-937707-54-6

Vol. 36-Planets and Pulsars
ed. J. A. Phillips, S. E. Thorsett, and S. R. Kulkarni ISBN 0-937707-55-4

Vol. 37-Fiber Optics in Astronomy II
ed. P. M. Gray ISBN 0-937707-56-2

Vol. 38-New Frontiers in Binary Star Research
ed. K. C. Leung and I. S. Nha ISBN 0-937707-57-0

Vol. 39-The Minnesota Lectures on the Structure and Dynamics of the Milky Way
ed. Roberta M. Humphreys ISBN 0-937707-58-9

Vol. 40-Inside the Stars, IAU Colloquium 137
ed. Werner W. Weiss and Annie Baglin ISBN 0-937707-59-7

Vol. 41-Astronomical Infrared Spectroscopy: Future Observational Directions
ed. Sun Kwok ISBN 0-937707-60-0

Vol. 42-GONG 1992: Seismic Investigation of the Sun and Stars
ed. Timothy M. Brown ISBN 0-937707-61-9

Vol. 43-Sky Surveys: Protostars to Protogalaxies
ed. B. T. Soifer ISBN 0-937707-62-7

Vol. 44-Peculiar Versus Normal Phenomena in A-Type and Related Stars
ed. M. M. Dworetsky, F. Castelli, and R. Faraggiana ISBN 0-937707-63-5

Vol. 45-Luminous High-Latitude Stars
ed. D. D. Sasselov ISBN 0-937707-64-3

Vol. 46-The Magnetic and Velocity Fields of Solar Active Regions, IAU Colloquium 141
ed. H. Zirin, G. Ai, and H. Wang ISBN 0-937707-65-1

Vol. 47-Third Decinnial US-USSR Conference on SETI
ed. G. Seth Shostak ISBN 0-937707-66-X

Vol. 48-The Globular Cluster-Galaxy Connection
ed. Graeme H. Smith and Jean P. Brodie ISBN 0-937707-67-8

Vol. 49-Galaxy Evolution: The Milky Way Perspective
ed. Steven R. Majewski ISBN 0-937707-68-6

Vol. 50-Structure and Dynamics of Globular Clusters
ed. S. G. Djorgovski and G. Meylan ISBN 0-937707-69-4

Vol. 51-Observational Cosmology
ed. G. Chincarini, A. Iovino, T. Maccacaro, and D. Maccagni ISBN 0-937707-70-8

Vol. 52-Astronomical Data Analysis Software and Systems II
ed. R. J. Hanisch, J. V. Brissenden, and Jeannette Barnes ISBN 0-937707-71-6

Vol. 53-Blue Stragglers
ed. Rex A. Saffer ISBN 0-937707-72-4

Vol. 54-The First Stromlo Symposium: The Physics of Active Galaxies
ed. Geoffrey V. Bicknell, Michael A. Dopita, and Peter J. Quinn ISBN 0-937707-73-2

Vol. 55-Optical Astronomy from the Earth and Moon
ed. Diane M. Pyper and Ronald J. Angione ISBN 0-937707-74-0

Vol. 56-Interacting Binary Stars
ed. Allen W. Shafter ISBN 0-937707-75-9

Vol. 57-Stellar and Circumstellar Astrophysics
ed. George Wallerstein and Alberto Noriega-Crespo ISBN 0-937707-76-7

Vol. 58-The First Symposium on the Infrared Cirrus and Diffuse Interstellar Clouds
ed. Roc M. Cutri and William B. Latter ISBN 0-937707-77-5

Vol. 59-Astronomy with Millimeter and Submillimeter Wave Interferometry
ed. M. Ishiguro and Wm. J. Welch ISBN 0-937707-78-3

Vol. 60-The MK Process at 50 Years: A Powerful Tool for Astrophysical Insight
ed. C. J. Corbally, R. O. Gray, and R. F. Garrison ISBN 0-937707-79-1

Vol. 61-Astronomical Data Analysis Software and Systems III
ed. Dennis R. Crabtree, R. J. Hanisch, and Jeannette Barnes ISBN 0-937707-80-5

Vol. 62-The Nature and Evolutionary Status of Herbig Ae / Be Stars
ed. P. S. Thé, M. R. Pérez, and E. P. J. van den Heuvel ISBN 0-937707-81-3

Vol. 63-Seventy-Five Years of Hirayama Asteroid Families: The role of Collisions in the Solar System History
ed. R. Binzel, Y. Kozai, and T. Hirayama ISBN 0-937707-82-1

Vol. 64-Cool Stars, Stellar Systems, and the Sun, Eighth Cambridge Workshop
ed. Jean-Pierre Caillault ISBN 0-937707-83-X

Vol. 65-Clouds, Cores, and Low Mass Stars
ed. Dan P. Clemens and Richard Barvainis ISBN 0-937707-84-8

Vol. 66- Physics of the Gaseous and Stellar Disks of the Galaxy
ed. Ivan R. King ISBN 0-937707-85-6

Vol. 67-Unveiling Large-Scale Structures Behind the Milky Way
ed. C. Balkowski and R. C. Kraan-Korteweg ISBN 0-937707-86-4

Vol. 68-Solar Active Region Evolution: Comparing Models with Observations
ed. K. S. Balasubramaniam and George W. Simon ISBN 0-937707-87-2

Vol. 69-Reverberation Mapping of the Broad-Line Region in Active Galactic Nuclei
ed. P. M. Gondhalekar, K. Horne, and B. M. Peterson ISBN 0-937707-88-0

Vol. 70-Groups of Galaxies
ed. Otto G. Richter and Kirk Borne ISBN 0-937707-89-9

Vol. 71-Tridimensional Optical Spectroscopic Methods in Astrophysics
ed. G. Comte and M. Marcelin ISBN 0-937707-90-2

Vol. 72-Millisecond Pulsars—A Decade of Surprise, ed. A. A. Fruchter
M. Tavani, and D. C. Backer ISBN 0-937707-91-0

Vol. 73-Airborne Astronomy Symposium on the Galactic Ecosystem: From Gas to Stars to Dust
ed. M. R. Haas, J. A. Davidson, and E. F. Erickson ISBN 0-937707-92-9

Vol. 74-Progress in the Search for Extraterrestrial Life,
ed. G. Seth Shostak ISBN 0-937707-93-7

Vol. 75-Multi-Feed Systems for Radio Telescopes
ed. D. T. Emerson and J. M. Payne ISBN 0-937707-94-5

Vol. 76-GONG '94: Helio- and Astero-Seismology from the Earth and Space
ed. Roger K. Ulrich, Edward J. Rhodes, Jr., and Werner Däppen ISBN 0-937707-95-3

Vol. 77-Astronomical Data Analysis Software and Systems IV
ed. R. A. Shaw, H. E. Payne, and J. J. E. Hayes ISBN 0-937707-96-1

Vol. 78-Astrophysical Applications of Powerful New Databases
ed. S. J. Adelman and W. L. Wiese ISBN 0-937707-97-X

Vol. 79-Robotic Telescopes: Current Capabilities, Present Developments, and Future Prospects for Automated Astronomy
ed. Gregory W. Henry and Joel A. Eaton ISBN 0-937707-98-8

Vol. 80-The Physics of the Interstellar Medium and Intergalactic Medium
ed. A. Ferrara, C. F. McKee, C. Heiles, and P. R. Shapiro ISBN 0-937707-99-6

Vol. 81-Laboratory and Astronomical High Resolution Spectra
ed. A. J. Sauval, R. Blomme, and N. Grevesse ISBN 1-886733-01-5

Vol. 82-Very Long Baseline Interferometry and the VLBA,
ed. J. A. Zensus, P. K. Diamond, and P. J. Napier ISBN 1-886733-02-3

Vol. 83-Astrophysical Applications of Stellar Pulsation, IAU Colloquium 155
ed. R. S. Stobie and P. A. Whitelock ISBN 1-886733-03-1

Vol. 84-The Future Utilisation of Schmidt Telescopes, IAU Colloquium 148
ed. Jessica Chapman, Russell Cannon, Sandra Harrison, and Bambang Hidayat ISBN 1-886733-05-8

Vol. 85-Cape Workshop on Magnetic Cataclysmic Variables
ed. D. A. H. Buckley and B. Warner ISBN 1-886733-06-6

Vol. 86-Fresh Views of Elliptical Galaxies
ed. Alberto Buzzoni, Alvio Renzini, and Alfonso Serrano ISBN 1-886733-07-4

Vol. 87-New Observing Modes for the Next Century
ed. Todd Boroson, John Davies, and Ian Robson ISBN 1-886733-08-2

Vol. 88-Clusters, Lensing, and the Future of the Universe
ed. Virginia Trimble and Andreas Reisenegger ISBN 1-886733-09-0

Vol. 89-Astronomy Education: Current Developments, Future Coordination
ed. John R. Percy ISBN 1-886733-10-4

Inquiries concerning these volumes should be directed to the:
Astronomical Society of the Pacific
CONFERENCE SERIES
390 Ashton Avenue
San Francisco, CA 94112-1722
415-337-1100
e-mail asp@stars.sfsu.edu

ASTRONOMICAL SOCIETY OF THE PACIFIC
CONFERENCE SERIES

Volume 89

ASTRONOMY EDUCATION: CURRENT DEVELOPMENTS, FUTURE COORDINATION

Proceedings of an ASP Symposium held in College Park, MD, 24-25 June 1995

Edited by
John R. Percy

TABLE OF CONTENTS

PREFACE

The symposium "Astronomy Education: Current Developments, Future Coordination" was held at the University of Maryland, College Park, on June 24-25, 1995, as part of the Astronomical Society of the Pacific's 107th annual meeting. Its stated goals were: to review current research programs, projects and other developments in astronomy education; to publicize these programs and projects through poster papers; to discuss how the theoretical and practical results of these programs and projects can be shared; to plan for the next decade of progress in astronomy education; to create networks of astronomy education resources – both materials and people; to bring together about 200 of the most active and knowledgeable people in North American astronomy education today; to disseminate the results of the symposium through a high-quality conference proceedings. These stated goals were rather ambitious for a two-day meeting, but it exceeded most participants' expectations in most respects. And it was a good start. There was overwhelming agreement that such meetings should be held on a regular basis. There have been numerous exciting developments in astronomy education, some of them due to interest and funding by NASA and NSF, some of them due to the ongoing programs of organizations like the ASP and AAS, and many of them due to the enthusiasm and imagination of local individuals and groups.

The primary sponsor and organizer of the symposium was the ASP. Since its inception in 1889, the ASP has grown into the largest general astronomy society in the world, with members in all fifty states, and over sixty-five other countries. The Society is unique in bringing together educators, professional and amateur astronomers, and many thousands of others who are intrigued by the universe, and eager to share the wonder of astronomy. The goal of the ASP is to enhance public understanding of astronomy through a variety of programs and resources. The ASP also makes a substantial contribution to astronomical research through its technical journal (Publications of the ASP), its Conference Series (in which this volume is published), and its scientific symposia. The present symposium was co-sponsored by: American Association of Physics Teachers (Astronomy Division), American Astronomical Society, American Association of Variable Star Observers, Canadian Astronomical Society, International Planetarium Society, and Royal Astronomical Society of Canada. We thank all these organizations for their support.

Special funding for this symposium was provided by National Aeronautics and Space Administration (IDEA Grants Program), National Science Foundation, Space Telescope Science Institute, and the V.M. Slipher Fund. We thank Laura Danly, Gerhard Salinger, and Dennis Schatz for their part in facilitating our grant applications. The sponsoring societies have also supported the symposium in various ways. My own involvement in this and other education activities was made possible by the enthusiastic support of my employer - Erindale College, University of Toronto.

The Organizing Committee consisted of Lonny Baker, Andrew Fraknoi, Robert Havlen, Mary Kay Hemenway, Jeff Lockwood, Janet A. Mattei, George Mumford, John Percy (chair), Dennis Schatz, John Trasco, and Don Wentzel (chair: Local Organizing Committee). The Local Organizing Committee (especially Don Wentzel) deserves our special thanks, as does the ASP headquarters staff and volunteers who were involved in planning the meeting, and making it go so smoothly. We should not forget that they were organizing several other streams of the meeting at the same time!

It is fitting that this "first" ASP symposium on astronomy education should be held at the University of Maryland, and be hosted by Don Wentzel. Don has made outstanding contributions to astronomy education over the years, both at the national and especially the international level.

This volume is obviously the product of many minds. It contains the invited review papers, the dozens of poster papers, the reports of eight small-group discussions, and the comments of the audience - in all, contributions by almost everyone at the symposium. The audience consisted of a wide spectrum of astronomy educators: teachers at levels from primary to postgraduate, professional astronomers, leaders of astronomical institutions and societies, staff of planetariums and science centers, and even a graduate student or two. In retrospect, there were several under-represented groups: teachers (especially those dealing with primary students, and with disadvantaged students), graduate students, amateur astronomers, representatives from publishing companies, industry, and the media, among others. This provides us with a challenge for the next meeting in this series. Nevertheless, we were exposed to a wide spectrum of interesting programs, projects, and people. We made new contacts and friends, and came away from the meeting inspired to continue our important work.

The camera-ready copy for this book was produced by Elizabeth Kobluk, Erindale College, University of Toronto, from a large number of diverse submissions which (despite the editor's instructions) were often unreadable by human or by machine. I thank Liz for her skill, energy, dedication, and high standards, and for cheerfully working away on this massive project on top of a "normal" heavy workload.

John R. Percy
Erindale College
University of Toronto

December 4, 1995

LIST OF PARTICIPANTS, AND NAME INDEX TO PAPERS

The numbers after each participant name/address are the page numbers for their papers. Names without addresses are non-participant authors.

Central Park W. at 81st St., New York NY 10024-5192: 73

Canizo, Thea L.: Vail Middle School, 8875 E. McClellan St., Tucson AZ 85710: 235

Carney, Bruce W.: Department of Physics and Astronomy, University of North Carolina, CB 3255 Phillips Hall, Chapel Hill NC 27599-3255; [bruce@sloth.astro.unc.edu]

Chaisson, Eric: Wright Center for Science Education, Tufts University, Medford MA 02155; [echaisso@pearl.tufts.edu]

Chapman, Stu: Harford County Public Schools, Southampton Middle Moores Mill Rd., Bel Air MD 21014; [schapman@umd5.umd.edu]: 178

Cheung, Cynthia Y.: NASA Goddard Space Flight Center, 301 Natick Ct., Silver Spring MD 20905-2780; [ccheung@nssdca.gsfc.nasa.gov]

Christian, Carol: 215

Chippindale, Suzanne: Santa Fe Community College, 1387 Lopez Ln., #8, Santa Fe NM 87505

Clemmett, Sarah: 224

Cochrane, Kathleen: Our Lady of Ransom School, 10411 Dearlove Rd., Niles IL 60714; [kathy@newton.dep.anl.gov]: 180

Comins, Neil F.: Department of Physics and Astronomy, University of Maine, Orono ME 04469; [galaxy@maine.maine.edu]: 116, 181

Cooper, P.R.: 245

Craine, Eric: 183

Crawford, David L.: NOAO, PO Box 26732, Tucson AZ 85716; [crawford@gnat.org] 183

Dagenais, Gisele: 94 Marmora St., Trenton ON Canada K8V 2J3; [gisele@astro.uwaterloo.ca]: 186

Danby, Arlene: Herndon High School, 6521 Palisades Dr., Centreville VA 22020

Danly, Laura: Space Telescope Science Institute, 3700 San Martin Drive, Baltimore MD 21218; [danly@stsci.edu]: 103, 188, 227, 228

Deans, Paul: Pacific Space Centre, 1100 Chestnut St., Vancouver BC Canada V6J 3J9

DeGioia-Eastwood, Kathy: Department of Physics and Astronomy, Northern Arizona University, Flagstaff AZ 86011-6010; [kathy.eastwood@nau.edu]: 140, 190, 283

Descoteaux, Denise: Addison-Wesley Interactive, 4 Robin Rd., West Newbury MA 01985

DeVore, Edna: SETI Institute, 2035 Landings Dr., Mountain View CA 94043; [edna_devore@setigate.seti-inst.edu]: 228

Donahue, Megan: Space Telescope Science Institute, 3700 San Martin Drive, Baltimore MD 21218; [donahue@stsci.edu]

Donoho, David C.: 1107 Wood Duck Hollow, Jacksonville FL 32259

Dow, Kimberly: Harvard-Smithsonian Center for Astrophysics, 60 Garden St., Cambridge MA 02138; [kdow@head-cfa.harvard.edu]: 192

Dubick, Thomas C.: Charlotte Latin School, 9502 Providence Rd., Charlotte NC 28277

Duncan, Douglas: The Adler Planetarium, 1300 South Lake Shore Dr., Chicago IL 60605; [duncan@oddjob.uchicago.edu]: 266

Dukes, Robert J.: Department of Physics and Astronomy, The College of Charleston, Charleston SC 29424; [dukesr@cofc.edu]: 195

Eastlick, Pam: Planetarium Co-ordinator, University of Guam, UOG Station, Mangilao, Guam 96923; [pameastl@uog9.uog.edu]: 197

Eckroth, Charles: St. Cloud University, 740 4th Ave. S., Saint Cloud MN 56301

Eder, JoAnn: Arecibo Observatory, PO Box 995, Arecibo PR 00613-0995; [eder@naic.edu]: 156

Edwards, Suzan: Astronomy Department, Smith College, Northampton MA 01063; [sedwards@smith.smith.edu]: 88

Fentress, Steve: 230

Field, J.R.: 289

Fierro, Julieta: Instituto de Astronomia, UNAM, Apartado Postal 70-264, CP 04510 DF Mexico; [julieta@astroscu.unam.mx]: 94, 199

Fleming, Steve: Melbourne Planetarium, PO Box 24, Kinglake, Victoria 3763, Australia

Fraknoi, Andrew: Department of Astronomy, Foothill College, 12345 El Monte Rd., Los Altos Hills CA 94022; [fraknoi@admin.fhda.edu]: 9, 134, 200, 202, 302

French, Linda M.: Wheelock College, 135 Conant Rd., Boston MA 02215; [linda@annie.wellesley.edu]: 129, 204

Friend, David B.: Department of Physics and Astronomy, University of Montana, Missoula MT 59812; [pc_dbf@lewis.umt.edu]: 206

Gage, Barbara: Prince George's Community College, 9485 Marshall Corner Rd., Pomfret MD 20675

Gallagher, Amie: Hayden Planetarium, 114 Hidden Valley Dr., Edison NJ 08820

Gardner, Michael A.: Longway Planetarium, 1310 E. Kearsley St., Flint MI 48503

Garrett, Thomas L.: Tidewater Community College, 4704 Haywood Dr., Portsmouth VA 23703

Gillespie, C. Jr.: 228

Ginder, John: Royal Astronomical Society of Canada, 2459 Colter Ct., Mississauga ON Canada L5L 3K7

Gittler, Leonard H.: Syracuse Astronomical Society, 78 Cherry Tree Cir., Liverpool NY 13090

Good, R.F.: 245

Graham, Mary: Grijalva Elementary School, Tucson AZ 85746: 208

Greer, Christian: The Adler Planetarium, 1300 South Lake Shore Dr., Chicago IL 60605; [cgreer@midway.uchicago.edu]

Grice, Noreen: Charles Hayden Planetarium, Boston Museum of Science, Boston MA 02114-1099; [grice@A1.mos.org]: 210

Gutsch, William: 25 The Crossway, Smoke Rise, Kinnelon NJ 07405; [102417.2073@compuserve.com]

Hall, Donald E.: Department of Physics and Astronomy, California State University, 6000 J St., Sacramento CA 95819-6041; [hallde@ccvax.ccs.csus.edu]: 212

Harding, Russell: 187 Noroton Ave., Darien CT 06820

Haskin, John: Frost Valley Environmental Education Center, 2000 Frost Valley Rd., Claryville NY 12725

Havlen, Robert J.: Astronomical Society of the Pacific, 390 Ashton Ave., San Francisco CA 94112; [rhavlen@stars.sfsu.edu]: 214

Hawkins, Isabel: Center for EUV Astrophysics, University of California, 2150 Kittredge St., Berkeley CA 94720-5030; [isabelh@cea.berkeley.edu]: 131, 215

Hayden, M.B.: 245

Heinzman, Tony: Apple Valley Middle School, 15191 Nokomis Rd., Apple Valley CA 92307

Hemenway, Mary Kay: AAS Education Officer, Astronomy Department, University of Texas, Austin TX 78712-1083; [marykay@astro.as.utexas.edu]: 143, 218, 283

Hennig, Lee Ann A.: Thomas Jefferson High School for Science, 7714 Lookout Ct., Alexandria VA 22306: 297

Henry, Todd: 272

Hildreth, Scott: 200

Hill, Lon Clay Jr.: Department of Physical Sciences, Broward Community College, Central Campus, Ft. Lauderdale FL 33314; [lchj@alpha.acast.nova.edu]

Hoff, Darrel B.: Department of Physics, Luther College, Decorah IA 52101; [hoffdarr@martin.luther.edu]: 298

Hollow, Robert: Blue Mountains Grammar School, Private Mail Bag 6, Wentworth Falls NSW 2782, Australia; [rhollow@harpo.nepean.uws.edu.au]: 221, 289

Honywood, M.A.: 289

Howell, David C.: Peter Andrew Planetarium, Deerfield Academy, Deerfield MA 01342

Hull, G.: 228

Imhoff, Catherine L.: Science Programs, Computer Sciences Corporation, 10000-A Aerospace Rd., Lanham-Seabrook MD 20706; [imhoff@iuegtc.gsfc.nasa.gov]: 224

Inglis, Neil L.: 4998 Battery Ln., #502, Bethesda MD 20814

Jacoby, Suzanne: NOAO Education Officer, NOAO, PO Box 26732, Tucson AZ 85726; [sjacoby@noao.edu]: 225

Jarman, A.J.: 289

Jensen, Sally Jean: AASTRA, Holderness Central School, RFD #3, Box 197C, Plymouth NH 03264; [sallyj@oz.plymouth.edu]

Johanson, Heather: 262

Jones, C.: 192

Jones, Paul A.: 289

Kardel, W. Scott: 167

Katsu, Carl F.: Science Department, Fairfield School District, PO Box 245, Fairfield PA 17320; [ckatsu@aol.com]

Kernohan, James C.: Milton Academy, 170 Centre St., Milton MA 02186; [jck@world.std.com]

Kinney, Anne: Space Telescope Science Institute, 3700 San Martin Dr., Baltimore MD 21218; [kinney@stsci.edu]: 227, 228

Knappenberger, Paul H.: The Adler Planetarium, 1300 South Lake Shore Dr., Chicago IL 60605; [pknappen@midway.uchicago.edu]: 140

Koch, David: NASA Ames Research Center, Moffett Field CA 94035; [koch@chandon.arc.nasa.gov]: 228

Kolena, John A.: Physics Department, Duke University, Durham NC 27706; [kolena@ncssm-server.ncssm.edu]

Kubinec, William R.: 195

Landis, Rob: Space Telescope Science Institute, 3700 San Martin Dr., Baltimore MD 21218; [landis@stsci.edu]: 230

Lark, Neil L.: Physics Department, University of the Pacific, Stockton CA 95211; [nlark@uop.edu]: 233

Lebofsky, Larry A.: Lunar and Planetary Laboratory, University of Arizona, Tucson AZ 85721-0092; [lebofsky@lpl.arizona.edu]: 235

Lebofsky, Nancy R.: 235

Lee, Kim: Charlotte Latin School, 9502 Providence Rd., Charlotte NC 28277

Leiker, P. Steven: Center for Astrophysics, 60 Garden St., Cambridge MA 02138; [leiker@cfa.harvard.edu]

Levine, Joel M.: Orange Coast and Coastline Community College, 1 Belaire, Laguna Niguel CA 92677; [joel=levine%mathscience%occ@banyan.cccd.edu]: 237

Lightner, G. Samuel: Physics Department, Westminster College, New Wilmington PA 16172

Linton, David A.: Parkland Community College, 3103 Valleybrook Dr., Champaign IL 61821

Lockwood, Jeffrey: 1234 S. Camino Seco, Tucson AZ 85710; [jlockwood@astro.as.arizona.edu]: 125

Lopez, Ramon E.: The American Physical Society, One Physics Ellipse, College Park MD 20740; [lopez@avl.umd.edu]: 110

Lovan, Shea A.: Department of Physics, University of California, Santa Barbara CA 93106; [shea@rot.ucsb.edu]: 238

Lubin, Philip M.: 238

Luehrmann, M.K.: 245

Lynds, Beverly T.: Project SkyMath, UCAR, 3244 6th St., Boulder CO 80304; [blynds@hidata.ucar.edu]: 241

Maciolek, Ardis: Grosse Pointe North High School, 707 Vernier, Grosse Pointe MI 48236; [maciolek@cirrus.sprl.umich.edu]: 243

Malina, Roger: 215

Mandel, E.: 192

Manning, James G.: Museum of the Rockies, Montana State University, Bozeman MT 59717-0272; [ammjm@gemini.oscs.montana.edu]: 80

Marschall, Laurence A.: Department of Physics, Gettysburg College, Gettysburg PA 17325; [marschal@gettysburg.edu]: 148, 245

Mascotti, Larry: Mayo High School, 1420 S.E. 11th Ave., Rochester MN 55904

Mattei, Janet A.: American Association of Variable Star Observers, 25 Birch St., Cambridge MA 02138-1205; [jmattei@aavso.org]: 131, 247, 249

Mechler, Gary: Physical Sciences Department, Pima Community College, 2202 W. Anklam Rd., Tucson AZ 85709-0001; [gmechler@pimacc.pima.edu]

Mendez, Flavio: Space Telescope Science Institute, 3700 San Martin Dr., Baltimore MD 21218; [mendez@stsci.edu]: 188

Moche, Dinah L.: Queensborough Community College, PO Box 440, Mamaroneck NY 10543

Moore, Marla: NASA Goddard Space Flight Center, Code 691, Greenbelt MD 20771; [ummhm@lepvax.gsfc.nasa.gov]

Mosley, John: Griffith Observatory, 2800 East Observatory Road, Los Angeles CA 90027; [jmosley@cello.gina.calstate.edu]: 134, 251

Motta, Mario: 249

Muratore, Mary: ARTIST-TUCSON, 9600 Baber Ln., Tucson AZ 85747: 208

Murray, S.: 192

Mumford, George S.: Department of Physics and Astronomy, Tufts University, Medford MA 02155; [gmumford@pearl.tufts.edu]: 116, 253

Nagaraj, Arati M.: Addison-Wesley Interactive, 1 Jacob Way, Reading MA 01867

Nations, Harold L.: 195

Nokleby, Lois A.: Farmington Middle School, 4200 208th St. W., Farmington MN 55024: 255

Novacek, Greg: 167

Osell, Fritz: University of Hawaii, 96-045 Ala Ike, Pearl City HI 96782

Pack, S. Hughes: Northfield Mt. Hermon School, PO Box 706, Mt. Herman MA 01354; [hughes_pack@nmh.northfield.ma.us]: 158

Pasachoff, Jay M.: Hopkins Observatory, Williams College, Williamstown MA 01267; [jay.m.pasachoff@williams.edu]: 66

Pennypacker, Carl: Lawrence Berkeley Lab, University of California, Berkeley CA 94720; [pennypacker@csa2.lbl.gov]: 61, 158

Percy, John R.: Erindale College, University of Toronto, Mississauga ON Canada L5L 1C6; [percy@astro.utoronto.ca]: 1, 143, 247, 256

Pompea, Stephen M.: Pompea & Associates, 1321 E. 10th St., Tucson AZ 85719-5808; [spompea@as.arizona.edu]: 259

Poole, William: Addison-Wesley Publishing Company, 7 Windsor St., Chelmsford MA 01824

Procter, Alison: Department of Astronomy, Yale University, PO Box 208101, New Haven CT 06520-8101; [aprocter@astro.yale.edu]: 262

Quackenbush, R.L.: 101 Northampton Terr., Chapel Hill NC 27514-5724

Radzilowicz, John G.: Christa McAuliffe Planetarium, 3 Institute Dr., Concord NH 03301; [j_radzil@tec.nh.us]

Ratcliff, Stephen J.: Department of Physics, Middlebury College, Middlebury VT 05753; [ratcliff@middlebury.edu]: 264

Ratcliff, Martin: 230

Rest, Carole: Space Telescope Science Institute, 3700 San Martin Dr., Baltimore MD 21218; [crest@stsci.edu]: 227

Reynolds, Susan: Onondaga Cortland Madison Planetarium, 213 Hulbert St., Minoa NY 13116-1908

Richard, Michael: Weymouth High School, 1051 Commercial St., E. Weymouth MA 02189; [mrichard@pcix.com]

Richter, Jessica: Project ASTRO, ASP, 390 Ashton Ave., San Francisco CA 94112; [jrichte@cello.gina.calstate.edu]: 200

Roettger, Elizabeth E.: The Adler Planetarium, 1300 South Lake Shore Dr., Chicago IL 60605; [eroettge@midway.uchicago.edu]: 143, 266

Rosendhal, Jeffrey D.: Office of Space Science, NASA Headquarters, Washington DC 20546; [jrosendhal@smtpgmgw.ossa.hq.nasa.gov]: 99

Roy, Sharmi: RTI, Box 329, Elkins WV 26241; [sbr@euclid.dne.wvnet.edu]

Rudolph, Alexander: Department of Physics, Harvey Mudd College, Claremont CA

91711; [rudolph@hmc.edu]

Ruiz, T.: 192

Sadler, Philip M.: Harvard-Smithsonian Center for Astrophysics, 60 Garden St., Cambridge MA 02138; [sadler@cfa.harvard.edu]: 46

Safko, John L.: Department of Physics and Astronomy, University of South Carolina, Columbia SC 29208; [safko@psc.psc.sc.edu]: 148, 270

Sajor, Andrew: Plattsburgh State University College, 41 Trafalgar Dr., Plattsburgh NY 12901

Saken, Jon: Space Telescope Science Institute, 3700 San Martin Dr., Baltimore MD 21218; [saken@stsci.edu]: 272

Sakimoto, Philip J.: NASA Headquarters, Code EV, Washington DC 20546; [psakimoto@oeop.hq.nasa.gov]

Saladyga, Michael: 247, 249

Salinger, Gerhard: National Science Foundation, 4201 Wilson Blvd., Room 885, Arlington VA 22230; [gsalinge@nsf.gov]: 96

Sampson, Gary E.: Wauwatosa West High School, 11400 W. Center St., Wauwatosa WI 53222; [gsampson@omnifest.uwm.edu]: 273

Schatz, Dennis: Pacific Science Center, 200 2nd Ave. N., Seattle WA 98109-4895; [schatz@pacsci.org]: 33, 129, 275

Schreier, Ethan J.: Space Telescope Science Institute, 3700 San Martin Dr., Baltimore MD 21218; [schreier@stsci.edu]

Schreur, Eric: KVCC - Hans Baldauf Planetarium, 2324 Bronson Blvd., Kalamazoo MI 49008-2409

Seacord, Andrew W. II: 4117 Woodhaven Ln., Bowie MD 20715

Segall, Linda: Physics Department, University of California, Santa Barbara CA 93106

Shankar, Anurag: Department of Astronomy, Indiana University, Bloomington IN 47405; [anurag@astro.indiana.edu]

Shawl, Stephen J.: Department of Physics and Astronomy, University of Kansas, Lawrence KS 66045; [s-shawl@ukans.edu]: 276

Singel, Amy L.: 266

Skurzynski, Gloria: National Geographic Society, 2559 Spring Haven Dr., Salt Lake City UT 84109-4032

Slavsky, David: 283

Smith, Dale W.: BGSU Planetarium, Department of Physics and Astronomy, Bowling Green State University, Bowling Green OH 43403; [dsmith@newton.bgsu.edu]: 279

Snow Theodore P.: CASA, University of Colorado, Campus Box 389, Boulder CO 80309; [tsnow@casa.colorado.edu]

Snyder, G.A.: 245

Strom, Stephen E.: Astronomy Program, University of Massachusetts, Amherst MA 01003; [sstrom@donald.phast.umass.edu]: 88

Teames, Sallie: Texas Society of Astronomy Educators, 1033 Reed St., Hurst TX 76053; [steames@tenet.edu]: 281

Toler, Dale: 158

Trasco, John: Astronomy Program, University of Maryland, College Park MD 20742-2421; [jtrasco@astro.umd.edu]: 283

Trowbridge, Patsy: Space Center Museum, PO Box 555, Mesilla NM 88046; [ptrowbri@nmsu.edu]

Tucker, George F.: Sage Junior College of Albany, 140 New Scotland Ave., Albany NY 12208; [tuckeg@zeus.sage.edu]: 112

Tyson, Neil D.: Hayden Planetarium, American Museum of Natural History, Central Park W. at 81st St., New York NY 10024-5192; [tyson@amnh.org]: 73

Urry, Meg: Space Telescope Science Institute, 3700 San Martin Dr., Baltimore MD 21218; [cmu@stsci.edu]

Vanstone, Kirsten: Ontario Science Centre, 770 Don Mills Road, North York ON Canada M3C 1T3; [Kirsten.M._Vanstone@fcgate1.osc.on.ca]: 285

Voit, G. Mark: Space Telescope Science Institute, 3700 San Martin Dr., Baltimore MD 21218

Waller, William H.: 40 Ridge Rd., #7, Greenbelt MD 20770; [waller@achamp.gsfc.nasa.gov]

Wasiluk, Elizabeth S.: Berkeley County Planetarium - H.H.S., 800 West Addition St., #2, Martinsburg WV 25401

Watson, K.: 289

Wentzel, Donat G.: Astronomy Program, University of Maryland, College Park MD 20742; [wentzel@astro.umd.edu]

Wesney, Joseph C.: AASTRA, Greenwich High School, 6 Orchard St., Cos Cob CT

06807; [wesney@lx.netcom.com]

West, Mary Lou: Math/CS/Physics, Montclair State University, Upper Montclair NJ 07043; [west@apollo.montclair.edu]: 287

Wetzel, Marc: McDonald Observatory, PO Box 1337, Fort Davis TX 79734

White, Graeme L.: Physics Department, UWS Nepean, Kingswood NSW 2747 Australia; [g.white@lancelot.st.nepean.uws.edu.au]: 221, 289

Wilder, Christina: Educational Consultant, 377 Mandarin Dr. #208, Daly City CA 94015-2746: 291

Wiley, Sheila: Astronomical Unit, Project ASTRO, 3765 Torino Dr., Santa Barbara CA 93105

Williams, Harold Alden: Planetarium Director, Montgomery College, Takoma and Fenton, Takoma Park MD 20912-4197; [haroldw@umd5.umd.edu]: 292

Williams, Karen: 610 W. Meadowbrook Dr., Midland MI 48640

Worek, Thaddeus F.: Department of Physics and Physical Sciences, Community College of Allegheny County, 808 Ridge Ave., Pittsburgh PA 15212; [tworek@ccac.edu]

Yorka, Sandra: Department of Physics and Astronomy, Denison University, 172 Monroe Ln., Westerville OH 43081; [yorka@cc.denison.edu]

Young, Donna L.: AASTRA, The Maine School for Science and Mathematics, Limeston ME 04967; [youngd@eagles.lcs.k12.me.us]

Young, T.M.: 289

Zeilik, Michael: Department of Physics and Astronomy, University of New Mexico, Albuquerque NM 87131-1156; [zeilik@chicoma.la.unm.edu]: 295

Astronomy Education: An International Perspective

John R. Percy
Erindale Campus
University of Toronto
Mississauga, Ontario
Canada L5L 1C6

Internet: percy@astro.utoronto.ca

Introduction

Education is important to astronomers because it affects the recruitment and training of future astronomers, and because it affects the awareness, understanding and appreciation of astronomy by taxpayers and politicians. Indeed, astronomers have an obligation to share the excitement and the significance of their work with the general public. Unfortunately, education is often neglected by the scientific and professional community, and by many research universities. The Astronomical Society of the Pacific plays an important role by bridging education and research.

There are other reasons why astronomy should be part of our education system and our culture. Astronomy is deeply rooted in the history of almost every society, as a result of its practical applications and its philosophical implications. It still has everyday applications to timekeeping, seasons, navigation and climate, as well as to longer-term issues such as climate change and biological extinctions. Astronomy not only contributes to the development of physics and the other sciences, but it is an important and exciting science in its own right. It deals with the origin of the stars, planets, and life itself. It shows our place in time and space, and our kinship with other peoples and species on earth. It reveals a universe which is vast, varied and beautiful, and promotes curiosity, imagination, and a sense of shared exploration and discovery. It provides an enjoyable hobby for millions of people, whether they be serious amateur astronomers, armchair astronomers, or casual skygazers. In a school context, it demonstrates an alternative approach to the "scientific method" - the observational/theoretical mode. It can attract young people to study science and engineering, and can increase public interest and understanding of science and technology - both of which are important in all countries, both developed and developing.

Why, then, is astronomy not taught in more schools? Why are there so many misconceptions about astronomical topics? Astronomy educators all over the world have discovered "universal" barriers to the effective teaching and learning of astronomy (and some of them have discovered solutions!):

1. Students have misconceptions about astronomical topics (such as the causes of the seasons), which are not overcome by the teaching methods which are commonly used. [In some countries, of course, seasonal effects are less pronounced than in others!]

2. Teachers have misconceptions about the teaching of astronomy (and perhaps about astronomical content as well) which discourage them from teaching it well, or teaching it at all.

3. Teachers (especially in elementary schools) have very little knowledge about astronomy or astronomy teaching, while astronomy has progressed by leaps and bounds since these teachers were in school themselves.

4. The most effective teaching tools are simple, inexpensive, hands-on activities, and these are not widely used, or not widely available. These must get around the problem that "the stars come out at night, but the students don't".
5. Teachers are not always aware of the materials available. There is not a single national or international journal which deals specifically with astronomy education, or an effective network for informing teachers of materials available. In Europe, astronomy educators have recently formed an association. Although there are language barriers to overcome, ideas and experiences can transcend language.
6. The best or most fortunate students may receive good education, but the others - girls, minorities, the disabled, inner-city or rural students - may be left out.

We need to take a wider look at astronomy education, and its problems. We cannot afford to re-invent the wheel, ignoring the experiences of others. There are innovative projects and programs in many countries, in many cases well-tested. By taking an international perspective, we achieve a deeper historical and cultural understanding, which is especially important in our multicultural societies. We take some pleasure in knowing that, as astronomy educators, we have "kindred spirits" around the world and, in the case of the developing countries, we take some satisfaction in knowing that we can help scientists and educators less fortunate than ourselves.

Systems of Education

There are two basic *systems* of education - the traditional or European system, and the US system. In the traditional system, there is usually a national curriculum, and standardized examinations which select the best students for high schools and universities. In this system, astronomy tends to be taught in high school and university as a rigorous science. Teachers, on the average, are well-trained. In this system, astronomy is rarely taught to non-science students.

In the US system, there is no national curriculum (though there are several projects to develop national standards for the mathematics and science curricula). There are no standardized content examinations. About half of all high school graduates go on to some form of post-secondary education. Astronomy is most often taught as a unit in an earth science course in junior high school. Some schools offer an astronomy course, often taught by a teacher who is an amateur astronomer. In university, astronomy is most often taken by non-science students who are required to take a "science option" as part of their graduation requirements; about 250,000 students a year take such courses. Good textbooks and teaching material are readily available. Some science students take astronomy courses, and some, of course, major in astronomy.

One of the advantages of this system is that politicians and business leaders, who tend to be non-scientists, may well have taken an astronomy course as a "science option". If they were well taught, they may be interested in, and sympathetic to astronomy.

There are also two (or more) *methods* of education. At one end of the spectrum is the method based on memorization and regurgitation of the material presented in the lectures and textbooks. This method can be reasonably successful if the students are highly motivated and disciplined, and if the material is well presented but, in most contexts, it does not result in effective learning. At the other end of the spectrum are the inquiry-based methods involving hands-on activities and problem-solving. This method appears to be much more effective in the long-term, according to research studies from many parts of the world.

The Needs of the Developing Countries

Whichever of these systems and methods are used, the developing countries face problems not encountered by astronomy educators elsewhere. Their special needs include: an understanding and interest on the part of astronomy educators elsewhere; opportunities to gain experience in more developed countries, and be visited by astronomers and educators from abroad; access to appropriate books, journals, and data; access to equipment. Many of the programs and projects of the International Astronomical Union (described elsewhere in this book) are intended to satisfy these needs.

In many developing countries, there is only one "lone astronomer" (or at most a small group) to do all the educational, research and administrative work which is shared among many in the more developed countries. The accomplishments of these individuals are remarkable.

Elementary School

The astronomical topics most often taught around the world are: geometry and motions of Earth, moon and sun (day and night, seasons, eclipses, moon phases, and maybe – even in landlocked countries – tides); general information about the planets, and constellation identification are also often taught. There are occasional exceptions: in the state of Parana, in Brazil, astronomy makes up a third of the school science curriculum at all levels! Unfortunately, in almost every country, elementary school teachers are poorly trained and poorly equipped for the teaching of astronomy. Teachers need to be aware of students' misconceptions, to develop appropriate curriculum, appropriate means of feedback and evaluation, and simple, inexpensive, hands-on demonstrations and activities.

"Inexpensive", of course, is relative. In South Africa, for instance, science education must be developed in the black regions, but the cost of even the simplest imported material is prohibitive. So local solutions must be found. In China, simple equipment can be cut-out from mass-produced templates, assembled by the students, and taken home to keep.

Secondary School

There are several contexts in which astronomy is taught - if at all - in secondary school. In several countries, there are optional courses in general astronomy which can be taken. Uruguay and Greece are the only countries which I know of which have a compulsory astronomy course in high school. Teachers in Uruguay have developed simple hands-on equipment and activities to accompany this course. Elsewhere, astronomy can be introduced to illustrate concepts in gravitational physics, or spectroscopy. Often, the only astronomy in secondary school is a review of Earth motions, at a higher level, as part of a physical geography course. There are other possibilities: in one state of Australia, cosmology is available as a "distinction course" which senior students can take. There are science or space magnet schools in many countries including the US and Canada, and even an "astronomy high school" in Sweden. In China, high school students can take part in evening and weekend astronomy activities at "children's palaces" of culture - many of which have observatories and planetariums.

In the US, generous funding from the National Science Foundation supports curriculum renewal, and the development of activities and research projects based on real astronomical data, such as variable star measurements, and digital images.

Two related problems which become apparent at the secondary school level are: (i) attracting and retaining women in physical science courses, and (ii) attracting and retaining *any* students in *any* science courses. The latter problem is particularly apparent in countries such as China and Russia, in which business and economics are providing lucrative careers. There is also the perception that physical sciences are more difficult than other fields. The women-in-science problem has some interesting geographical variations: while the proportion of women astronomers is high in France, Italy, Spain and Latin America, it dwindles to zero in some countries in northern Europe, UK, Australia, New Zealand and South Africa.

College and University

The situation in the US is outlined in detail in the other papers in this volume. At some large universities such as Maryland and Texas, the number of students enrolled in astronomy courses (mostly for non-science students) is so large that there can be a diversity of courses, with ample resources (instructors, teaching assistants, equipment ...). At the other end of the scale are the two-year colleges which typically offer one or two courses, often staffed by overworked instructors with little background in astronomy. Fortunately, the US National Science Foundation, NASA, and professional and educational societies such as AAPT, AAS, and ASP provide strong support for undergraduate astronomy education. This includes the development of curriculum and lab activities, provision of real data such as images for lab experiments or research, funding of telescopes and instruments for undergraduate research, and enhancement of instructor skills through workshops and education sessions at society meetings.

In Sweden, where the weather is often cloudy, several universities have co-operated in developing a "standard" radio telescope and receiver for use in undergraduate teaching.

There may also be astronomy courses for science students not majoring in astronomy. In some countries, academic programs are so rigid that is is almost impossible for students to schedule these courses, except as "extras". This problem could possibly be solved by "marketing": informing both students and departmental administrators of the benefits of having students take these courses - the additional perspective, knowledge and skills provided.

Distance Education

In some countries, distance education is much better developed than in North America. The Open University, in the UK, is probably the best example. Its one astronomy course has a larger enrolment than all other astronomy courses in the UK put together! The University of South Africa (UNISA) has 150,000 students across Africa and beyond. These universities can often put a lot more thought and effort into their courses than most of us can - simply because of the economy of scale. They must also use a form of self-paced learning - which is probably more effective than the lock-step approach which is usually used in the traditional lecture mode.

The Training of Astronomers

Undergraduate astronomy major programs, and MSc and PhD programs, are quite similar from one country to another, though there are some differences which reflect political and economic realities. Astronomers may be trained in astronomy departments or in physics departments; in either case, they receive strong training in physics and mathematics as well as astronomy. The fact that some physics graduates have done their research work on astronomical topics raises the interesting question "how does one define 'an astronomer'"? The training in physics and mathematics, along with laboratory and computing skills, obviously qualifies the graduate for a wider range of jobs. And "jobs" is a major concern for astronomy graduates in almost every country. Even in the most industrialized countries, government budgets are shrinking, and many academic positions are filled by tenured faculty hired in the "boom years" of the 1960's and 1970's. In some countries, there is a feeling that astronomy (and other science) training should be broadened even more - to include teaching and communications skills, computing, and business. On the other hand, in other countries, employers outside astronomy are reluctant to hire graduates without specialized training in their field.

Another issue is research training. In some countries, students receive little or no experience in research until they enter a PhD program. Even then, they may take on a research problem suggested by their supervisor, rather than being forced to develop their own. In the US, there is extensive government support for undergraduate research, in the form of grants for telescopes and instruments, and for summer salaries for students and their supervisors. At my own university (Toronto), there are new first-year seminar courses, and second-year "research opportunity" courses, as well as the traditional "senior thesis" courses.

In many countries, the barriers to undergraduate research are (i) shortage of books and journals; (ii) shortage of research data, and (iii) shortage of faculty willing and able to supervise undergraduates. Problems (i) and (ii) can be partly alleviated by electronic access to publications and data. Problem (iii) is an attitude problem which needs to be changed.

For students from the developing countries, there is a special need to broaden the perspective, since most of the students' experience will come from one or two astronomers. The IAU's International Schools for Young Astronomers, and the Vatican Observatory Summer Schools, play an important role in this regard.

Training Teachers

Pre-service training of teachers usually takes place in a university setting. Lack of teacher training in astronomy is a universal problem. There is now adequate research on student learning processes, and ample material on activities and resources. The problem is in exposing pre-service and in-service teachers to this material. In many countries, there is a structural problem in the universities, namely that astronomy and education departments do not interact. In-service teacher workshops must then provided by educational and scientific societies (like the ASP), outside the educational system. The French Comite de Liaison Enseignants et Astronomes has played a leading role in teacher training for almost two decades, by organizing summer workshops, developing and testing activities and equipment, and providing a follow-up magazine "Les Cahiers Clairaut" for over 1000 teachers. Astronomers and teachers have also been very active in Spain, where there have been a series of five international conferences on the teaching of astronomy.

Informal Education

Informal education (sometimes called popularization, or public education) has a major impact on both the public and on students. Any astronomy instructor knows that their students are interested in the latest news reports on astronomical topics. Popularization is one way of reaching politicians and other decision-makers. Surveys in many countries (such as Canada, UK, US) show that, despite (or because of) the impact of the news media, the level of public scientific literacy is rather low. Some scientists disapprove of those colleagues who work to popularize science. We must change this attitude by explaining to our students that popularization is not only important, but it is also fun. We must give them training and experience in working with schoolchildren, teachers and the public, and assure them that education is an honourable profession. Popularization requires some ability, and lots of enthusiasm, initiative and patience. It demonstrates clearly that science is a human endeavour. There are always opportunities for popularization, especially in smaller centers, and in less developed settings. Amateurs frequently play an important role.

Mass Media

The highest-impact forms of popularization are TV, films, newspapers, newsmagazines and (in some countries) radio. Patrick Moore, a self-proclaimed amateur astronomer from the UK, has hosted a monthly TV program for almost 40 years; the average viewing audience is 4,000,000! Terence Dickinson, another amateur astronomer, from Canada, writes a weekly astronomy column in a newspaper with a readership of over a million. Yet these media are largely ignored by the astronomy education community. They were not present at this symposium, and I, as the chair of the Organizing Committee, am directly to blame. We must find out how to make contacts with these media, what they want from us, and how we can provide it. We can issue short and simple press releases, provide the media with names of willing contacts from the astronomical community, and make use of the media to capitalize on events such as eclipses and comets, and on major astronomical discoveries.

Specialized Media

Specialized magazines such as *Astronomy*, *Mercury*, and *Sky and Telescope* are always looking for photos, short articles, and news notes. Local club newsletters are another vehicle for getting started in popularization. In the developing countries, there are often simple science magazines, and astronomy is a "must-have" in these - sky charts, upcoming astronomical events, and recent discoveries. You are guaranteed a readership which is interested and reasonably knowledgeable.

Planetariums, public observatories, and science centers and museums, play a special role in popularizing astronomy. They may be the major astronomy facility in their region or country. They often come about as a result of a sustained campaign by local professional and amateur astronomers and teachers. Their visibility provides an excellent naming opportunity for a government, corporate or private donation. Occasionally these facilities are installed in a shiny new building but, as often as not, they are installed in a converted building, in a school or university, or in a simple, inexpensive building. Imaginative sources of funding must often be found.

Planetariums come in all sizes. Their purposes are to educate, enlighten and entertain. They present public and school shows, lectures and courses, publications and other information. They may train teachers, and planetarium staff. They may provide a home for the local astronomy club. They may have exhibits, or a small observatory. Planetariums may reflect their local geography or culture. There is a planetarium in Norway which specializes in programs about the northern lights. In India, where astronomy is deeply rooted in culture, planetariums make use of the cultural connections in developing their shows. Despite their key role in astronomy education, planetariums do not always enjoy close links with the rest of the astronomy community. That deficiency must be overcome at every level.

Public observatories are dedicated facilities for research by staff, students and amateurs, for skygazing by amateurs and the public, and for education of people of all ages. They serve as a focus for astronomical activity, as do planetariums. They are a cultural facility which is common in countries where science is a part of culture - Europe, and increasingly in Japan - but not so much in North America. Public observatories can come about as a result of strong marketing. In Brazil, a group of amateur and professional astronomers mounted a successful campaign to convince several municipal governments to build public observatories in their cities.

Science centers also come in many sizes and forms. There are major facilities such as the Exploratorium in San Francisco, the Ontario Science Centre in Toronto, and the Deutschesmuseum in Munich. There are also small science centers, such as one which I visited in the Department of Physics, at the University of Pretoria. It featured simple hands-on exhibits, with a staff of one enthusiastic teacher who presented programs to thousands of schoolchildren each year. There are science centers in old observatories (as in Sydney, Australia) and railway stations (as in Christchurch, New Zealand). There are even travelling science centres (as there are travelling planetariums). In Mexico, the science centre has taken some of its exhibits into the subway stations, where they are experienced by millions of travellers. Interactive exhibits are now the norm (though the old-fashioned static exhibits should not be overlooked), and leading science centers now work in partnership with their communities to reach a wider audience, including minorities, the disabled, and other under-served groups.

Books are a traditional means of reaching the public. In North America, Carl Sagan's *Cosmos*, and Stephen Hawking's *A Brief History of Time* immediately come to mind. In the UK, Patrick Moore has published over a hundred books. In virtually every other country I have visited (even the developing ones), someone has taken it upon themselves to write astronomy books - often with a special local flavor.

Public lectures are often thought of as a rather low-impact form of popularization, but they have the advantage of showing that science is done by real people, who can communicate their enthusiasm to the public. The great Indian cosmologist and educator Jayant Narlikar once gave a series of public lectures (in the local language, not in English); the last lecture attracted over 10,000 people! Public lectures can be organized at schools or universities, astronomy clubs, planetariums or science centres. Service clubs and seniors' clubs are especially receptive and satisfying. It is easy to give a public lecture if you follow a few simple rules: start from basics; keep it clear and simple (no jargon and equations); use models, analogies, pretty pictures and legible slides or viewgraphs; be enthusiastic; keep the discussion under control.

Non-credit courses can also be influential (because the "students" are often professionals from other fields) and satisfying (because the students are interested, receptive, interactive and appreciative). Courses for retired people, like the North

American ElderHostel program, will be increasingly in demand.

The Role of Amateurs

Amateur astronomers are those who do astronomy as a hobby. They range from armchair astronomers, through recreational skygazers, to those who make "serious" contributions to astronomy. Many are professionals in other fields. Their talents and accomplishments are significant. They also have a desire to learn. Amateur astronomers in Hong Kong, for instance, have a training program for their club leaders and, at one time, the leading university in Hungary had a special program of courses for leaders of amateur groups.

In almost every aspect of education, amateur astronomers are enthusiastic and effective partners. Professionals should cultivate them and work with them, rather than - as is sometimes the case - look down upon them. Amateurs form groups which interested people can join; hold public lectures; visit schools to help teachers, talk with students, and demonstrate telescopes; hold "star parties" in the parks; organize International Astronomy Day programs; mount travelling displays and programs in libraries and shopping malls; participate in teacher training workshops; lobby for planetariums, science centers and observatories; provide information to the media; write books and articles, and produce TV shows. These are things which my local amateur group (Toronto Centre of the Royal Astronomical Society of Canada) does. Perhaps other groups do even more. We should remember that an amateur astronomer - Frank Bateson - was the "father" of all astronomy, both amateur and professional, in New Zealand. Amateur astronomers also discover comets which excite public interest in astronomy and (in the case of the AAVSO's "Hands-On Astrophysics" project described elsewhere in this volume) generate data which teachers can use for educational purposes. The direct contributions of amateurs to research are well known, of course, but the educational contributions are often overlooked.

It is difficult to do justice to the whole range of international astronomy education in one short article. I refer you to "The Teaching of Astronomy", edited by J.M. Pasachoff and J.R. Percy (Cambridge University Press, 1990), the proceedings of an International Astronomical Union colloquium held at Williams College, Williamstown MA in 1988. This is still the best overview of the topic.

Discussion

Bisard.

I feel the title "Teaching Astronomy" for ASP or IAU future conferences should be altered to "Learning Astronomy." This would be more appropriate to student-centered learning research and best practice strategies coupled with cognitive structure research results.

Crawford.

One useful response is a list of addresses, such as for developing country organizations needing journals.

The State of Astronomy Education in the U.S.

Andrew Fraknoi
Chair, Astronomy Department
Foothill College
12345 El Monte Rd.
Los Altos, CA
USA 94022

E-mail: fraknoi@admin.fhda.edu

Introduction

I would like to set the scene for the discussion that will follow in this book and bring up a few disturbing points which I believe any discussion about astronomy education will ultimately have to address.

Let me begin by posing the following question: where does astronomy education take place in the United States? Those readers who teach will probably say that it takes place in classrooms like theirs (anywhere from first grade through the university.) But I want to argue that astronomy education happens in many other places besides the formal classroom.

It happens in hundreds of planetaria and museums around the country; it happens at meetings of amateur astronomy groups; it happens when someone reads a newspaper or in front of television and radio sets; it happens while someone is engrossed in a popular book on astronomy, or leafs through a magazine like *Sky & Telescope*; it happens in youth groups taking an overnight hike and learning about the stars; and it happens when someone surfs the astronomy resources on the internet. When we consider astronomy education, its triumphs and its tribulations, we must be sure that we don't focus too narrowly on academia and omit the many places that it can and does happen outside the classroom.

Some Disturbing Statistics

Certainly, the classroom is a vital and important part of the picture. After all, formal education is a big business in our country. At all levels, from kindergarten to college, the country enrolls some 60 million students and has 4.6 million teachers and other staff. The total cost of educating our young is $466 billion, about 8% of our gross domestic product (National Center for Education Statistics 1994). Are we getting our money's worth?

In 1988, the Public Opinion Laboratory at Northern Illinois University conducted a survey of a representative sample of 2,041 American adults to get a sense of their scientific literacy. Among the 75 basic science questions was a question about the motion of the Earth. Respondents were asked, "Does the Earth go around the Sun or the Sun around the Earth?" 21% got it wrong and 7% said they did not know. The 72% who got it right were then asked what period of time the trip took. Only 45% got it right, 17% said 1 day, 2% said one month, and 8% said they did not know.

This means that 94 million people in our country could not correctly say that the Earth went around the Sun AND that it took a year to do so! And astronomy is not alone in being a field about which Americans know little. The Carnegie Commission on Science, Technology and Government reported in 1991 that 47% of US 17-year olds could not convert 9 parts of ten to a percentage and that (in a multiple choice survey) 63% of adult Americans thought that lasers work by focusing sound waves (Beardsley 1992).

In September 1993, the Department of Education issued the report of a survey on adult literacy in the U.S. Those in the survey were given a calculator, told the cost of a carpet per square yard, and were given the size of a room. The question was, how much should carpet cost to cover the entire room. Even with a calculator, 96% of the representative sample of American adults could NOT do it correctly (Barber 1993).

And that's only one part of our problem. In June 1990 the Gallup organization conducted a national survey of beliefs among adult Americans. About one in four said they believed in the basic premise of astrology, and 74% read their horoscopes at least occasionally. 47% think that UFO's are real, and 27% think that aliens have actually touched down and visited the Earth.

You may laugh at the notion of taking this kind of belief seriously. But when funding for the SETI program was being debated in Congress, at least one of the people's representatives said on the record that there was no need for a radio astronomy search because UFO records revealed that aliens were already here and could thus be contacted at no expense to the government. More seriously, if a large number of our citizens (and even our leaders) believes that our lives are governed by and decisions can be made by magic and superstition, will they feel the same urgency we do about the need for greater understanding of science and technology to solve the difficult problems of our age?

And if you don't believe that such fiction sciences have an influence on public policy, I should perhaps just remind you of the revelation that came out of the Reagan administration – that a San Francisco astrologer named Joan Quigley was given almost total control of the President's schedule during much of his time in the White House. The White House chief of staff, Donald Regan, wrote later:

> "Virtually every major move and decision the Reagans made during my time as White House chief of staff was cleared in advance with a woman in San Francisco who drew up horoscopes to make certain that the planets were in favorable alignment for the enterprise...Although I had never met this seer,...she had become such a factor in my work, and in the highest affairs of the nation, that at one point I kept a color coded calendar on my desk as an aid to remembering when it was propitious to move the president of the United States from one place to another, to schedule him to speak in public, or commence negotiations with a foreign power" (Regan 1988).

What makes the widespread belief in pseudoscience and widespread ignorance about science possible in our country? Greedy and ignorant media, cynical publishers, and many scientists sticking their heads in the sand all bear some of the responsibility, but our system of education is certainly a main culprit. Many teachers, especially at the elementary level, just don't receive adequate training in science and the scientific method. Thus it is a lot easier and less frightening for them to teach as little science as possible or to teach science out of the textbook. And many high schools in this country have relaxed their requirements and offerings in science to the point where we might say they are not just relaxed, they're ASLEEP.

In 1986, the National Science Teachers' Association surveyed the high school teaching scene. To take one example, of the 24,000 high school in the US, about

1/3 offered no physics course at all. Many of the courses that *are* offered in physics are taught by teachers whose training is in some other field. And of the physics teachers, 82% teach only two or one physics classes in a school year!

That same survey showed that in the year of the survey, only 57% of US high school students were enrolled in any science class! 43% were taking no science at all that year! Of the 1990 graduating class, fewer than 50% took a chemistry class, and about 20% took physics. Contrast that with other countries where students take three science classes every year of high school.

A 1985 survey by the Stanford School of Education revealed an interesting fact about the roots of the problem. A typical elementary school student in the US spends about 25 hours per week in instructional activities (that by itself is sobering). But of those 25 hours, how much time does a student spend on science? What would you guess...a fifth, an eighth? The answer turned out to be 44 minutes or about 3% – and much of that on learning vocabulary rather than discovering ideas.

But I want to end my depressing statistics with a positive thought: survey after survey reveals that when students are asked what topic in science is most interesting to them, the top two winners are consistently dinosaurs and outer space. The fascination of astronomy is a powerful tool for engaging the intellect and imagination of our youngsters, and this is what encourages us to make it part of the positive school experience of every child.

Nor do I want to imply that the public at large is necessarily uniformly hostile toward astronomy (or even science). There is by all indications a tremendous hunger among many people in this country to share in the excitement of scientific discovery and to know more about the fruits of scientific exploration. And astronomy seems to be near the top of the list of topics the public is eager to learn more about. The problem is more that this hunger is so often left unsatisfied, that the meager meal of science gruel set before the public by our educational system and the media, leaves them, like Oliver Twist, dreaming of a richer repast.

So let us turn from this very quick taste of the problem to the places where solutions might be expected. I've divided the places where astronomy education takes place in the United States into six broad (and somewhat arbitrary) areas:

1. Graduate Education

According to the American Institute of Physics, there were 125 PhD's in astronomy awarded in the U.S. in 1994, plus about 71 physics degrees in astrophysics, for a total of 196. This was up significantly from the combined figure of 125-150 per year that the AIP has been reporting for some time. (There were also 34 terminal masters degrees awarded.) What kind of job are we doing in training these graduates of our programs?

The answer to that question depends on what you expect these graduates to be doing. If all they will be asked to do is astronomical research, then by all measures we are doing an excellent job. All over the world, students in the sciences long to come to the U.S. and Canada to get a top research education. Certainly, the culture of most of our astronomy departments is based on this expectation: research skill is what is sought out, research skill is what is rewarded. It is true that many of these researchers will wind up in academia and will also be asked to teach, but many (not all, but many) graduate programs manage to convey to research students that having to teach is a minor irritant that any smart person can learn to put up with.

On the other hand, if we expect teaching to be a significant part of the students' future responsibilities, then our picture of graduate training is a bit more somber. A few astronomy departments do make a serious effort to help their students become better teachers, but most follow the old prescription for how you teach a kid to swim: throw a graduate student into a pool of lukewarm students and let him fend for himself – he'll soon pick it up!

Now in many large introductory undergraduate courses around the country, the first person non-science students have close contact with in astronomy is a graduate student teaching assistant. Given the emphasis on research skills, this teaching assistant often turns out to be someone pretty unprepared for making much of this "first contact." The resulting experience is thus often an unsatisfying one for each side.

If we are going to change the culture of the astronomy departments, we must first foster a sense that education has value in the training and work of astronomers commensurate with the value they place on research. In some astronomy departments (as they are currently structured), this may be harder than finding a snowball on Venus. But in others, the slow winds of change are starting to blow.

A number of universities are now starting or considering a masters program that will combine astronomy and education, and specifically prepare students for teaching, planetarium education, or science communication. As an example, both Sonoma State University near San Francisco and the University of Arizona in Tucson have had experimental programs in this direction. And at the Center for Extreme Ultraviolet Astrophysics, a research institution in Berkeley, it is now official policy that any staff member may spend up to 10% of his or her time on pursuing science education projects in or out of the office.

I think much more can be done in this area, and it can be done without endangering the research excellence for which the astronomical community in the United States has justly been renowned. Indeed, one needs only listen to some of the talks at astronomical meetings to know that stronger teaching and communications skills could even benefit those astronomers who never see the inside of a classroom.

2. Undergraduate Education

Here I refer to teaching non-science majors about astronomy, as part of some general education program, not to the much smaller area of teaching specialized courses for undergraduate astronomy majors. To appreciate the growth of this part of the astronomy education enterprise, we must bear in mind that there has been a tremendous increase in the number of people getting college degrees in the U.S. in recent decades.

Before World War II, only 8% of Americans went to college; today more than half will get a bit of college education, and about a third will actually graduate from college. Even in 1950, only about 500,000 college degrees were awarded in the country: 432,000 bachelors, 58,000 masters, 6,000 PhD's. By 1995, projections are for 1.7 million degrees, among them almost 1.2 million bachelors, 75,000 medical, legal and other professional degrees, 377,000 masters, and 41,000 PhD's (National Center for Education Statistics 1994).

All this has meant a large increase in the number of college professors and instructors in many fields, not just astronomy. (This increase has been very dependent on federal and state government funds, and there may be significant changes in the college and university systems as the pressure of decreasing government funding begins to hit home in the next decade.)

In 1993, there were 2,157 4-year colleges and universities in the US plus about 1,500 2-year colleges (but some of these are technical training schools, so the number of 2-year colleges as we would think of them may be closer to 1200.) In 1995, the U.S. Dept of Education projects that about 15 million students will be enrolled in some institution of higher education, from community college to research university. And in the United States, we keep learning beyond our younger years: In 1990-91, out of a total US adult population of 182 million, the U.S. National Center for Educational Statistics reported that 57 million (about 32%) had taken some sort of adult education or training course in the last 12 months. How many of these students are taking an introductory astronomy course?

I wish we had a more precise answer to this question. No one seems to know how many introductory astronomy students there are per year in the U.S. In preparing this summary, I surveyed textbook authors and publishers who keep track of these sorts of statistics. The best estimate I can come up with is that about 200,000 students take astronomy in the US per year. The last survey I am aware of was done by Darrel Hoff in 1980 (Hoff 1982), in which he sampled about 20% of the colleges that had even one astronomy faculty member. He estimated that the general education astronomy enrollment in 1980 was about 300,000; this fits in with the general sense everyone has that the numbers have declined in the last decade or so.

When we think of who teaches such astronomy courses, those of us from four-year university backgrounds tend to picture a person trained in astronomy, who may teach one or two classes per semester. But in real life, many of the courses are taught at smaller and two-year colleges, by people with training in other sciences and perhaps a less than complete familiarity with the results of modern astronomy. In some of these institutions, a teaching load might be 15 units per semester: five 3-unit classes each term (sometimes even the same class offered in 5 different sections.)

At such colleges, the instructors serving large numbers of students generally work in isolation from the research community and rarely have the opportunity to upgrade their skills by coming to conferences such as this one. They rely on the textbook and the ancillary materials that publishers provide for much of their information, and rarely have a chance to interact with other teachers of astronomy. Many teach the way they were taught, and even the best teachers find themselves unable to maintain the enthusiasm of their early years. The astronomical community (and the AAS and ASP in particular) really do need to come up with creative ways to include these instructors in our programs and activities and help them to revitalize their teaching from time to time.

3. K-12 Education

Currently, the U.S. enrolls about 46 million K-12 students per year and tries with varying degrees of success to prepare them for life in the 21st century. In many ways, the school system in America reminds an observer of Saturn's ring system: from far away, it looks ordered and beautiful, a testament to the organizational powers at work in the system. But the minute you get closer, what you see is the chaos of countless individual pieces, some colliding, some moving together, some all knotted up. And what are all the pieces doing ultimately? Going in circles of course!

I've already mentioned some of the serious problems in science education in US schools – those problems are, of course, only a tiny subset, of the much larger problems confronting the entire educational system. For example, one legacy of our frontier past is that the US education scene consists of many fiercely independent "empires". All fifty states have different rules and requirements; and there are 16,000 school districts in the country, in many of which local school boards maintain their own priorities and regulations with a grim determination.

Our society expects our schools to do much more than merely teach our students basic liberal knowledge. Today, we expect our schools to fulfil many of the functions that were the province of the extended family, of religious institutions, and of the community at large just a few decades ago. As Peter Schrag wrote recently in *The New Republic* magazine, "No country has ever done, or even tried, what this country is now trying: To take such a diverse population of children – 20% of them from below the poverty level, many of them speaking little English, many from one parent or no-parent families – and educate each child at least through the 12th grade..." (Schrag 1991).

In 1990, Bruce Alberts, now the head of the National Academy, wrote poignantly about high school science teachers in California: Such a teacher might typically teach 5 classes a day, with three different preparations and a total of more than 160 students. A dedicated teacher, with labs, setup, preparation, grading, student conferences, and all the paperwork schools require, will work something like 70 hours a week, and have an average starting salary of $22,000 per year. Who would want this job? Or who, once having this job, would not be tempted to cut corners, give lots of rote assignments, skimp on hands-on lab experience, and just try to survive? (Albers 1990)

If we really felt that education is important in the US, would we pay teachers so much less compared to lawyers, accountants, business leaders? Why, in a culture that glorifies money, are we surprised when students (especially students from low-income families) get the message early on – teaching and learning must not be so important to adults, or film stars and basketball players would not earn outrageously more than a mentor teacher.

But, despite our expectations of them, schools in our country cannot by themselves solve all the problems that threaten the education of young people. If we really want to make changes in the ways our children are educated, we must be prepared to examine the other influences on their lives as well. For some children, these include poverty, violence, drugs, neglect, ill health, and lack of proper housing. But even for children growing up in economically and emotionally stable environments, there is a pernicious influence that was much discussed as a problem when I was growing up, but today seems to be treated mainly with resignation – television.

A typical student in the US spends about 900 hours a year in school and between 1200 and 1800 hours in front of a television set (Barber 1993). In 1993, the average US household had a television set on for about 8 hours per day (Wright 1995). And what does a youngster learn from commercial television – often the exact opposite of what we try to teach them in school:

- That the only thing that matters is the instant gratification of every urge ("just do it," as the popular slogan says);
- That what counts in America is money, celebrity, and fun;

- That anything can be true and can happen (and that anecdotal evidence is enough to prove any assertion.)

Think of how teachers are shown on most television programs on the commercial services: mostly as objects of derision. And how are scientists usually portrayed (when they are shown at all)? Mostly as evil villains, or naive agents of serious catastrophe that result from their experiments getting out of control. These "lessons" of television viewing are rarely lost on our students!

Before leaving the disquieting arena of K-12 education, let me turn just briefly to the places where so many of our teachers are trained – our schools of education. In the 20th century, the US has evolved a whole slew of specialized schools and departments for preparing our teachers. In many of these programs, you do not need to take any science (and certainly any physical science) to become an elementary level teacher. It is one of the few majors that do not have such a requirement, which means it actually selects out those people who are afraid of science. This is one reason why so many elementary teachers transmit a fear, a hostility, a shudder at the very mention of science to their young students, at just the time when kids' minds are seeking not just information but values.

Estimates are that more than 2/3 of math and science teachers in the US today do not meet the recommended proficiency standards of their own professional associations (Beardsley 1992). But, especially in the higher grades, the problem is not that teachers do not know enough science. The problem is more often that the schools of education, despite their many courses on teaching technique, rarely prepare teachers to present science the way it should be taught – in ways the students, at their own stages of reasoning, are ready to understand. So science teachers often model their teaching of science on how they learned science, in big college lecture courses, from textbooks full of facts. No wonder 11-year olds get turned off.

Now some schools of education in this country do make a valuable effort to train teachers in the value of science and its most effective presentation. But as long as schools of education remain separate "fiefdoms" from the departments of science at our universities, and as long as these programs continue to allow teachers to graduate without proper grounding in the method and teaching of science, our children will continue to have role models whose own attitudes toward science would be charitably described as luke-warm.

In 1983 the National Commission on Excellence in Education published *A Nation at Risk,* a report that decried the state of education in the US It called the "rising tide of mediocrity" a "peril to our very future as a nation and a people." It compared what was happening in our schools to an act of war, except it was an act we had declared on ourselves. The report (and others like it) made many recommendations, but so far there have been few effective follow-ups and few additional resources for waging this hidden war.

You can read about a few really good initiatives to reform science teaching later in this volume. But no amount of teacher training in science will address the fundamental problems of low pay, low morale, overcrowded classrooms, boring and often outdated textbooks, and masses of students overwhelmed by life that confronts teachers in so many schools today. The entire country must begin to deal with these issues, and must face them with adequate resources to do the job. Thus far, I am very pessimistic about the will or the ability of our elected leaders to do so.

4. Informal Education Institutions

Much of the learning of astronomy in this country is done outside the classroom, in institutions and through organizations that sometimes interact with the schools (through class visits, for example) and sometimes act independently. These include planetaria, science museums, observatory visitor centers, NASA facilities, youth groups, and many similar organizations. Let us examine each category briefly, with the understanding that you can find more about each of them in other parts of this volume.

A. Planetaria

There are approximately 1,100 planetaria in North America, visited by millions of people each year (see the paper by Manning later in this volume.) About 30% of these serve school groups only, while about 60% do both school and public shows. For many youngsters, a planetarium visit is their first (and in some cases only) introduction to astronomy. The quality of this introduction can vary widely, depending on the skill and background of the presenter and how the planetarium environment is used. Nevertheless, most children have a fond recollection of their planetarium experience, and for many children in cities, it may be the only time they really experience a *dark* night sky.

Planetarium educators are organized into a number of regional organizations, and into the International Planetarium Society (although not everyone belongs to these groups.) My main observation of the field is that planetarium educators tend to be somewhat isolated from the astronomy research and even college education community; this can occasionally lead to some problems in keeping up with current science, but I don't think these are a major cause for concern. More important is a sense that planetarium educators get of being peripheral to astronomy, despite the large numbers of people for whom they serve as primary contact with the world of astronomy. I think it would be very useful for the main astronomical societies to make more of an effort to involve and get involved with the planetarium enterprise; the AAS has recently taken some first steps in this direction.

B. Museums

Many science museums have astronomy exhibits, where visitors can read or participate in activities relating to astronomy or at least space exploration. Many museums also sponsor youth and education programs after school or on weekends. As in planetaria, there are many museum visitors who are first exposed to modern astronomy through such programs. Science museum educators also have an organization, called the Association of Science and Technology Centers (ASTC), with offices in Washington. The same comments I made above for planetaria would apply to astronomy staff at many science museums as well.

C. Observatory Visitor Centers

A number of the major observatories are expanding or putting in visitors centers which have an educational component. Some of these centers accommodate a large number of visitors; for example, Kitt Peak has 150,000 visitors per year. An especially good center has been built at the Lowell Observatory, where you can experience a computer simulation of a night at the observatory. Of course, there are budget problems that prevent much expansion in this area, but my feeling is that more cooperation among observatories and more communication among those of us working in astronomy education would have a salutary effect on these efforts.

Another aspect of the work of observatories is responding to requests for information from the public. Some observatories have developed excellent materials of their own, others use the materials developed by such groups as the Astronomical Society of the Pacific. Again, it would be helpful to have a list of available resources for those whose responsibility is responding to public inquiries.

D. NASA Education Efforts

NASA has extensive efforts to help in science education, but it is difficult for a mere mortal (including personnel at NASA) to know and understand all the programs they have and all the decisions they make. Astronomers and astronomical organizations have had a lot of trouble finding ways to work with the NASA Education Division, despite much recent effort on both sides. The organizers of this conference, for example, were unable to get a representative of the NASA education side to participate. In addition NASA has also had a reputation of being much stronger in the area of education about human space flight, than about astronomy.

As a result of the frustration some NASA scientists have had in getting the word out about NASA results in astronomy, a few years ago NASA's Astrophysics Division decided to hire its own education staff to produce educational materials and activities. The first two leaders of this effort were Jeff Bennett (now at the University of Colorado) and Cherilynn Morrow. The current person in this position is someone with a great deal of senior policy experience at NASA, Jeff Rosendhal (see his article later in this volume.) As a result of recent changes at NASA Headquarters, much of the Astrophysics Division education effort will now actually be carried out through the Space Telescope Science Institute.

One of the things we hope to do through Project ASTRO at the Astronomical Society of the Pacific is to compile a list of the activities at all the NASA Offices and Centers that relate to astronomy education. Some of these can already be found in the Appendix to this volume which lists national astronomy education projects.

E. Others:

Other astronomy education programs that take place outside the formal school systems include the Challenger Centers with their space flight simulators for kids, the Young Astronauts clubs around the country, which occasionally do astronomy activities, scout and other youth groups, which have inspired many youngsters through astronomy merit badges and similar programs, and astronomy camps, some of which you will read about in this volume. In response to some of the crises in K-12 education that I discussed above, a number of interesting grass-roots efforts are springing up to supplement science in the schools with after-school and summer activities, sometimes in connection with a local science center, youth group, or amateur club. At the present time, there is unfortunately no central clearing house that would keep track of and disseminate the results of such efforts.

5. Amateur Astronomers

The U.S. has a large population of amateur astronomers, people whose hobby is astronomical observing or following astronomical developments in a serious sort of way. I like to divide the amateur community into three categories: *Research-level amateurs* are those who have sophisticated telescopes and detectors, or who carry out serious observing programs. These amateurs are usually members of such specialized organizations as the American Association of Variable Star Observers, Association of Lunar and Planetary Observers, or the International Amateur-Professional Photoelectric Photometry Group, or they are working in conjunction with a professional astronomer in their community. There are probably not more than a few hundred of this group in our country.

Observing amateurs are those who have a telescope and regularly take it out for observing the sky, either for their own amusement or with a community or school group. These amateurs are frequently members of some of the more than 200 amateur clubs in the U.S., many of which are, in turn, members of the umbrella organization called the Astronomical League. The League currently has a combined membership of almost 13,000 people (Beaman 1995).

Armchair amateurs are those who mainly prefer to read about astronomy and may do some casual observing from time to time. Some of these amateurs are members of local clubs, but many are not, and pursue their interest in astronomy through magazines or books they read, programs they watch on television, and lecture series they may attend. Some are members of such national organizations as the Astronomical Society of the Pacific or the Planetary Society. New converts to this group these days can come from those browsing the many interesting astronomy rest-stops on the information superhighway.

Many members of the three groups are tied together by the two main magazines for amateurs, *Sky & Telescope* and *Astronomy*. This latter has a circulation of about 170,000, while *Sky & Telescope's* is over 120,000. Estimates of the total number of amateurs in the U.S. range from 200,000 to 500,000, often depending on how exactly you define the term. This figure, 40 to 100 times the number of professional astronomers, represents a tremendous population with potential in astronomy education.

Many amateurs are already involved in education, by going for occasional visits to local schools or putting on neighborhood star parties, where youngsters get their first look through a telescope. The amateur community organizes a National Astronomy Day each spring, where they make a special effort to bring telescopes to where people are and show them the night sky. The Astronomical League has a number of educational programs and publications, although they are limited by being purely volunteer efforts with no budget to support them.

But much more could be done. Many amateurs have time, knowledge, energy, and enthusiasm, which could much more actively be harnessed in the service of education. Some professional astronomers and educators worry that amateurs will tell students erroneous things; but at the level of a 5th grade class, the physics of quasar energy mechanisms isn't really a relevant topic. The phases of the Moon, why telescopes are needed to observe celestial objects, or the joys of hunting comets are much more relevant to the reasoning level of the youngsters.

This was the thought behind Project ASTRO, a two year program we have been piloting at the Astronomical Society of the Pacific with support from the National Science Foundation: to set up ongoing partnerships between amateur (and professional) astronomers and 4th to 9th grade teachers in 45 sites around the state of California. Astronomers visit "their" classroom not once, but at least four times, and work with the teacher to put together age-appropriate hands-on classroom and after-school activities.

We found that, with proper training, and when they are provided with a suite of good activities and teaching resources, amateurs (and professionals) can do an excellent job in helping students get excited about astronomy and science in general. We now hope to extend the project to five other states around the country and to develop training workshops for interested partners at both AAS and ASP meetings (see the paper later in this volume.)

Project ASTRO has also produced *The Universe at Your Fingertips,* an 815-page loose-leaf notebook of exemplary activities, resource lists, and teaching suggestions that incorporate the best ideas from our project and many others around the country. It is available through the A.S.P. Catalog.

6. The Astronomy Interpretation Community

In some ways, the interpreters to the public are the most far-reaching part of the astronomy education community, because they include the media. It is sobering to remember that one episode of "Unsolved Mysteries" on television is seen by more people than all the students all of us in the symposium will ever teach during our entire careers.

The astronomy interpretation community includes editors and reporters at daily newspapers and magazines, producers and writers on radio and television, the authors of introductory books on astronomical topics, the writers of children's books, and the authors of astronomy software. Let's look at this world briefly:

A. Magazines

If you examine a list of the top 100 magazines by circulation, you do find some rays of hope amidst the gathering darkness of gossip and entertainment magazines. In 1994, there was one in the top 5 US magazines that regularly features very high quality astronomy articles: can you guess which magazine that is? .... *National Geographic* (with a circulation of about 9.5 million).

In the top 20, we have *Time* and *Newsweek*, both of which have had excellent physical science reporting. The top 30 includes *Smithsonian* and the top 40, *Popular Science*. And one of the largest circulation periodicals in the country, the Sunday newspaper supplement called *Parade* (which is mostly pap), regularly features wonderful essays by Carl Sagan which extol the scientific perspective and debunk popular pseudo-sciences.

In the category of smaller circulation special interest magazines, we have a number that do an excellent job of reporting astronomy to their readers: In addition to *Sky & Telescope* and *Astronomy* (which we have already mentioned), there are *Discover, Scientific American, American Scientist,* and *Air and Space*. Excellent and regular coverage also appears in the news pages of such magazines as *Science, Nature,* or *Science News*. Plus many of the astronomical and space interest societies issue their own magazines, such as *The Planetary Report* from the Planetary Society, *Mercury* from the Astronomical Society of the Pacific, or *Ad Astra* from the National Space Society. Here, although specialists occasionally complain about a subtle point being missed, the reporting is very, very good indeed, and astronomy stands out among sciences as receiving and offering the best coverage for readers with a serious interest in the field. Of course such readers are relatively small in number compared to the population of the country.

B. Daily Newspapers (and Radio and TV News)

If you follow astronomy reporting in U.S. newspapers, you are probably used to reading the syndicated copy of some of the very best science reporters (from such newspapers as *The New York Times, The Boston Globe,* or *The Los Angeles Times*. But these reporters, many of whom have some training in science, and who generally belong to a trade organization called the National Association of Science Writers, are really the cream of the profession. Under that cream comes a much larger group of reporters, editors, and radio and TV people whose training and judgment have become a cause for national concern.

In the old days, many journalists were trained in some academic subject (or self-trained) and went out to report the news. Today, there are over 400 schools or departments of journalism in the US (although many are slowly changing their names to include words like communication and media – in part because many of their graduates hope to earn large salaries reading the news from cue cards on television.) These graduates, armed with courses as unspecific and muddled as anything our schools of education can come up with, are beginning to fill the ranks of the radio, television, and print media in the country. Many have minimal or no training in science and a good fraction share the larger public's distrust of and sense of intimidation by scientists.

Even many of the elite journalists in the first group don't do as much digging and reporting these days as you might imagine. They have a kind of symbiotic relationship with the public information officers at universities, research labs, and scientific societies, whose job it is to get out the news from their institution and have it be as widely disseminated as possible. (The two groups are actually so interwoven that journalists frequently move from one to the other and back again with ease.) Because no reporter can keep up with all that is happening in every science, many have come to rely on public information officers to identify noteworthy stories for them. And those are usually the stories that are then developed. There are exceptions – often stories with a local angle or a whiff of scandal – but many journalists are happy to pluck the fruit from the low hanging branches of the information tree; it is rare to see them do much climbing on their own.

In San Francisco for example, a city that has three of the very best science journalists on the staff of its two newspapers, it has nevertheless been true that the majority of astronomy stories the public got to read or hear about in the past two years have been determined by only two processes: 1. what stories Steve Maran, the Public Information Officer of the American Astronomical Society, decides to feature at the society meeting press conferences (which are attended by many top science reporters whose stories are frequently syndicated around the country); and 2. what stories NASA and Space Telescope Science Institute public information officers decide to hold news conferences or issue illustrated news releases about.

When the non-elite group of journalists cover some science story (especially on radio and TV), they often do so reluctantly and frequently make a muddle of it. Or they wind up simply rewriting a press release or wire service copy, or focusing on a local scientist who can comment in 20 seconds on the story. And, often it turns out that most of what these journalists call science news is actually news about medicine or applied technology.

It is these second-tier journalists who often get confused (to make the most charitable interpretation) between science and pseudoscience. They were the ones who reacted to the Nancy Reagan astrology revelations by interviewing local astrologers (who made it all sound like the most natural thing in the world that the president's schedule should be determined by astrological forecasting) and by doing puff pieces on all the movie stars that also guided their lives by astrology.

But there is another part of modern journalism, which more rarely gets discussed: the *gatekeepers of the media*. Decisions about what gets on the evening news are usually not made by the people you watch on your screen. Decisions about what gets significant coverage in the newspapers are not made by the people whose bylines you know. Decisions about what topics are discussed on the radio are often not made by the people whose voices you hear.

Making these decision is the role of the gate-keepers, editors and producers who filter the news, select the stories to be covered, and direct the tone of the material that will be read on the air. These gate-keepers are often young, superficially educated, and lack any serious knowledge of science. Yet they are the ones who assign the stories to reporters, determine the length of articles or broadcast pieces, and decide where these pieces will appear. They have been, many of them, raised on a steady diet of the kind of journalism we now have, and so accept the current system without question.

Furthermore, it is this group, more than the reporters, that is actually charged with minding the corporate bottom line at their institutions, and is thus most likely to pander to the worst instincts of the public instead of educating them. (Just count the number of news stories about ghosts around Halloween or the predictions of psychics around the new year.) Their sense of the public trust of journalism is founded much more on an entertainment than an educational model, and the trends in what gets reported and how it gets reported clearly bear out their influence. Small wonder that serious science and skeptical inquiry, which is the best thing the scientific method has to teach us, get such short shrift.

C. Radio and Television Programming

Now we get further into the media wasteland, where oases are harder to find. You will hear about some of the oases today, such as the *Stardate* and *Earth & Sky* radio programs, the *Nova* television series on PBS, the new telecourse on astronomy from Coast Community College, an occasional special somewhere on a cable channel, etc. The most successful astronomy TV program was *Cosmos,* which I believe is still the most watched program in PBS history. It is estimated to have been seen by 500 million people in more than 60 countries since 1980. But it is a sad exception to the general rule of what is seen on television on an average night.

Overall, the television picture is very dim indeed, with vast quantities of pseudo-science being served up by most of the commercial networks. (This is being exacerbated by the tremendous growth of what is called "tabloid TV" – shows that generally imitate the contents and approach of the tabloid newspapers.) Science on many stations is often limited to ten-second newsbites on the evening news which make little sense to the uninitiated (and are generally what I call Guinness Book of World Records stories – the farthest galaxy, the biggest black hole, the most distant comet.)

By the way, if you are one of the people in this country who watches little or no television, and therefore thinks the problem of so little worthwhile contents on the air is not a major concern, you should know that in 1993, the average American household had a television set on for 7 hours and 51 minutes each day (Barber 1993)!

That's only the average! It boggles the mind to think of it (and rots the mind to do it.)

As a country, we need to recognize the enormous power that the media have, and pay more attention to how decisions are made in the media. By this I don't mean censorship, but rather undertaking a long-term educational program to make sure that those who have the responsibility for deciding what goes on the air (or what is printed) have the education and the informed background to make sensible judgments. One step would be to introduce and require excellent science overview courses developed specifically for students going into journalism. Another would be for everyone who is dissatisfied with science coverage on the media to make his or her voice heard locally or nationally. (In San Francisco, the work we have done at the ASP has increased the amount of astronomy on local radio and television significantly and, although it took a while, it was not a very painful exercise.)

The argument the media always make in response to such concerns is that they are merely giving the public what it wants. But this argument is specious. Someone like me, who began life in Communist Eastern Europe, never got to taste a mango or an avocado as a child. Thus I would never have known to ask for a mango or an avocado or that I wanted one. But after we came to America, I was introduced to a much wider range of foods: my palate was educated...and now I ask for mango and avocado regularly. Similarly, I would suggest that when the media tell us that people are much more interested in the Bermuda triangle that the Great Attractor, they are merely confessing our joint failure at educating both the gatekeepers of the media and the public.

In focusing so strongly on the entertainment aspect of their mission over the educational aspect, the media lose sight of the fact that there is an alternative to pandering to the lowest taste. It is the far more difficult and long term job (and the almost forgotten pleasure) of elevating the tastes and desires of their audience to new levels of perception and understanding.

I should mention that there is a national organization which is making a creditable effort to help the media sort out science from fiction science. The group is called the Committee for the Scientific Investigation of Claims of the Paranormal (CSICOP), a mouthful of a name, but a group with their hearts in the right place. It consists of scientists, educators, magicians, philosophers, lawyers, and other skeptics whose interest is to get the rational, skeptical perspective about fiction science out to the media and the public. Among astronomers, Carl Sagan, David Morrison, Ed Krupp, Don Goldsmith, Steve Shore and I have all been very active in CSICOP projects relating to astronomical pseudoscience.

Through their superb magazine, *The Skeptical Inquirer,* their meetings and workshops, and their information releases to the media, CSICOP has managed to alert and educate a significant number of reporters to stop and consider what they are doing when they file a story involving pseudoscience. They also work to bring together such reporters with skeptical spokespeople before a story is written. (They can be contacted at: P.O. Box 703, Amherst, NY 14226; I urge everyone in astronomy education to get to know their work and support what they are doing.)

D. Nontechnical Books for Adults

In 1993, roughly 45,000 new books were published in the U.S. [*Publishers Weekly,* Mar. 7, 1994, data from R.R. Bowker Co.] Of these, about 2,000 were classified as science, although I suspect that the librarians who do this classifying may well include some pseudoscience in this category. Such books, written by both scientists and science journalists, can be an important way that educated laypeople learn about new developments and ideas in science.

When a really excellent book comes along, such as *Lonely Hearts of the Cosmos* or *First Light,* it can give laypeople marvelous insight into how astronomy is really done today. A best seller, such as *Cosmos* or *A Brief History of Time* (which by the end of 1993 had sold over 5.5 million copies worldwide), can turn more people on to astronomy than thousands of astronomy courses.

On this front, there is good news and bad news: the good news is that fine books in popular astronomy are still finding a publisher; the bad news is that these publishers rarely promote or advertise such books with any energy or enthusiasm. Instead, they reserve their advertising budgets for the books they consider "guaranteed big sellers", a phrase which then becomes a self-fulfilling prophecy. As a result, some of the best astronomy books of the last decade have sunk without a trace in the vast murky ocean of modern publishing.

E. Children's books

One area of astronomy education that has received very little attention are children's books, despite the fact that these can have a strong influence on youngsters. Several astronomers and science writers have produced whole series of astronomy books for kids, among them the late biochemist Isaac Asimov, the former director of the Hayden Planetarium Franklyn Branley, British astrophysicist David Darling, the current director of the Griffith Observatory Ed Krupp, the associate director of the Pacific Science Center Dennis Schatz, journalist Seymour Simon, and NASA aerospace specialist Gregory Vogt. Alas, mixed in with these excellent and reliable books -- often on specific single topics in astronomy – are a host of muddle-headed, error-filled books written by people with little science background and published by organizations and publishers who should – in many cases – know better.

At the ASP, we have collected and reviewed hundreds of children's books on astronomy and will be publishing recommendations for the best of them in *The Universe at Your Fingertips* Resource Notebook that will come out of Project ASTRO.

Conclusion

In this quick overview of astronomy education, I have purposely focused on the problems and challenges, because many of the other papers in this volume will focus on the various solutions being proposed. Throughout the country, scientists, engineers, and educators are becoming concerned about these problems and looking for ways we can change the cultures we live in to improve the nation's understanding and appreciation of science.

In astronomy this effort is just beginning (on all but the most local of scales), and those of us involved in educational reform, whichever of the communities that I've discussed we belong to, still feel a bit isolated and untethered. There is a great need for a more organized and coordinated effort in astronomy education, and I very much hope this conference can be the beginning of such coordination and organization.

The American Astronomical Society has formed an Astronomy Education

Policy Board, which is looking at how to bring about institutional changes in the worlds of astronomical research and university education that will encourage and reward work in astronomy education. More about the plans of this Board can be found in the paper by Suzan Edwards later in this volume.

In the appendix to this volume is a catalog we have compiled at the Astronomical Society of the Pacific of some of the national projects in astronomy education. (Many of these are also described in contributed papers throughout this volume.) We intend to continue to update and expand this catalog, and would welcome sugggestions and additions from readers.

Let me conclude with the following more general thoughts: Everyone who attended this conference is already dedicated to astronomy education, or else you would not be sacrificing a pleasant summer weekend attending a symposium on the subject. But we make up only a tiny fraction of the total astronomical community. I hope as part of your dedication you will share your enthusiasm for education with colleagues, students, amateurs, and all the other categories I've discussed.

I'd like to make the simple suggestion that everyone in astronomy ought to devote a minimum of 1 percent of their time to improving astronomy education for nonscientists: if we take a 40 hour work-week and 52 weeks in a year, 1 percent works out to 21 hours a year. That's enough for seven three-hour sessions with a local teacher and 6th grade class; for preparing or giving a workshop for local astronomy graduate students and postdocs about recent developments in education and things they can do to help; for organizing a meeting of a local amateur astronomy club on what they can do to get involved in the problems I've outlined; or for spending several afternoons with the staff of the local planetarium.

I urge everyone who works in or loves astronomy to do this not only out of a sense of duty or compassion. The ominous news from recent developments in our nation's capitol is that (as Jeff Rosendhal has been pointing out) the unwritten compact through which the federal government has supported science and universities for the sake of their value to national security shows signs of breaking down. New generations of our leaders (like new generations of our students) do not feel immense loyalty or warmth toward science. And their apathy may well allow very serious cuts in the budgets that support so much of our scientific enterprise.

Our crisis in the public understanding of science is like a disease that has slowly attacked our country over the years. If it is allowed to fester untreated for much longer, the body politic will surely send its own antibodies to the source of irritation, and the resulting "cure" may be worse than anything scientists can now imagine. Thus the work we do today in science education may well turn out to be the single most important thing that can be done to assure the continued good health of the science of astronomy in the United States.

References:

Albers, B. "Agenda for Excellence" in *Journal of National Institutes of Health Research,* Apr. 1990, p. 19.

Barber, B. "America Skips School" in *Harper's Magazine*, Nov. 1993, p. 39.

Beaman, B. "Notes from the President" in *The Reflector,* May 1995, p. 4.

Beardsley, T. "Teaching Real Science" in *Scientific American,* Oct. 1992, p. 98.

Fraknoi, A., ed. *The Universe at Your Fingertips.* 1995, Astronomical Society of the Pacific.

Hoff, D. "Astronomy for Nonscience Students: A Status Report" in *The Physics*

Teacher, Mar. 1982, p. 175.
Menke, D. "Status of the Planetarium Profession" a paper given at the 1994 Meeting of the International Planetarium Society (private communication).
National Center for Education Statistics. *Digest of Education Statistics.* 1994, U.S. Dept. of Education.
Regan, D. *For the Record.* 1988, Harcourt Brace Jovanovich.
Schrag, P. in *The New Republic,* Dec. 16, 1991, p. 20.
Wright, J., ed. *The Universal Almanac.* 1995, Andrews & McMeel.

Discussion

Bisard.

I strongly believe the ASP must conduct a national survey of "Learning Astronomy" rather than "Teaching Astronomy" at all levels of K-16. This could be a small grant supported project but one which answers who? what? where? when? with regard to the learning of astronomy.

Crawford.

Building awareness and interest are critical, and the younger the better. IDA's Star Watching Program has had some excellent student feed back, for example:
"I never saw a star before, now they're my friends."
"I never looked up at night before." etc.

Mechler.

Tabloids comment and call for more on science itself as a process in astronomy textbooks.

Hoff.

I made a comment on the fact that H.S. students spend only about 1.5 hours out-of-doors per week. This is both a fact to be concerned with and a challenge for those who develop astronomy activities.

Fraknoi.

I agree with the concern. With the overwhelming role that television and video/computer games now play in the evening lives of our students, some rarely get to observe the night sky and thus become aware of its beauties and patterns.

Better Rather Than More Astronomy Education

Andrew Ahlgren
Associate Director, Project 2061
American Association for the Advancement of Science
1333 H Street NW
Washington DC
USA 20005
Professor Emeritus
Department of Curriculum and Instruction
University of Minnesota

E-mail: aahlgren@aaas.org

Getting more astronomy taught in the schools is not the way in which astronomy educators can make the most progress in fostering general understanding of astronomy. There are ample subjects that are taught even more than they should be (e.g., chemistry, algebra) which students don't understand when they study and remember only isolated scraps of later. The astronomical community would do better to work out what ideas are (a) most important for everyone to know and (b) reasonable to expect that everyone could learn. In short, teach less but teach it well enough that students actually understand it. (OK, maybe in the long run students can understand more than we try to teach now, but the place for astronomy educators to start is with learning how to teach at least some basics well.)

All astronomers owe it to themselves to see the videotape *A Private Universe* which shows that a majority of fresh Harvard graduates can't explain lunar phases, or seasons, either. (The 18-minute tape is available from the Annenberg Collection for $40 phone 1-800-LEARNER.) For the Harvard graduates it isn't a matter of not being exposed. And it isn't just that they are ignorant. Children form naive ideas about phenomena, sometimes even as a result of incompletely understanding what they have been taught, that are very resistant to change. (*A Private Universe* illustrates that, as well.) A familiar saying in the educational research community is that college students have the same misconceptions about the natural world that they had in elementary school – except that they express them in longer words.

Many teachers – and, a fortiori, professors – are appalled at the misconceptions of the students seen in the videotape, but are confident that their own students do not have such difficulties. These optimists should be encouraged to probe their own students' knowledge a few months after the final exam. But probing means asking for thorough explanations as evidence of understanding, not accepting mere pronunciation of a few phrases. For example, as *A Private Universe* demonstrates, saying "The tilt of the earth's axis" or "indirect rays" is not adequate evidence of understanding the seasons. (Try to avoid anger in doing this. Many teachers find themselves shouting something like, "Don't you remember that I told you...?")

Science for All Americans (AAAS, 1989) was an attempt to identify the important and learnable ideas in astronomy as part of science literacy. Astronomy content appears most obviously in the sections The Universe and The Earth in Chapter 4: THE PHYSICAL SETTING. Obviously, too, there are essential ideas in the Chapter 4 sections Motion (including waves and light) and Forces of Nature. Chapter 10: HISTORICAL PERSPECTIVES, includes the astronomical episodes Displacing the Earth from the Center of the Universe [Copernicus, Kepler, and Galileo], Uniting the Heavens and Earth [Newton], and Relating Matter & Energy and Time & Space [Einstein]. Less obviously, there are ideas relevant to astronomy in Chapter 1: THE NATURE OF SCIENCE, and Chapter 11: COMMON THEMES (sections Systems, Models, Constancy & Change, and Scale).

Better than just selecting important ideas, astronomy educators should consider sketching out some "progression of understanding" maps that depict what precursor ideas are necessary for more sophisticated ideas and when they might be learned. Making such a map forces consideration of what leads to what, what is necessary for what, and brings out the most essential ideas and their precursors. Figure 1 shows a draft of a progression-of-understanding map for Gravity; the boxed statements are abbreviations and paraphrases of actual benchmarks. When teachers have studied a progression-of-understanding map (or made their own), they have a much better feeling for where ideas have come from and where they are going next, which helps to shape their priorities and emphases in teaching.

A progression-of-understanding map is built in part on the logic of the ideas and partly on what students are capable of learning when. Astronomers are perfectly capable of thinking through the logic (e.g., the moon has to be thought of as being spherical before its phases can be explained), but they will need a lot of help in estimating what students can understand. In *Benchmarks for Science Literacy* (1993) the goals of *Science for All Americans* are elaborated, with the same chapter and section structure, to specify what progress toward them could be expected by grades 2, 5, 8, and 12. The grade-placement of ideas in *Benchmarks* reflected implicit progression-of-understanding-map kinds of thinking, but explicit maps good enough to publish are still under construction.

If possible, teachers at different levels should work together on planning how learning would progress over K-12 (or K-16). When possible, also, progression-of-understanding maps should draw on the growing body of research on how and when students learn ideas in, and relevant to, astronomy. For example, no matter how faithfully they may repeat the teacher's words, second graders do not know (and perhaps are developmentally incapable of knowing) what it means to say that the earth is a sphere with people living all over it, that down is toward the center, or what planets are. And they aren't sure what a thousand is, much less the temperature, time, and distance numbers found in astronomy. They can't understand the phases of the moon or the cause of the seasons. *Benchmarks* Chapter 15: THE RESEARCH BASE gives section-by-section summaries of the available educational research. The Appendix shows the summary for The Universe and part of the summary for The Earth.

This essay has included some points that although not likely to be popular, should be taken seriously in trying to promote broader and deeper understanding of astronomy:

- Teaching *more* astronomy should wait on learning how to teach it *better*.
- Expertise in astronomy is not enough to be an effective astronomy educator.
- Teachers at all levels overestimate what their students learn.
- Good teaching cannot be freely invented, but requires research.

This essay also has neglected the motivational value of gee-whiz, up-to-the-minute aspects of astronomy in favor of the more lasting value of understanding basics well. Both have their places in effective astronomy education. But actually understanding something in the science curriculum is rare enough that it has considerable motivational value in itself.

Appendix: Excerpts from Benchmarks Chapter 15: The Research Base

4. THE PHYSICAL SETTING

There is more research on student conceptions about The Physical Setting than in any other area. The Pfundt and Duit (1991) bibliography reveals that more than 70% of the published papers about students' conceptions in science were concerned with topics related to The Physical Setting benchmarks. Much research has focused on topics related to the Earth, Structure of Matter, Energy Transformations, and Motion. Topics related to The Universe and Forces of Nature have also received attention, but for the Processes That Shape the Earth, there is little research. Even in the frequently researched areas, relatively few studies report on long-term teaching interventions that try to improve students' ideas about the physical setting. The available literature on students' understanding of topics related to The Physical Setting has been reviewed in Driver, Guesne, & Tiberghien (1985). Conference proceedings on these topics include Driver & Millar (1985); Duit, Goldberg, & Niedderer (1992); Jung, Pfundt, & Rhoeneck (1981); Lijnse (1985); and Lijnse et al., (1990).

4a The Universe

Research available on student understanding about The Universe focuses on their conceptions of the sun as a star and as the center of our planetary system. The ideas "the sun is a star" and "the earth orbits the sun" appear counter-intuitive to elementary-school students (Baxter, 1989; Vosniadou & Brewer, 1992) and are not likely to be believed or even understood in those grades (Vosniadou, 1991). Whether it is possible for elementary students to understand these concepts even with good teaching needs further investigation.

4b The Earth

Shape of the earth. Student ideas about the shape of the earth are closely related to their ideas about gravity and the direction of "down" (Nussbaum, 1985a; Vosniadou, 1991). Students cannot accept that gravity is center-directed if they do not know the earth is spherical. Nor can they believe in a spherical earth without some knowledge of gravity to account for why people on the "bottom" do not fall off. Students are likely to say many things that sound right even though their ideas may be very far off base. For example, they may say that the earth is spherical, but believe that people live on a flat place on top or inside of it – or believe that the round earth is "up there" like other planets, while people live down here (Sneider & Pulos, 1983; Vosniadou, 1991). Research suggests teaching the concepts of spherical earth, space, and gravity in close connection to each other (Vosniadou, 1991). Some research indicates that students can understand basic concepts of the shape of the earth and gravity by 5th grade if the students' ideas are directly discussed and corrected in the classroom (Nussbaum, 1985a).

Explanations of astronomical phenomena. Explanations of the day-night cycle, the phases of the moon, and the seasons are very challenging for students. To understand these phenomena, students should first master the idea of a spherical earth, itself a challenging task (Vosniadou, 1991). Similarly, students must understand the concept of "light reflection" and how the moon gets its light from the sun before they can understand the phases of the moon. Finally, students may not be able to understand explanations of any of these phenomena before they reasonably understand the relative size, motion, and distance of the sun, moon, and the earth

(Sadler,1987; Vosniadou, 1991).

Figure 1 A progression-of-understanding map for Gravity
© Project 2061 – Work in Progress

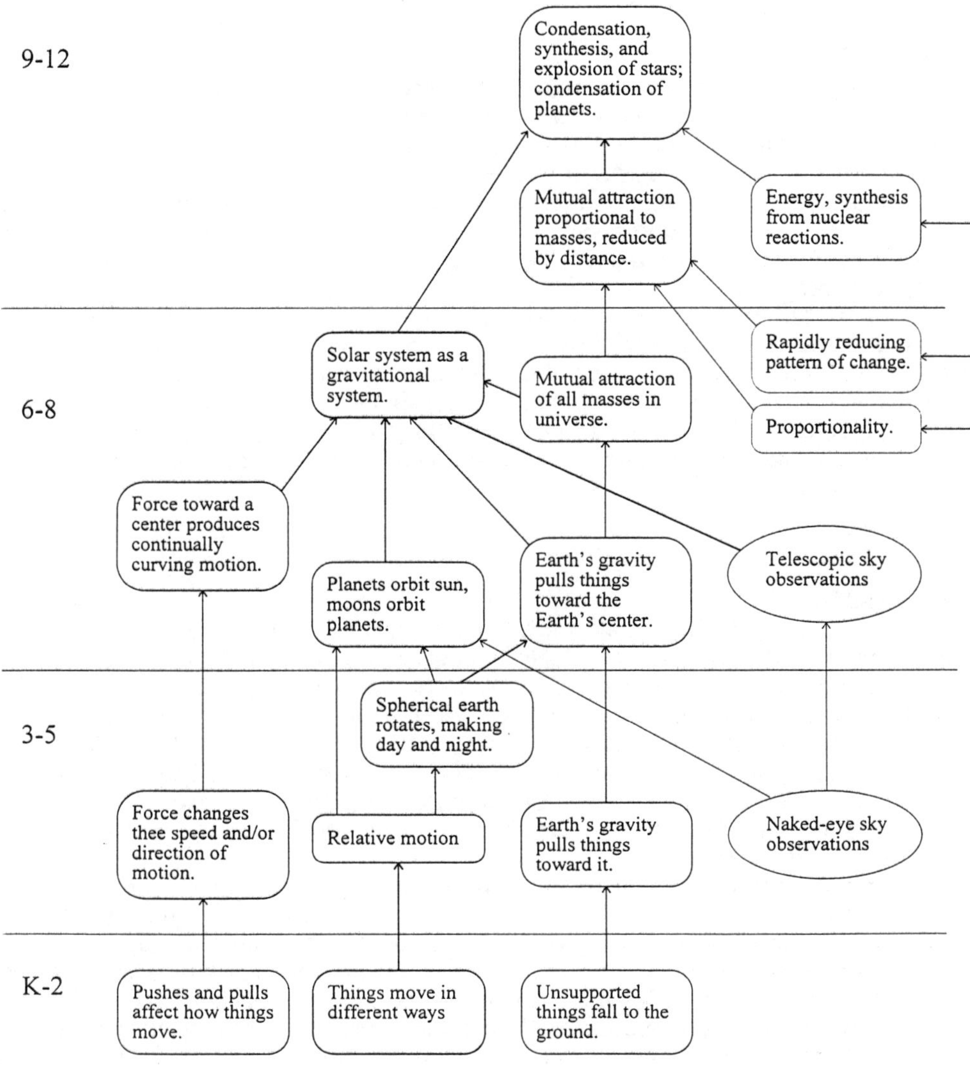

Discussion

Bisard.

2 Questions: (1) Are there any minima (in your project) written for prerequisites for the K-3 levels? (2) The cognitive structure approach is a very rich technique with regard to conceptually analyzing a topic (like density, gravity, or ...). Which audience is most responsive to this in your hundreds of presentations? (K-6, 7-12, post-secondary)?

Pasachoff.

I was very disappointed and upset to see the chart by Dr. Ahlgren, since it included very little more recent than 1906 (Einstein's special relativity) and nothing more recent than 1939 (Bethe's work on nuclear energy in stars). It is my view that astronomy is a current and vital field that continues to make important progress and to discover important and interesting things, and that it is inappropriate and misleading to give the impression to students that, in the words of Sir W.S. Gilbert in Gilbert & Sullivan's *Patience*, "art stopped short in the cultivated court of the Empress Josephine."

It is bad for students, bad for astronomy, and bad for physical science in general when so much of the quota for astronomy is taken up with basic things like phases and seasons, and so much else is basic items like Newton's law of gravity or Kepler's laws of motion. Though such things should be taught, it is also important to teach students that science is a contemporary vital enterprise. I see that either through ignorance or through conscious omission, contemporary astronomy is all but omitted from Project 2061's lists.

Ahlgren.

The omission was conscious.

Pasachoff.

I am sorry to hear that, since I think astronomy can be useful vehicle for bringing a wide variety of scientific concepts before students. and though Project 2061 and the National Academy of Sciences/National Research council's National Standards are not technically syllabi, it is my experience that such lists will be translated into syllabi by state education authorities and thus will be adopted by textbook companies to govern the content of their texts. Thus omitting or minimizing the last fifty years of advances in astronomy from the lists of Project 2061 and the National Standards project will lead to shortchanging the students of the future.

I think that there are many things from ongoing thought about astronomy that should be taught to students. For example, the notion that stars change and evolve over time, with such fascinating objects as white dwarfs, neutron stars (which we detect as pulsars) and black holes as end points, is one that interests students and that should be covered in junior-high school as well as in high school and in college. Indeed, when I lecture to second-grade and other elementary-school students, I often get excellent questions about black holes. Why shouldn't we teach fundamental aspects of such topics, especially when they coincide with things that students want to know. I stress that it is not just the vocabulary that I want to teach, but the fundamental ideas (a) that stars go through a life cycle, (b) that we can observe the stages of those life cycles in certain ways that verify for us scientifically that we have the true picture, and (c) that scientists (astronomers in particular) continue to participate in interesting and important research.

I think it is important for students to be excited by things and ideas in science and not just knowledgeable about everyday occurrences.

Ahlgren.

What you say may be for the top students, but there is a lowest quartile. And you are a college professor, aren't you?

Pasachoff.

I have also, with my wife (Naomi Pasachoff) and another colleague written Earth Science and Physical Science texts for junior-high and, with still more colleagues, science texts for elementary school.

I think that what I am saying goes for all quartiles, including the lowest. It isn't just the lowest quartile that doesn't come to comprehend phases and seasons; Phil Sadler's work in "A Private Universe" and elsewhere shows that even top students don't usually comprehend those things. Students of all levels go on to be voting citizens, and I think it is important for them to be introduced to the ongoing vitality of astronomy. Among other things, when it comes time for them to vote for Representatives or Senators, they may have some idea of the importance of scientific research. When the biology teachers teach even elementary students about DNA but when physical-science teachers stop with Einstein or before, the students mistakenly get the idea that it isn't important to fund physical science anymore. The Superconducting Super Collider may have been part of the fallout of our lack of teaching enough about the value of ongoing research in physical science.

It is the process of science that we must be teaching. I agree that it is important for students to learn more than the vocabulary involved; it is not enough for them to know scientific-sounding words without comprehending the concepts. but there are fundamental ideas of science that we can include for students of almost any age and ability. For example, the notion that I can see three times more clearly when I put on my eyeglasses, that today's big telescopes provide about a factor of 100 in clarity, and that the Hubble Space Telescope gives us another factor of 10 in clarity can be taught in a conceptual way. The notion that an exploding star, a supernova, that went off in 1987 in a nearby galaxy can tell us about the life stories of stars in a new and exciting way through study of the light and the particles we received from it also strikes me as an important idea to transmit. As curricula evolve over the next decade or so, some other ideas can replace these or add to them, so there must be space in Project 2061 and in the National Standards for those new developments to come into teaching. Otherwise we will find interested students asking their teachers in 4th-grade or in 8th-grade about important astronomical ideas and being rejected because "they aren't in the book" or, worse, "you couldn't understand them."

Ahlgren.

You are asking for more astronomy, but other scientists (geologists, and so on) would also want more of their fields.

Pasachoff.

The original committees that were set up for Project 2061 and for the NAS/NRC National Standards were deficient in representation of astronomers compared with geologists, physicists, and so on. I would be more than glad to make my points before a committee with a fair and widespread representation of the different fields of science. I invite you to set up such a committee and have other astronomers or me present along with scientists from a variety of fields. I think we will succeed in making a case that astronomy has been particularly badly served in Project 2-61 and the NAS/NRC National Standards.

Though I know that you represent the American Association for the Advancement of Science and its Project 2061 – and I was Chair of the Astronomy

Division of the AAAS when the first report of the Project 2061 came out, and spoke out then and was on a review committee – and so do not speak for the National Academy of the Sciences/National Research Council, let me mention some comments on their latest draft that I sent them.

Recent astronomy is almost entirely omitted from the Physical Science Standards. I asked that they add: K-4: Nature of the sun, moon and planets; 5-8: Stars and how they shine; 9-12: Nature of galaxies and the Universe, cosmology.

Whatever astronomy is present is represented in the Earth and Space Science Standards. I suggested that they change "Objects in the sky" for K-4 to "Nature and motion of objects in the sky," and that they add "The life stories of stars" to grade 5-8. I also suggested that they add "The nature of galaxies" to levels 9-12.

At the very least, these changes would give teachers an impetus for teaching about ongoing things, and would lessen the tendency to tell inquiring students that topics they ask about aren't in the syllabus. I am concerned with the many teachers – a majority – who do not have extra knowledge and who will be bound to the text and to the syllabus. for them, we need to include space in the texts and the syllabus for ongoing advances in physical science.

Ahlgren.

Remember I am describing a common core for all. You can add as much as you like for specially interested students.

Professor Pasachoff raised some important, and as yet unsolved, questions about how to differentiate the 'syllabi' for the students most and least interested in astronomy. I think that the fundamental understandings of which he has just spoken are good choices. And I hope that the parts of actual curricula relevant to astronomy will be interesting, even seductive, to students and as up to date as achieving the many other goals of the curriculum will allow. Yet the problems of student understanding of which I have spoken cannot be dazzled away. (Whether Professor Pasachoff lecturing to 2nd-graders is an example of that I cannot say, but I hope he listens to the children as well – and to all of them, not just those with their eager hands raised.) Given the time limitations of schooling, astronomy educators will individually have to decide on the relative values they place on up-to-date interest and underlying understanding – for example, on some 2nd-graders having 'excellent questions about black holes' vs. all 5th-graders understanding that the earth is round.

Shawl.

One interesting method I have used is to give oral final exams to students who have been doing "A" work. Such exams are revealing – sometimes enthralling – sometimes discouraging. Additional comment: An overall problem is that people do not feel a part of nature; we need to teach in a way that will make people identify with, and feel close to, nature.

Wasiluk.

I didn't get to ask: Isn't all this talk on curriculum and reform work and books and ideas just one way of pretending to be doing something when you are just shirking the issue?

Education Reform: Implications for the teaching of astronomy It's not WHAT we teach, but HOW we teach, that needs to change

Dennis Schatz
Pacific Science Center
200 2nd Avenue N.
Seattle, Washington
USA 98109

Introduction

Last week as I finished the preparations for my talk, I stacked up all of the material I have related to Education Reform in Science, Mathematics and Technology Education. This includes everything from the 2061 Benchmarks and National Science Education Standards, to various state outcomes and the New Standards Project initiative in assessment reform. The pile measures slightly over four feet - and staying abreast of Education Reform is far from my highest job priority.

In the short time that I have, I cannot cover all that is stated or implied in the emerging standards, compare their differences, and give definitive guidelines on what astronomy should be taught at each grade. I can only give a general sense of the content of astronomy being emphasized in the emerging standards, and more important the implications of how astronomy - really all subjects - should be taught.

What Astronomy Should Be Taught - The Content Standards

The astronomy content that should be taught is found in the Earth Science Area of the National Standards, or the Physical Universe Section of the 2061 Benchmarks. According to the National Standards, the astronomy topics to cover are:

Kindergarten through Grade 4 - Objects in the sky

- The sun, moon, stars, clouds, birds, and airplanes all have properties, locations, and movements that can be described and that may change.
- Objects in the sky have patterns of movement. The sun, for example, appears to move across the sky in the same way every day, but its path changes slowly over the seasons. The moon moves across the sky on a daily basis much like the sun. The shape of the moon seems to change from day to day in a cycle that lasts about a month.

Grade 5 through 8 - Earth in the solar system

- The Earth is the third planet from the sun in a system that includes the moon, the sun, eight other planets and their moons, and smaller objects such as asteroids and comets. The sun, an average star, is the central and largest body in the solar system.
- Most objects in the solar system are in regular and predictable motion. These motions explain such phenomena as the day, the year, phases of the moon, and eclipses.
- Gravity is the force that keeps planets in orbit around the sun and governs the rest of the motion in the solar system. Gravity alone holds us to the Earth's surface and explains the phenomena of the tides.

Grades 9 through 12 – The origin and evolution of the Earth system; The origin and evolution of the universe

- The sun, the Earth, and the rest of the solar system formed from a nebular cloud of dust and gas 4.6 billion years ago. The early Earth was very different from the planet we live on today.
- The origin of the universe remains one of the greatest questions in science. The Big Bang Theory places the origin between 10 and 20 billion years ago, when the universe began in a hot dense state and it has been expanding ever since.
- Early in the history of the universe matter, primarily the light atoms hydrogen and helium, clumped together by gravitational attraction to form countless trillions of stars. Billions of galaxies, each of which is a gravitationally bound cluster of billions of stars, now form most of the visible mass in the universe.
- Stars produce energy from nuclear reactions, primarily the fusion of hydrogen to form helium. These and other processes in stars have led to the formation of all the other elements.

Each content standard in the National Standards document also provides a general overview of how students should develop understanding of these subjects. For the K-4 content standard given above, the National Standards document states:

> By observing the day and night sky on a regular basis, children learn to identify sequences of changes and to look for patterns in these changes. As they observe sequences of changes, such as the movement of an object's shadow during the course of a day, and the positions of the sun and the moon, they find the patterns in these movements. They can draw the moon's shape for each evening on a calendar and then determine the pattern in the shapes over several weeks. Research studies make it clear that these understandings should be confined to observations, descriptions, and finding patterns.
>
> Attempting to extend this understanding into explanations or the use of models will be limited by the children's inability at this age to understand that Earth is spherical, little understanding of gravity, and misconceptions about the properties of light that allow us to see objects such as the moon. (Even though they verbalize that they live on a ball, probing questions will reveal that their thinking may be very different.)
>
> Emphasis in Grades K-4 should be on developing observation and description skills and the explanations based on these observations.

For comparison, the astronomy topics to cover according to the 2061 Benchmarks are:

Grades K through 2 – The Universe

- There are more stars in the sky than anyone can easily count, but they are not scattered evenly, and they are not all the same in brightness or color.
- The sun can be seen only in the daytime, but the moon can be seen sometimes at night and sometimes during the day. The sun, moon, and stars all appear to move slowly across the sky.
- The moon looks a little different every day, but looks the same again about every four weeks.

Grades 3 through 5 – The Universe, The Earth

- The patterns of stars in the sky stay the same, although they appear to move across the sky nightly, and different stars can be seen in different seasons.
- Telescopes magnify the appearance of some distant objects in the sky, including the moon and the planets. The number of stars that can be seen through telescopes is dramatically greater than can be seen by the unaided eye.
- Planets change their positions against the background of stars.
- The earth is one of several planets that orbit the sun, and the moon orbits around the earth.
- Stars are like the sun, some being smaller and some larger, but so far away that they look like points of light.
- Like all planets and stars, the earth is approximately spherical in shape. The rotation of the earth on its axis every 24 hours produces the night-and-day cycle. To people on earth, this turning of the planet makes it seem as though the sun, moon, planets and stars are orbiting the earth once each day.

Grades 6 through 8 – The Universe; The Earth

- The sun is a medium-sized star located near the edge of a disk-shaped galaxy of stars, part of which can be seen as a glowing band of light that spans the sky on a very clear night. The universe contains many billions of galaxies, and each galaxy contains many billions of stars. To the naked eye, even the closest of these galaxies is no more than a dim, fuzzy spot.
- The sun is many thousands of times closer to the earth than any other star. Light from the sun takes a few minutes to reach the earth, but the light from the next nearest star takes a few years to arrive. The trip to that star would take the fastest rocket thousands of years. Some distant galaxies are so far away that their light takes several billion years to reach the earth. People on earth, therefore, see them as they were that long ago in the past.
- Nine planets of very different size, composition, and surface features move around the sun in nearly circular orbits. Some planets have a great variety of moons and even flat rings of rock and ice particles orbiting around them. Some of these planets and moons show evidence of geologic activity. The earth is orbited by one moon, many artificial satellites, and debris.
- Large numbers of chunks of rock orbit the sun. Some of these that the earth meets in its yearly orbit around the sun glow and disintegrate from friction as they plunge through the atmosphere – and sometimes impact the ground. Other chunks of rocks mixed with ice have long, off-center orbits that carry them close to the sun, where the sun's radiation (of light and particles) boils off frozen material from their surfaces and pushes it into a long, illuminated tail.
- We live on a relatively small planet, the third from the sun in the only system of planets definitely known to exist (although other, similar systems may be discovered in the universe).
- Because the earth turns daily on an axis that is tilted relative to the plane of the earth's yearly orbit around the sun, sunlight falls more intensely on different parts of the earth during the year. The difference in heating of the earth's surface produces the planet's seasons and weather patterns.

- The moon's orbit around the earth once in about 28 days changes what part of the moon is lighted by the sun and how much of that part can be seen from the earth – the phases of the moon.

Grades 9 through 12 – The Universe

- The stars differ from each other in size, temperature, and age, but they appear to be made up of the same elements that are found on the earth and to behave according the same physical principles. Unlike the sun, most stars are in systems of two or more stars orbiting around one another.
- On the basis of scientific evidence, the universe is estimated to be over ten billion years old. The current theory is that its entire contents expanded explosively from a hot, dense, chaotic mass. Stars condensed by gravity out of clouds of molecules of the lightest elements until nuclear fusion of the light elements into heavier ones began to occur. Fusion released great amounts of energy over millions of years. Eventually, some stars exploded, producing clouds of heavy elements from which other stars and planets could later condense. The process of star formation and destruction continues.
- Increasingly sophisticated technology is used to learn about the universe. Visual, radio, and x-ray telescopes collect information from across the entire spectrum of electromagnetic waves; computers handle an avalanche of data and increasingly complicated computation to interpret them; space probes send back data and materials from the remote parts of the solar system; and accelerators give subatomic particles energies that simulate conditions in the stars and in the early history of the universe before stars formed.
- Mathematical models and computer simulations are used in studying evidence from many sources in order to form a scientific account of the universe.

Avenues for teaching astronomy also occur under other Benchmarks:

Grades 6 through 8 – Displacing the Earth from the Center of the Universe
Grades 9 through 12 – Motion
Grades 9 through 12 – Forces of Nature
Grades 9 through 12 – Displacing the Earth from the Center of the Universe; Uniting the Heavens and Earth; Relating Matter & Energy and Time & Space

The National Science Education Content Standards also provides other opportunities to include astronomy subjects under:

Grades 9 through 12 – The Structure of Atoms
Grades 9 through 12 – Interactions of Energy and Matter

Many educators ask which of these two major documents should be the one to consult when trying to identify the astronomy content to include in their curriculum. A recent comparison of the Benchmarks with the National Standards concludes, "...what is most impressive is the great similarity between the content of the two versions, not their differences."

Ultimately, it will probably not be necessary to choose between the two because education reform will be driven by the State Frameworks developed by each state. These frameworks will determine the assessment tools in each state that will control future dollar allocations and legislative mandates.

Washington State is near the compilation of its Essential Learnings in Science, which profited from the Benchmarks in final form and the National Standards in draft form. The crafters of the Essential Learnings decided to move away from the traditional discipline listing of standards used in both Benchmarks and National Standards. Instead, Unifying Concepts are used to organize the Essential Learnings. The unifying concepts, "...that organize the conceptual knowledge of the natural sciences are ideas that transcend science disciplinary boundaries, and prove fruitful in learning science in a holistic manner." The unifying concepts are:

1) Properties, Characteristics, Measurement and Scale

2) System, Order and Structure

3) Change and Interaction

4) Models

Benchmarks related to astronomy in the Washington State Essential Learnings are found under various unifying concepts:

Grades K through 4 – Properties

- Know a property of light is that it can travel in straight lines in all directions from any point on a light source until it strikes an object.
- Understand light is reflected by a mirror, refracted by a lens or absorbed by an object.

Grades 5 through 8 – Properties

- Understand how to use properties of a scientific nature to classify materials and objects (minerals, rocks, substances – metals, soil, etc.).

Grades 5 through 8 – Measurement

- Understand how to compare the relative distances, sizes and duration of large and small objects and events on the earth, and in the solar system.

Grades 9 through 12 – Measurement

- Understand how to use very large and very small numbers in terms of powers of 10 in order to think about and to compare things that vary significantly in size, distance or duration.

Grades K through 4 – Order

- Understand the relationship between seasons and weather.
- Understand light travels in straight lines from a source (unless it is reflected off a mirror, scattered by hitting other things or refracted by entering water, glass or plastic).

Grades K through 4 – Order

- Understand many aspects of the natural world behave in predictable ways such as the phases of the moon, patterns of stars, seasons and inherited characteristics of organisms.

Grades 9 through 12 – Order

- Understand the idea of order and regularity in the natural world is explained by laws such as Newton's Laws of Force and Motion, Kepler's Laws of Planetary Motion, Conservation Laws, Darwin's Laws of Natural Selection, Einstein's Special Theory of Relativity, the Theory of Plate Tectonics, Theory of

Electromagnetism, and Atomic Theory.

Grades K through 4 – Change and Interaction

- Recognize the characteristics of cyclic events such as animal and plant life cycles, phases of the moon and seasons.
- Describe the motion of the sun, moon and earth using relative position and position over time.

Grades 5 through 8 – Change and Interaction

- Understand how the inclination of the earth relative to the sun causes seasons.
- Understand the motion of the objects in the solar system and how these motions explain phenomena such as the day, year, eclipses and phases of the moon.
- Understand how gravitational, electrical and magnetic forces interact at a distance (gravitational attractive forces between masses, electrical attractive or repulsive force between unlike and like charges respectively, and magnetic attractive or repulsive forces between unlike and like magnetic poles).
- Understand the ways in which energy is transferred, including conduction, convection and radiation.
- Understand in what ways gravitational, magnetic and nuclear forces are at work in the universe.

Grades 9 through 12 – Change and Interaction

- Understand cyclic events such as seasons, length of day/night, tides, plant life cycles and phases of the moon.
- Understand how the earth, moon, sun, planets and comets interact in the solar system.
- Understand the effects of electrical, magnetic, gravitational, and nuclear forces (strong and weak) on objects in the universe.
- Understand how accelerated charges emit electromagnetic waves, including visible light, as well as longer and shorter wavelengths.

Grades K through 4 – Models

- Use a scale model and explain how it is similar to and different from the real object (map, drawing, physical scale model – examples: stuffed animal, etc.).

Grades 5 through 8 – Models

- Construct a physical model to scale
- Design a model that demonstrates the functioning of a natural phenomena or designed object e.g. airplane wing, volcano, plate tectonics, water flow, solar system, current electricity, etc.

A comparison done by Project 2061 staff of the content standards in both Project 2061 and the National Science Education Standards shows that, "Both visions of what is *essential* for *everyone* to know **exclude** a host of topics...that congest the traditional science curriculum," including Ohm's Law, series and parallel circuits, geometric optics, cloud type, ideal gas laws, simple machines and balancing chemical equations.

This comparison also states that, "Any one of these [topics] could nevertheless be studied by all students as a context in which to learn many other things that *are* specified." A close examination of the various astronomy related content standards show that a person trying to justify the teaching of a "pet" topic, can easily find a reason to teach it.

Thus, the content standards or Benchmarks – and most state standards – in themselves do not limit what an individual instructor can attempt to teach in an astronomy course. Those who have been waiting for the National Standards to identify a limited amount of content that is essential so one can implement the concept of "Less is More" will be disappointed. Instructors need to be prepared to deal with "More is More", because there are several aspects of the emerging standards and state guidelines that demand more teaching time, and will have the greatest implications on the teaching of astronomy because they will dramatically change How To Teach for many instructors.

How Astronomy Should Be Taught – Unifying Concepts.

Each of the frameworks cited in this paper emphasize the importance of using unifying concepts (sometimes called themes) as, "...ideas that transcend disciplinary boundaries and prove fruitful in explanation, in theory, in observation, and in design." (Project 2061: *Science for All Americans*);. Thus the National Science Education Standards ask schools to design their K-12 curricula so that:

As a result of activities in grades K-12, all students should develop understandings and abilities aligned with the following concepts and processes:

- Order and organization
- Evidence, models and explanation
- Constancy, change, and measurement
- Evolution and equilibrium
- Form and function

A similar set of unifying concepts that all students should know is provided in the 2061 Benchmarks, along with suggestions for how student understanding of them should build from kindergarten to 12th grade:

- Systems
- Models
- Constancy and Change
- Scale

Earlier in this document we saw that the Washington State Essential Learnings has its own set of unifying concepts":

- Properties, Characteristics, Measurement and Scale
- System, Order and Structure
- Change and Interaction
- Models

Many educators will decide to ignore the use of unifying concepts. Few curricula are organized this way, and the traditions of discipline-based teaching are hard to overcome. It will be easiest to ignore the unifying concepts in states where the State Framework is organized according to discipline, with unifying concepts as a secondary emphasis. Unifying Concepts will be harder to ignore in states like Washington where the standards are organized first by concept, and then by discipline. This will especially be true if the assessment tools developed to accompany the standards have a similar emphasis on the unifying concepts vs the discipline-based content.

Inserting the unifying concepts into the astronomy curriculum will demand time from other topics, time that some instructors will be reluctant to provide. But unifying concepts are important for fully understanding the nature of science, especially in today's world where the boundaries between disciplines are becoming fuzzier. We no longer study astronomy, biology, and physics. We study astrophysics, molecular biotechnology, and biophysics. The 2061 Benchmarks nicely sums up the reason for emphasizing unifying concepts:

> Some powerful ideas ... are not the intellectual property of any one field or discipline. Indeed, notions of systems, scale, change and constancy, and models have important applications in business and finance, education, law, government and politics, and other domains, as well as in mathematics, science, and technology. These common themes are really ways of thinking...

How Astronomy Should Be Taught – Science as Inquiry

The most dramatic changes in astronomy education will not be due to the Content Standards, or the Unifying Themes. It will be due to the emphasis placed on the Science as Inquiry Standards. The National Science Standards state:

> Science as inquiry is a basic and controlling principle in the ultimate organization of and activities in students' science education...Students at all grade levels and in every domain of science should have the opportunity to use scientific inquiry and develop the ability to think and act in ways associated with the processes of inquiry, including asking questions, planning and conducting an investigation, using appropriate tools and techniques, thinking critically and logically about the relationships between evidence and explanations, and communicating scientific arguments.

> The statement in the National Standards goes on to state:

> In the vision presented by the Standards, inquiry is a step beyond "science as a process," in which students learn skills, such as observing, inferring, and experimenting. The new vision includes the "processes of science" and requires that students combine those processes and scientific knowledge as they use scientific reasoning and critical thinking to develop their understanding of science. Engaging students in inquiry serves five essential functions:
>
> - It assists in the development of understanding of scientific concepts.
> - It helps students "know how we know" in science.
> - It develops an understanding of the nature of science.

- It develops the skills necessary to become independent inquirers about the natural world.
- It develops the dispositions to use the skills, abilities, and habits of mind associated with science.

The specific Inquiry Standards by grade level in the National Standards are:

Grades K through 4 – Understanding about scientific inquiry; abilities necessary to do scientific inquiry

- Ask a question about objects, organisms, and events in the environment.
- Plan and conduct a simple investigation.
- Employ simple equipment and tools to gather data.
- Use data to construct a reasonable explanation.
- Communicate investigations and explanations.
- Scientific investigations involve asking and answering a question and comparing the answer to what scientists already know about the world.
- Scientists use different kinds of investigations depending on the questions they are trying to answer. Types of investigation include describing objects, events, and organisms, classifying them and doing a fair test (experimenting).
- Simple instruments, like magnifiers, thermometers, and rulers, provide more information than scientists obtain using only their senses.
- Scientists develop explanations using observations (evidence) and what they already know about the world (scientific knowledge). Good explanations are based on evidence from investigation.
- Scientists make the results of their investigations public; they describe the investigations in ways that enable others to repeat the investigations.
- Scientists review and ask questions about the results of other scientists' work.

Grades 5 through 8 – Understanding about scientific inquiry; abilities necessary to do scientific inquiry

- Identify questions that can be answered through scientific investigations. Students should develop the ability to refine and refocus broad and ill-defined questions.
- Design and conduct a scientific investigation.
- Use appropriate tools and techniques to gather, analyze, and interpret data.
- Develop descriptions, explanations, predictions, and models using evidence.
- Think critically and logically to make the relationships between evidence and explanations.
- Recognize and analyze alternative explanations and predictions.
- Communicate scientific procedures and explanations.
- Different kinds of questions suggest different kinds of scientific investigations.
- Current scientific knowledge and what scientists understand guide scientific investigations.

- Technology used in data gathering enhances accuracy and allows scientists to manipulate and quantify results of investigations.
- Scientific explanations use evidence, logically consistent arguments, and propose, modify, or elaborate principles, models, and theories in science.
- Science advances through legitimate scepticism.
- Scientific investigations sometimes result in new ideas.

Grades 9 through 12 – Understanding about scientific inquiry; abilities necessary to do scientific inquiry

- Identify questions and concepts that guide scientific investigations.
- Design and conduct scientific investigations.
- Use technology to improve investigations and communications.
- Formulate and revise scientific explanations and models using logic and evidence.
- Recognize and analyze alternative explanations and models.
- Communicate and defend a scientific argument.
- Scientists usually base their investigating on existence questions or causal-functional questions.
- Scientists conduct investigations for a variety of reasons, such as exploration of new areas, discovery of new aspects of the natural world, confirmation of prior investigations, prediction of current theories, and comparison of models and theories.
- Scientists rely on technology to enhance the gathering and manipulation of data.
- Scientific explanations must adhere to criteria such as: a proposed explanation must have a logical structure; it must abide by the rules of evidence; it must be open to questions and possible modification; it must be based on historical and current scientific knowledge; and the methods and procedures that scientists used to obtain evidence must be adequately reported to enhance opportunities for further investigation.
- Results of scientific inquiry – new knowledge and methods – emerge from different types of investigations and public communication among scientists.

The 2061 Benchmarks have similar standards expressed under The Nature of Science and Habits of the Mind Chapters. The Washington State Essential Learnings dedicates an entire Essential Learning section to: Students Conduct Scientific Inquiry. The key elements in this section are:

Scientific Investigations

- Observing
- Hypothesizing
- Using Appropriate Resources
- Establishing Validity of Explanation
- Reporting
- Investigating safely
- Discovery motivates learning

Thinking Skills:

- Critical thinking

- Problem solving
- Creative thinking
- Thinking about thinking

Scientific Habits of Mind:
- Intellectual honesty
- Critical response
- Perseverance

Science Through Inquiry – Not Just a K-12 Phenomena

Many of you in this room may say this does not apply to me because I do not teach at the pre-college level. But most of you do professional development workshops for teachers. The National Standards provide guidelines for the nature of teacher professional development that makes it clear that using an inquiry approach is imperative:

> The standards are intended to inform everyone with a role in professional development. They speak to the science and education faculties of colleges and universities, who have the primary responsibility for the initial preparation of teachers of science...
>
> The current reform effort in science education requires a substantive change in how science is taught in our schools. Implicit in this reform is an equally substantive change in professional development practices at all levels, beginning with the undergraduate college experiences. Most current approaches to professional development involve traditional teaching of science content and emphasis on technical training – far short of what is needed. For example, undergraduate science courses typically communicate science as a body of facts and rules to be memorized, rather than a way of knowing about the natural world. Institutes and courses for teachers often do the same. Teacher preparation courses and inservice activities in science methods frequently emphasize technical skills rather than theory and reasoning.
>
> The view of science and how it is learned are difficult to convey to students in schools if teachers have never experienced it themselves. [Thus] All professional development, beginning with teacher preparation in the undergraduate years, [should] prepare teachers to understand and use the techniques and perspectives of inquiry.
>
> Standard A in the National Standards regarding Professional Development (Learning Science Content) states:
>
> Science learning experiences involve teachers in actively investigating scientific phenomena, interpreting results, and making personal sense of finding consistent with currently accepted scientific understanding.
>
> - Learning experiences address issues, events, problems, or topics significant in science and of interest to participants.
>
> - Teachers are introduced to scientific literature, media, and technological resources that expand their science knowledge and their ability to access further knowledge.

- Experiences build on the teacher's current science knowledge, skills, and attitudes.
- Learning experiences incorporate ongoing reflection on the process and outcomes of understanding science through inquiry.
- Teachers are (to) be encouraged and supported in efforts to collaborate.

Thus, the implication of the emerging education reform standards cannot be clearer. What needs to change is HOW we teach astronomy, not what we teach.

To help you consider effective teaching strategies that emphasize a more inquiry approach, examine the following methods for teaching about the H-R Diagram at the high school or college level. Which approach is most aligned with the emerging National Standards?

A. Show slides of H-R diagrams that plot the spectral class vs. absolute magnitude of stars in various types of star clusters. Point out the main sequence, white dwarf and red giant areas of the diagram. Explain how the age of a cluster can be determined by where the upper part of the main sequence ends.

B. Plot on the board the temperature and intrinsic brightness of the 20 apparently brightest stars in the sky, and the 20 nearest stars. As the students to help you list the differences between the two diagrams. Explain what these differences are.

C. Arrange a laboratory session where students measure their height and weight, and plot this information for the entire class. Students examine the results for general trends and conclusions. Students then plot the temperature and intrinsic brightness of 30-50 stars, and again examine the results for general trends and conclusions.

Now consider the same question for the following suggestions for introducing the study of galaxies.

A. Project a sequence of several slides of galaxies which show the characteristics of the generally accepted galaxy types (elliptical, irregular, and spiral - barred and non-barred). Point out these characteristics in each slide. Show a 21 cm map of the Milky Way and indicate how this shows the Milky Way's spiral structure.

B. Describe the characteristics of the major galaxy types: spiral (barred and non-barred), elliptical and irregular. Show examples of each of these using slides. Then go through a number of galaxy slides and state which galaxy type each is.

C. Simultaneously project nine galaxies which include a sampling of the major galaxy types. Have the students generate a list of general characteristics and classify the galaxies using these characteristics. Then show them several other galaxies and have them classify these galaxies.

D. Arrange a laboratory session in which the students examine a sample of 20 to 40 photographs from the Hubble Atlas of galaxies (the sample includes examples of all galaxy types). Students are asked to find similarities and differences in the galaxies and the classify the galaxies using a classification scheme they devise. Then give the students a number of other galaxy photographs to classify.

If you ranked them in the reverse order from how they are listed for both topics, then you understand how astronomy teaching needs to evolve. How we teach astronomy needs to change so that teaching goes beyond telling students what we know, or even how we know it. Students must experience for themselves how we know what we know.

Bibliography

1. National Science Education Standards, November 1994 Draft, National Academy Press, Washington, D.C., 1994.
2. Benchmarks for Science Literacy, AAAS Project 2061, Oxford University Press, New York, 1993.
3. Washington State Essential Learnings in Science, May 1995 Draft, Washington State Commission On Student Learning, Olympia, WA, 1995.
4. Summary Comparison of Content between November draft of National Science Education Standards and Project 2061 Benchmarks for Science Literacy, AAAS Project 2061, 1995.
5. Effective Astronomy Teaching and Student Reasoning Ability, Schatz, Fraknoi, Robbins and Smith, Lawrence Hall of Science, University of California, 1978.

Discussion

Bisard.
Do you feel there is a need for an ASP book or guide on Learning & Teaching Astronomy which is not activity-centered but cognitively and pedagogically centered?

Hill.
Follow through to answer to my question. I agree that we need to make better models (more appropriately scaled, etc.), but we need to keep in mind the notion that by its very nature astronomy extrapolates some relationships that cannot be literally diagrammed.

Zeilik.
I am concerned that all the work on national standards and state reform have very little impact on us – professors of astronomy at the university level. I fear we have fallen into a trap of "we teach astronomy, but do our students learn science?"

Schatz.
This is a key problem. College professors can feel they can ignore the essence of the new reform movement, because it applies only to the K-12 grades. But college professors are one of the greatest barriers to changes on how we teach science. They are a model for the future K-12 teaching and provide much teaching in-service. The universities also provide no incentive to teach effectively. Until there is, there will be little change.

One pressure to change the teaching technique will be for college professors requesting Teacher Enhancement funds for teaching in-service. Reviewer should only fund projects that emphasize an enquiry approach.

An outlandish idea would be to have the professional organization endorse the concepts in the National Standards, and encourage astronomers – especially graduate students – to obtain appropriate training at professional meetings.

Astronomy's Conceptual Hierarchy

Philip M. Sadler
Harvard-Smithsonian Center for Astrophysics
60 Garden Street
Cambridge, MA
USA 02138 Tel.: (617) 495-9798 E-mail: psadler@cfa.harvard.edu

I. Introduction

Why are the fundamental concepts of astronomy so difficult for our students? Many teachers of astronomy at pre-college and college levels view students as ill-prepared for their courses. Students do not understand fundamental concepts on which teachers hope to build. Students are not familiar with the motions in the heavens, the moon's phases, the earth's seasons, the nature of light, or the sizes and scale of astronomical systems. The graphical and visual representations that we use throughout our courses (spectra, light curves, H-R diagrams) appear foreign and strange. Equations and simple order-of-magnitude estimations are within the experience of our students, but are rarely useful to them or reproducible by them.

Teachers of astronomy may wonder whether our own teaching and curricula are to blame for our students' difficulties. We change demonstrations, assignments, the order of topics, and textbooks anticipating improvement in student understanding. Others hypothesize that students' prior courses in science are at fault. This frequently takes the form of a cascade of blame from professors to high school teachers to junior high teachers to elementary teachers to parents of pre-schoolers. Three explanations arise repeatedly concerning student preparation, that: students do not have adequate exposure to astronomy prior to our own courses, teachers at lower levels are ineffective because they lack astronomical knowledge, and, even if astronomical topics are taught, fundamentals are never mastered, because the pace of instruction is so rapid. Examining each of these claims individually is an instructive start to understanding the problem.

First of all, astronomy is not a sparsely taught, uncommon subject, but a popular topic throughout the pre-college curriculum. In elementary school, students first are exposed to seasons, moon phases, and the planets. In junior high school, students begin their study of the solar system and stars, usually as a part of their earth science course. Somewhere between 10%-15% of US high schools offer semester or yearlong elective courses in astronomy (Sadler 1992). Introductory college courses in astronomy are popular and often convenient ways to fulfill science requirements. Enrollments in introductory science courses in which astronomy has a major role are estimated (Figure 1) drawing upon the results of several surveys (Hoff 1982, Weiss 1987, Welch 1985). The majority of our high school and college students in astronomy have had substantial exposure to astronomy in grades 7-9. This early instruction in junior high general science and earth science courses dwarfs later enrollment in introductory astronomy courses.

Second, a belief held by many science teachers is that student learning is highly dependent upon the subject matter knowledge of their teachers. In most states, changed certification requirements, requiring a major in a subject matter field and disallowing education degrees, seem to reflect that belief. Research does not support the conclusion that such conditions make a difference. Teachers with heavy coursework and higher degrees do not either appear to have students that score higher on standardized tests or are held out as exemplars by their supervisors (Yager 1988). Nor are the difficulties that students have in astronomy the result of being taught by teachers without adequate subject matter knowledge. Recent advances in cognitive psychology show that learning is a process of *constructing* knowledge, not simply of transmission of knowledge from expert to novice. Although it can be of benefit for teachers to know the subject that they are teaching, advanced coursework does not guarantee more effective teaching. Nevertheless, while many elementary school teachers may lack courses in the physical and earth sciences, at the junior high school and high school level teachers have much more substantial backgrounds. Among teachers who teach a separate astronomy course at the high school level (roughly 15% of schools), many own telescopes and most subscribe to magazines such as *Sky & Telescope* or *Astronomy*. An amazingly high 77% consider themselves amateur astronomers (Sadler and Luzader 1988).

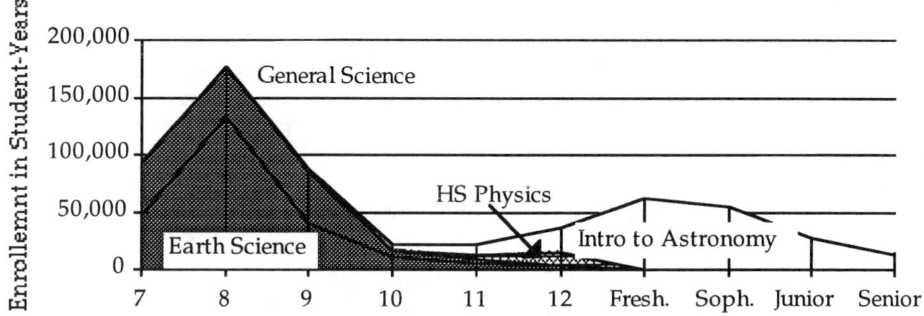

Figure 1, Enrollment in Courses Teaching Astronomy at the Introductory Level.

Thirdly, what then is the content in pre-college astronomy courses? Don't introductory courses cover the fundamentals? The answer is that the key concepts are rarely covered to the extent that promotes mastery. Students tested at the start and end of introductory astronomy courses have the same mistaken ideas about how the natural world behaves. For example, many do not believe that light propagates as particles with differing energies, that gravity is not a result of air pressure, or that the Sun is the only star in our solar system (Lightman and Sadler 1992). Astronomy is no different from other sciences in this regard. This circumvention of fundamental concepts is exacerbated in high school astronomy, where most teachers use popular college level textbooks, crammed with information from archeoastronomy to cosmology. Junior high earth science texts cover fewer of the same difficult concepts, but emphasize the field's specialized vocabulary. Elementary curricula are more selective, but often present concepts that are beyond the developmental level of most students. Rare is the fifth grader who can take the "God's-eye" view of our solar system, while living in a geocentric world. For these students, explaining seasons as resulting from "the tilt of the earth's axis" rather than the apparent position of the sun in our sky, is profoundly unproductive, since most of them do not view two-dimensional drawing as representing a three-dimensional model. Many explain artists' renditions of orbits as evidence for pictures taken from space by astronauts (Touger 1985).

Courses at every level appear to be collections of interesting topics with associated vocabulary. There is precious little thought put into how students structure what they learn. Historical (from Hipparcus to modern astronomy) or structural perspectives (progressing from the earth outward) appear logical and helpful to us, as astronomical cognoscenti. There is little evidence that students find these ways of organizing knowledge beneficial. A case in point is the unpopularity of the "new math" in the 1960's. Although the set-theoretic foundations of mathematics are useful to professional mathematicians, they appear to be a disastrous way of organizing mathematics for youngsters. What, then, is a productive way to structure an introductory astronomy course to optimize student learning? Which topics are prerequisites for others or can they all be taught without regard to order? Is there a natural progression in which most students come to understand astronomy or is it haphazard, idiosyncratic, and unpredictable? These are the questions which I and many of my colleagues have set for ourselves and which are discussed in this paper.

II. Studies of children's ideas

The educational psychologist, David Ausubel, was first to recognize the importance of students' prior knowledge. He posited "the unlearning of preconceptions might very well prove to be the most determinative single factor in the acquisition and retention of subject-matter knowledge (Ausubel 1978)." Ausubel makes a clear distinction between "meaningful" and "rote" learning. Meaningful learning denotes the incorporation of new concepts and facts into a student's scheme of the world that results in the restructuring of the student's knowledge, while rote learning is a process in which a student's existing scheme is unaffected. New information is never fully processed and connected to prior knowledge, but stored in isolation (and promptly forgotten when the semester is over).

Ausubel's ideas are at the root of much of the research on children's understanding of science. Probes of children's ideas have taken the form both of student interviews and written instruments. Each has been fruitful. Open ended interviews have served to expose domains that are fertile for more extensive investigation and have revealed the presence, but not the relative popularity, of a multitude of student ideas. These ideas are often quite fantastic and often seen as humorous by adults. As an example, I recently overheard a wonderful question directed by a six year old to a seven year old after noticing the moon in the daytime sky, "Is the sun ever out at night?" A lively discussion ensued. It was clear that this youngster had an alternative view of the reason for day and night which did not involve the sun.

Written instruments have aided in examining the beliefs of larger populations and have helped to establish which ideas appear to be shared by many students and those that are more idiosyncratic. Often these individual ideas are part of a much larger and structured "alternative framework (Driver and Easley 1978)." A good example is the Aristotelian framework, refuted by Newton, in which objects come to rest unless a constant force is applied (Caramaza, McCloskey and Green 1981, Brown and Clement 1986, Clement 1986). In astronomy, a common alternative framework is one in which the distance between objects is roughly the same as the size of the objects. For these students, lunar eclipses are moon phases, stars are crowded into our solar system and our spaceships can easily visit other galaxies. This view provides no context for the elaborate methods of the astronomer used to establish sizes and distances. Why use spectroscopy to measure composition if we can travel to and sample stars?

Several studies with application to the learning of astronomy have been carried out. They deal with a range of topics. It is not the purpose of this paper to review them, but the reader may wish to consult them for in depth treatments of student difficulties in each of these areas:

- cosmography (the study of the earth in space) (Nussbaum 1979, Sneider and Pulos, Vosniadou and Brewer 1987)

- gravity (Gunstone and White 1981, Ogar 1986)
- the solar system (Klein 1982, Treagust and Smith 1986, Touger 1985)
- Moon phases (Dai 1990, Cohen and Kagan 1979,
- cosmology (Lightman, Miller, and Leadbeater 1987)
- light (Andersen and Karrqvist 1983, Anderson and Smith 1983, Klein 1982)
- earth science (Schoon 1988)

The author's own experience interviewing students about their astronomical ideas began as the initial phase in the development of Project STAR, a high school astronomy curriculum funded by the National Science Foundation. Students were found to have very well developed beliefs about the reasons for day and night, the cause of the seasons, and the moon's phases that were at odds with those of their teachers (Sadler 1987). Student ideas were documented in the award-winning video *A Private Universe* (Schneps and Sadler 1988 [Available from the ASP]). Later interviews were carried out by the author, the staff of Project STAR, and teachers consulting to Project STAR. I am particularly indebted to Dr. Anne Young of the Rochester Institute of Technology, Dr. Linda Shore, now of the Exploratorium, and the dynamic team of Jenny and Paul Hickman, now of Belmont High School and Boston University Academy, for their work in this area.

Multiple choice tests can be constructed using the results of interviews and open-ended tests. In these instruments, items are constructed which capture the most prevalent ideas of students along with a scientifically correct answer (Freyberg and Osborne 1985). These tests appear quite normal to teachers until they let their students take them. Since popular notions are included, students inevitably find them more attractive than the scientifically correct answers, even after extensive instruction. This has commonly resulted in two disparate interpretations by teachers. Either they begin to question what their students actually understand or they attack the construction of the item. The latter always proves unproductive since interviews of students produce nearly identical results. These multiple-choice tests can then be given to large populations of students and scored by computer. The target population for the study which I will discuss drew students from grades eight through twelve who were in earth science or astronomy courses. The 1,200 subjects were the students of 22 teachers and took the test at both the start and end of their courses. The full description of the generation and validation of the test have been thoroughly examined in an earlier publication (Sadler 1992).

III. Excursion into Psychometrics

Much has been written about the use of tests as a way to measure student knowledge. A construct that is useful for our analysis is that of *difficulty*. This does not refer to a teacher's view of how hard or easy a question is, but the assignment of a numeric value to a particular test item. For a multiple choice item, item difficulty is often described as the fraction of students who chose the correct answer for a particular item within the test population. Items can then be compared based on these values. For example, the following two items were constructed from interviews with children. In a population of high school students taking astronomy or earth science, .66 of the students answered Item 1 correctly:

1. What causes night and day?

A. The Earth spins on its axis.
B. The Earth moves around the Sun.
C. Clouds block out the Sun's light.
D. The Earth moves into and out of the Sun's shadow.
E. The Sun goes around the Earth.

Item 34 was included on the same test. Only .28 of the students answered it correctly.

Figure 2, The Big Dipper Illustration from Item 34

34. The Big Dipper would have a noticeably different shape to the unaided eye:

A. if viewed from another star.
B. if viewed from Pluto.
C. if you looked at it a year from now.
D. if you viewed it from China.
E. never, it would always look the same.

Of course, the difficulty of a particular test item depends on the population taking the test. One would expect all graduate students in astronomy to get Item 1 right (as indeed Harvard students do), while few kindergartners will select the correct answer. Item 34 might prove more difficult for this population.

How can a test made up of misconception questions be used to help teachers understand how their students come to understand astronomical concepts? The key is to use the variation in a large population of students to discover the different stages of understanding. This requires a way to measure item difficulty that is independent of the level of the test population, so that novices can be compared to experts. Over 40 years ago, Lord (1952) introduced the idea of an item response model as a way to measure characteristics of test items that is independent of the examinee group. In his model, *item characteristic curves* are generated which illustrate the relationship between an underlying "latent trait" or ability and the probability of correctly answering the item, P(ability). By comparing student performance on the test as a whole (using a non-linear transformation of students' total test scores) to student performance on individual items (the fraction of students correctly answering a particular item), one can measure an item's intrinsic difficulty.

The model can be expressed as the probability of getting an item correct as a function of the ability of the student. Student ability is measured in units of standard deviation from the mean ability in the idealized, normally distributed population. More recently a model was developed that estimates item difficulty for both correct and incorrect answers (Thissen and Steinberg, 1984). This technique is extremely useful in analyzing misconception-based multiple-choice tests, where the "wrong" answers are of equal interest as those that are scientifically correct. Using such models, same item difficulties are generated from testing quite different populations. Many test populations can also be added to the dataset to seamlessly increase the statistical power of the model. In this way, it is possible to stretch our students out into a spectrum of ability and begin to understand the variation in comprehension of science concepts.

Let us take as an example the relatively easy question above (Item 1) concerning day and night. A graph of the probability of students selecting each answer as a function of their ability is shown below.

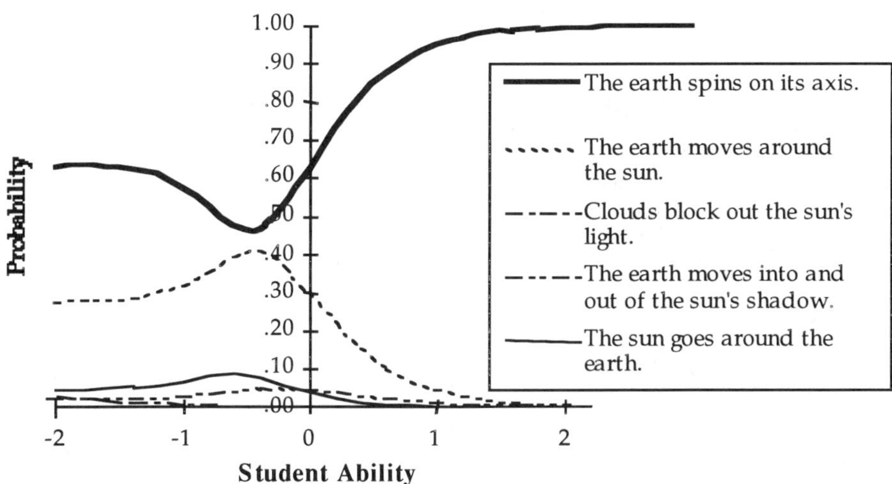

Figure 3, Item 1. What causes night and day?

This plot shows that students of the lowest ability (SD = -22) answer this question correctly about .65 of the time, while students of the highest ability answer the question correctly all of the time. Surprisingly, students of moderate ability (SD = -50) have only a 47% probability of answering this question correctly. Students of low ability (low overall test scores) appear to be at some advantage compared to students of average ability, who are more likely to believe that day and night are caused by the earth's revolution about the sun. This analysis shows that the "fact" that the earth's rotation makes day and night is not a fact at all. For many students, the idea is in conflict with the earth's orbit about the sun. It is the integration of such "facts" with the world that is misunderstood, not the existence of these facts by themselves. When called upon to predict student responses to this question, teachers inevitably choose that many students think that the sun goes around the earth. Although this seems reasonable given the historical development of the field, few students in grades 8-12 believe it, as shown in figure 3.

The problem of integrating specialized scientific information into children's frameworks is also evidenced in the results of the National Assessment of Educational Progress (figure 4). While students at all ages appear to know everyday science facts (e.g. the earth orbits the sun), their ability to integrate information into their world view is meager. It is not the learning of scientific facts and principles, per se, that is difficult, but coming to terms with their implications in all their variety. That is the challenge in learning and teaching science.

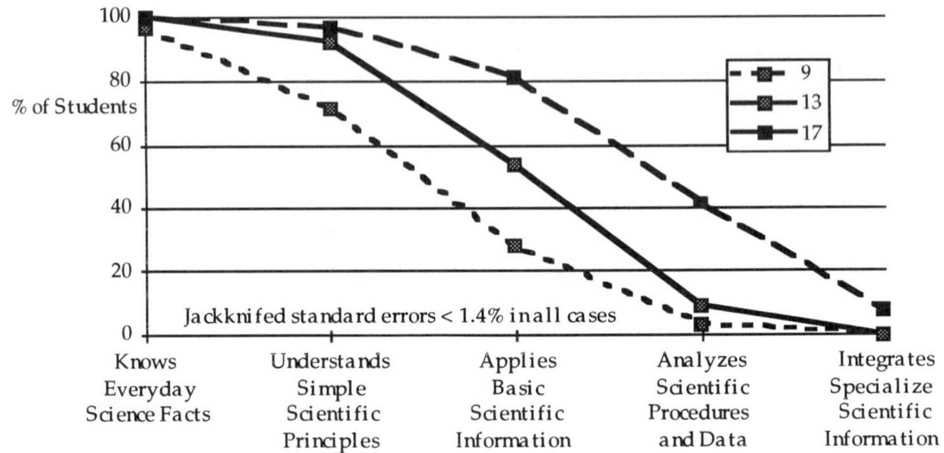

Figure 4, Proficiency in Science 1986 National Assessment of Educational Progress.

One can go beyond a straightforward interpretation of the item response graph of Item 1 (figure 3) simply as a static model to a description of how students progress though their learning of this concept. Although these graphs are drawn from large populations, students can be thought of as moving from the left to the right of the ability axis as they come to master the subject. This means that as student knowledge climbs as a result of taking a course in astronomy, their ability to answer a question correctly may fall for a period. Many researchers have found this to be the case at all academic levels. The reason for this surprising phenomenon is that instruction often reinforces and strengthens student misconceptions. In figure 3, a corresponding increase in student belief that it is the earth moving about the sun that is cause of day and night is reflected in a decline in the scientific explanation.

One can compare the relative difficulty of items by finding the student ability that corresponds to a probability of .50 of answering the question correctly. Even though there are two values that meet this criterion for some items, we will choose the one with the positive slope, so $P_{item\ 1}(.50)=-.30$. Let us take as an example a more difficult question, that of astronomical scale (figure 5). Several items on this test examine student understanding in this area. The most revealing has to do with the shape of a constellation when viewed from another star.

In this case, students of low ability answer the question correctly 25% of the time and the convergence of most of the curves to the left is indicative of student guessing at this level. Students of high ability answer correctly over 90% of the time. However, again students of moderate ability do worse on this problem, with only about 12% answering correctly at the -.5 level. This corresponds to an increase in the belief that the pattern of stars that we see in the sky is invariant and remains the same from any viewpoint. This item has a much higher level of difficulty, $P_{item\ 4}(.50)=.90$.

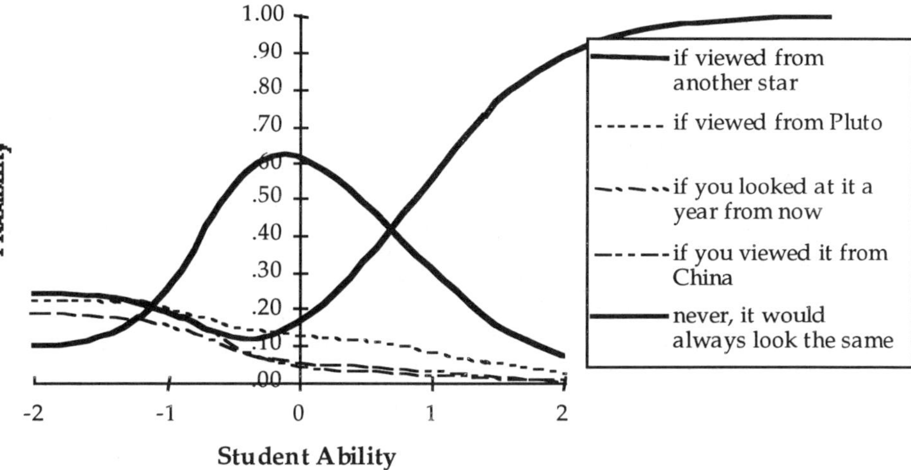

Figure 5, Item 34. The big dipper would have a noticably different shape to the unaided eye.

Looking at the test as a whole, one can measure the progress of students in units of standard deviations from the mean. In the test population, the average student had a pretest ability of 0.2 SD and a posttest ability of 0.4 SD in conventional classrooms, for a gain of 0.2 SD. This is in contrast to average teacher predictions of a 1.2 SD increase in performance. In experimental classrooms using Project STAR materials, students started at the same level of ability which increases to 0.8 SD, a gain of 0.6 SD. A future paper will examine this experiment in more detail.

Using this technique of calculating item response curves, one can establish the relative difficulty of many astronomical concepts. The majority of the 47 items in the Project STAR Misconception Instrument show similar profiles to those above. They are characterized by:

- A decrease in performance for moderate level students compared to low level students. Students who are engaged in the process of mastering a concept appear to be at a disadvantage to those who are clueless.
- An increase in belief in misconceptions before their eventual decline. Much teaching appears to reinforce students' prior knowledge even though the teacher's explanation may be correct.
- Many questions have difficulties higher than the average student ability. The concepts that teachers select to teach appear too difficult for many students. Ideas that we think students have mastered prior to entry to our classes may not really be understood to the extent that they can be applied.
- Teachers' overestimation of student gains. Students do not reach the levels of performance that their teachers predict on tests constructed by others. It is only on tests that teachers create for their own students that they appear to perform up to expectation.
- Mastery of basic concepts takes time and effort. It often entails a period of confusion. "Coverage" without an extended opportunity to dispel misconceptions and build true understanding appears to be detrimental to student learning.

IV. Hierarchy of Concepts

With the examination of student misconceptions completed let us turn to the surprisingly low student gains on what we as teachers consider relatively easy questions. Slow student progress argues for a reduced conceptual scope of our courses. An attempt to cover too much content appears to leave many students with reinforced misconceptions and a decreased ability to answer many astronomical questions. If we reduce scope, how do we choose which ideas should be covered? Should we only teach the easiest concepts, "dumbing down" our courses? Should we have high expectations for our students and teach the more difficult concepts, hoping they'll straighten out their misconceptions in the process? There is another approach that should be considered, i.e. choosing concepts on the basis of their structural relationship to learning other concepts. In this way, concepts that are required for understanding more difficult ideas should be taught first. When mastery of these ideas is achieved, one can move to harder topics that depend on these prerequisite ideas.

How then can we identify these prerequisites? Measuring item difficulty is not much of a help. Mastery of some randomly chosen easy concept will probably have no influence on learning a difficult one. Yet, within a population of students it is possible to determine which concepts appear to be prerequisite for others. The tool for establishing these relationships is the contingency table. As an example, let us consider the breakdown of women and men in two hospital wards. There are 100 patients, 60 in Ward A and 40 in Ward B. Seventy are male and 30 are females. Using these numbers alone, without knowing who is in which ward, we can calculate the number of females and males in each ward by assuming gender does not affect the probabilities of being in either ward. This is accomplished by treating the marginal totals as probabilities and simply multiplying them together (and dividing by the total) to fill in the interior cells of the table. These are the *expected values* (in italics within table 1A).

Table 1A, Expected Values

	Ward A	Ward B	Total
Male	*12*	*18*	30
Female	*28*	*42*	70
Total	40	60	100

Table 1B, Observed Values

	Ward A	Ward B	Total
Male	*30*	*0*	30
Female	*10*	*60*	70
Total	40	60	100

If we later carry out a census of the hospital patients, we can find the gender of patients in our wards. These are the *observed values* (italicized and underlined within table 1B). If the expected and observed values are the same, we can safely assume that the probability of being in a ward is independent of gender (we may wish to use a statistical test to be even more certain). However, if the distribution is quite different from expected, a prerequisite relationship may exist. Consider our hospital census result. There are no men in Ward B. The zero in one cell (Male, B) and nowhere else indicates a probable prerequisite relationship between being female and entering Ward B (for example, if it is a maternity ward). Note that what is outlined is not necessarily a correlation (in which diagonal cells in one direction are small compared to the other diagonal). Statistical measures would allow us to calculate the probability of the actual value being significantly different from the expected value with a chi-square test.

This technique was developed and applied to logical and mathematical knowledge by Airasian and Bart in 1973. In the same fashion, we can compare expected with observed values for the answers to two different test items which examine students' conceptual understanding. A cell with a zero value would indicate that one concept may be a prerequisite for another. Of course, the multiple choice format would change the minimum value of the key cell if the items are not similar in difficulty. Fisher's exact test is useful for calculating the probabilities in this case (Bart and Read 1984). By comparing every question on the test with every other, one can establish which are the statistically significant prerequisite relationships in the population of students. This means that 47x47 contingency tables must be tested and evaluated, a job that is straightforward, although time consuming, on a microcomputer.

An example is given in tables 2A and 2B for the contingency table relating the cause of night and day (Item 1) and the knowledge of how long it takes the earth to turn on its axis (Item 21). One can use these numbers to settle the question which comes first, knowing the reason for day and night, or knowing that the earth spins in 24 hours? There are only 30 students in the population (1%) who know the earth turns in 24 hours who do not know the cause of day and night. There are 406 who know the cause of day and night who do not know that the earth turns in 24 hours. There are many who both know or do not know both. One can argue that knowing the cause of day and night is a prerequisite for knowing that the earth spins in 24 hours.

Table 2A, Expected for Items 1 and 21

	Item 21		
Item 1	Incorrect	Correct	
Incorrect	*151*	*272*	423
Correct	*648*	*1165*	1813
	799	1437	2236

Table 2A, Observed for Items 1 and 21

	Item 21		
Item 1	Incorrect	Correct	
Incorrect	*393*	*30*	423
Correct	*406*	*1407*	1813
	799	1437	2236

Even though all of the concepts covered on the Project STAR test reveal student misconceptions and can be placed on a continuum of difficulty, a smaller set are actually related to each other in a prerequisite fashion. Below is a current analysis of items that pass muster (figure 6). This diagram can be interpreted on the following basis. The difficulty of concepts increases vertically. Items that are pointed to by arrows have prerequisites in that there are very few students who hold this concept who do not already understand the prerequisite concept. Yet, students may have prerequisite knowledge and not have the more difficult concept.

This diagram shows that certain elementary knowledge appears prerequisite for more difficult concepts. The understanding of day and night and the earth's yearly revolution about the sun appears to be key to mastery of the sun's motion in the sky, an understanding of seasons, and many other concepts. More importantly, this chart represents the great difficulty students have mastering advanced concepts without processing the prerequisite knowledge. This type of analysis suggests that understanding of science may be constructed much like astronomy's cosmic distance scale; accurate measures of more and more distant objects are dependent on the ways in which we measure closer objects. It may be impossible for students to acquire powerful scientific ideas without great attention to the basics.

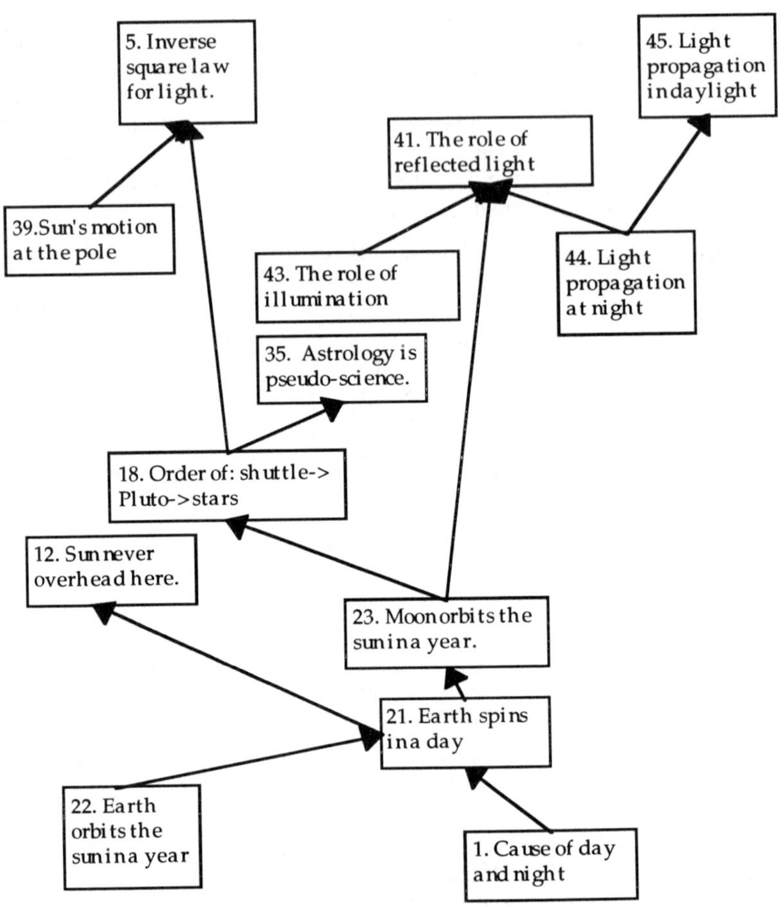

Figure 6, Prerequisite Relationships Between Astronomical Concepts.

V. Conclusion

Student difficulty in introductory astronomy courses has many causes. The interviews and formal testing carried out by Project STAR has revealed two major difficulties. Students hold preconceptions when they enter their courses and learning many astronomical concepts is dependent on mastery of easier concepts.

Students' ideas are quite different from those of their teachers, in spite of the fact that their teachers had to learn these ideas at one time. Many of these ideas are well-developed misconceptions that have been constructed by students from their own experiences and thoughts. These ideas are very resistant to change, but once changed are almost impossible to remember. Teachers rarely can recall their prior, non-scientific conceptual frameworks and tend to teach without attending to their students' prior knowledge. Several studies have shown that these misconceptions can be changed, but conventional courses do little. In many cases, teaching simply strengthens student misconceptions. The testing of their own ideas by students appears to be a key step in the changing of misconceptions.

Powerful ideas in science are often hierarchical in nature; they build upon one another. Students who do not have the foundation stones firmly in place will find it very difficult to construct an understanding of astronomy. It is important that teachers realize the difference between exposure to an idea and the mastery of it. While many students know that the earth orbits the sun, fewer know how it relates to cycles here on earth.

This research was supported by grants from the National Science Foundation and the Smithsonian Institution.

References

Airasian, Peter W. and William M. Bart. Ordering Theory: A New and Useful Measurement Model, Educational Technology, May 1993:56-60.

Anderson, B. and C. Karrqvist. "How Swedish pupils, aged 12-15 years, understand light and its properties." *European Journal of Science Education* 5 (4 1983):387-402.

Anderson, C. W. and E. L. Smith. "Teacher behavior associated with conceptual learning in science." Paper presented at *American Educational Research Association* in Montreal, 1983.

Ausubel, D. P., J. D. Novak and H. Hanesian. *Educational Psychology: A Cognitive View*. New York: Holt, Rinehart and Winston, 1978.

Bart, William M. and Sherry A. Read. A Statistical Test for Prerequisite Relations, Educational and Psychological Measurement 1984:44, 223-227.

Bouwens, Robert E. A. "Misconceptions among pupils regarding geometrical optics." In *GIREP - Cosmos* - an *Educational Challenge* in Copenhagen, European Space Agency, 369-370, 1986.

Brown, David E. and John Clement. "Misconceptions concerning Newton's law of action and reaction: The underestimated importance of the third Law." In *GIREP - Cosmos* - an *Educational Challenge* in Copenhagen, European Space Agency, 1986, pp.39-53.

Caramazza, A., M. McCloskey and B. Green. "Naive beliefs in sophisticated subjects: Misconceptions about trajectories of objects." Cognition 9 (1981):117.

Clement, John. "Overcoming students' misconceptions in physics: The role of anchoring intuitions and analogical validity." In *GIREP - Cosmos* - an *Educational Challenge* in Copenhagen, ed. by J. Hunt. European Space Agency, 1986, pp.84-97.

Cohen, Michael R. and Martin H. Kagan. "Where does the old Moon go?" *The Science Teacher* 46 (1979):22-23.

Dai, Meme F. "Misconceptions about the Moon held by fifth and sixth graders in Taiwan." National Science Teachers Association, 1990.

Driver, R. and J. Easley. "Pupils and paradigms: A review of literature related to concept development in adolescent science students." *Studies in Science Education* 5 (1978):61-84.

Freyberg, Peter and Roger Osborne. "Constructing a survey of alternative views." In *Learning in Science, The Implication of Childrens' Science*, ed. by Roger J. Osborne and Peter Freyberg. Auckland, New Zealand: Heineman, 1985, pp.166-167.

Gunstone, Richard F. and Richard T. White. "Understanding of gravity." *Science Education* 65 (3 1981):291-299.

Hoff, D. "Astronomy for the non-science student – A status report." *The Physics Teacher* March (1982):175.

Klein, Carol A. "Children's concepts of the Earth and the Sun: A cross cultural study." *Science Education* 65 (1 1982):95-107.

Lightman, Alan P., Jon D. Miller and B. J. Leadbeater. "Contemporary cosmological beliefs." In *Misconceptions and Educational Strategies in Science and Mathematics*, ed. by Joseph Novak. Ithaca, NY: Cornell University Press, 1987, pp.309-321.

Lightman, Alan and Philip Sadler. "Teacher Predictions versus Actual Student Gains." *The Physics Teacher*. College Park, MD: American Association of Physics Teachers, 31:3, 1993, pp.162-167.

Lord, F. M. "A Theory of Test Scores", Psychometric Monograph No. 7, Psychometric Society, 1952.

Nussbaum, Joseph. "Childrens' conception of the Earth as a cosmic body: a cross age study." *Science Education* 63 (1 1979):83-93.

Ogar, J. "Ideas about physical phenomena in spaceships among students and pupils." In *GIREP - Cosmos* - an *Educational Challenge* in Copenhagen, European Space Agency, 1986, pp.375-378.

Sadler, Philip M. and William M. Luzader. "The Teaching of Astronomy." In *International Astronomical Union, Colloquium 105* in Williams College, Williamstown, MA, ed. by Jay M. Pasachoff and John R. Percy. New York: Cambridge University Press, 1988, 257-276.

Sadler, Philip M. "High School Astronomy: Characteristics and Student Learning," *Proceedings of the Workshop on Hands-on Astronomy for Education*, E. Carlton Pennypacker, Ed., River Edge, NJ: World Scientific, 1992

Sadler, Philip M. "Misconceptions in Astronomy." In 2nd International Seminar on *Misconception and Educational Strategies in Science and Mathematics* in Ithaca, NY, ed. by Joseph D. Novak. Ithaca, NY: Cornell University Press, 1987, pp.422-425.

Sadler, Philip M. "The Initial Knowledge State of High School Astronomy Students." Ed.D. dissertation, Harvard University, 1992.

Schneps, M and Philip M. Sadler. "A Private Universe." Pyramid Films, 1988.

Schoon, Kenneth J. "Misconceptions in Earth and Space Sciences, A Cross-Age Study." Ph.D. dissertation, Loyola University, 1988.

Sneider, Cary and S. Pulos. "Childrens' cosmographies: understanding the Earth's shape and gravity." *Science Education* 67 (2 1983): 205-222.

Touger, J. S. "Students' conceptions about planetary motion." Paper presented at *American Association of Physics Teachers* in 1985.

Treagust, David F. and Clifton L. Smith. "Secondary students understanding of the solar system: implications for curriculum revision." In *GIREP - Cosmos* - an *Educational Challenge* in Copenhagen, ed. by J. Hunt. European Space Agency, 1986, pp.363-369.

Vosniadou, Stella and William F. Brewer. "Theories of knowledge restructuring in development." *Review of Educational Research* 57 (1 1987):51-67.

Weiss, Iris R. *Report of the 1985-86 National Survey of Science and Mathematics Education*. Research Triangle Institute, 1987.

Welch, W. W., L. J. Harris and R. E. Anderson. "How many are enrolled in science?" *The Science Teacher* 51 (9):1984.

Yager, R. E. (1988). "Features Which Separate Least Effective From Most Effective Science Teachers." *Journal of Research in Science Teaching*, 41, 210-222.

Discussion

Bisard.

The progression of a student through a set of frameworks from their "initial" ideas to those accepted as "correct" takes several steps. This might be called "Learning." What do you think of this?

Eckroth.

[after talk by Philip Sadler and several other later speakers – below not delivered orally, but written out later]

There are many, widespread misconceptions that are difficult to eradicate – but not impossible. None of us at this meeting, for example, still believes that the seasons are caused by a changing distance to the sun – though we probably did in the lower grades. How did we come to give up that misconception? In the same way that some students did in *A Private Universe* – by confronting contradictions. Probably in this example, by hearing that travellers to the southern hemisphere find the seasons are reversed there – a clear conflict with the erroneous "explanation." There are many easy observations that clearly conflict with common misconceptions. We should know of them and use them, and disseminate them to teachers at all levels. [Also, nobody in Albuquerque could ever measure the sun at 90° overhead!]

Fraknoi.

Based on these research findings, if you were not merely dean but czar of a school of education, how would you change the way they train teachers to teach science better?

Sadler.

Your hypothetical is quite a fantasy. No one really has such power at the university. But, what future teachers may need most are courses in which they learn science in the same way that we hope they will teach our children. After all, we naturally all teach in the way we've been taught. Lecture/recitation courses at the college level tend to beget themselves at the pre-college level. So, teaching without lecturing, with lots of activities and observations, and with attention to students' prior conceptions would be of great benefit to future teachers.

Pasachoff.

Some of the computer simulations – diskette or CD-ROM – can give students views of solar-system objects from a variety of different aspects, which could help students visualize the reasons for day/night, phases, seasons, and other positional and projective ideas. For example, a student using the *Red Shift* CD-ROM could look at the Earth from above the pole, from Mars, from the Sun, or from elsewhere and make the Earth (with continents visible) rotate to display the day/night cycle. This individual empowerment has the potential, if properly used, to change day/night, phases, and seasons from their unexpectly abstract nature to concrete objects in their virtual but apparently actual positions and thus break the logjam in student understanding of these topics.

I have long been struck by your wonderful work on the difficulty that students have with phases and seasons, and have shown *A Private Universe* to college classes and remediated their deficiencies. Jeanne Bishop is also talking (this meeting) about difficulties students have with "projective" concepts.

Let me ask why it is that we must teach phases and seasons in the K-12 years. Sure, it is common if not universal to do so. But is it just habit? Why is it important that students of any age understand the causes of phases and seasons? At the youngest ages, it may simply be too early in their intellectual development to hope for widespread understanding. Maybe we should be teaching fundamental ideas about modern science – such as the idea that telescopes help us see more clearly, and that telescopes in space can help us still further; or the idea that stars are born, live an ordinary life for a long time, and then die and become one of several end-products – in the early grades instead of phases and seasons. We can get to phases or seasons later, or not at all.

I know that the usual argument for teaching phases and seasons is that these are everyday phenomena. But divers don't have to know about angular momentum to dive into swimming pools, and cats don't have to know these laws of physics to land on their feet when jumping from great heights. We can deal with and appreciate the phases of the moon and the seasons without necessarily understanding the details. And if we are dealing with the moon, I would much rather have students comprehend, as they can at any age, that there are many moons of planets in the solar system that are bigger than our own moon, and that we can see surface features like craters or mountains on many of them by going close to them with spacecraft. Further, I like them to appreciate that ongoing scientific research is discovering more of these moons out in space and showing us drastically new things about the properties of the ones we know. These are also basic concepts and not just vocabulary. And these are concepts that bring students the idea that science is interesting and exciting to study, for new things are being found.

I would be interested in setting up a discussion on the fundamental matter as to whether these are indeed important matters to teach students in the elementary grades, worth the time we commonly take to overcome difficulties caused by the phase of their intellectual development.

Zeilik.

I have a research result that should make you happy. In my introductory astronomy course for non-science majors, I use Project STAR-like questions on pre- and post-tests. The gains are enormous: about 3 standard deviations of gain with an effect size of about 2!

Sadler.

It is wonderful to hear of successes like yours. No doubt, it is the result of your emphasis on both cooperative learning and explicit learning objectives. I think it is very valuable to use others' tests in assessing one's own students. Students can quickly learn to answer the type of questions that the teacher normally asks in class. By using others' test items, especially those that measure changes in misconceptions or apply astronomical concepts to terrestrial situations, we better serve our students. I am happy to send a report on such tests to anyone who contacts me.

Hands-On Astronomy for Education

Carl Pennypacker
Lawrence Berkeley Lab and Space Sciences Lab
University of California
Berkeley, CA
USA 94720

and

Jodi Asbell-Clarke
TERC

Abstract We review some of the key features of hands-on astronomy as applied in secondary and college classrooms. This field has experienced rapid growth, diversification, and numerous successes. We attempt to identify common features of successful programs, review hands-on activities in a few specific programs, identify obstacles to success, and finally as an example, review recent successes of the Hands-On Universe astronomical exploration program.

I. Introduction

The field of hands-on astronomy has experienced rapid growth in diverse areas. Many hands-on programs are succeeding by a number of metrics. These metrics of success include clear evidence that hands-on astronomy:

- engages students in learning,
- encourages students to consider careers in science and technology,
- helps students learn math and science by enabling them to see the relevance of math and science in genuine settings,
- helps them learn key computational skills of the 21st century.

A key feature of most of the hands-on activities is that learning is not conducted in a lecture/recitation session, but is largely learner-centered, involving computer activities, kinesthetics, simulations, scale model building, and use of astronomical instruments. Many of these programs are acknowledged by students to be fun and engaging; many are constructivist; some emulate and form partnerships with basic researchers; and most take important steps to meet national educational goals and standards. Other important characteristics of the hands-on successes are:

- the departure from memorization of facts,
- a slowing of the pace to avoid superficial coverage of a large number of topics,
- honouring students' naive models and model building,
- encouraging students to work in collaborative settings,
- integration of physics, math, astronomy, biology, creative arts, language arts, and history.

Although astronomy per se is not offered as a separate course in most schools, many do use astronomy in some of their curriculum. For example, Hands-On Universe (HOU) has been used in physics classes, earth science classes, principles of technology classes, after school programs, and dedicated astronomy courses. The road to success is often tenaciously navigated, and in the following sections, we outline some of the reasons we believe astronomy is in a special position with respect to other sciences in its power, its breadth, and its appeal. Appropriate hands-on astronomy, if well thought-out, delivered and supported, is a golden opportunity for teachers, students, researchers, and eventually our entire nation. But the hands-on paradigm and imprimatur is not enough--appropriate use of this powerful paradigm is key to long term sustainability.

II. Crisis in Secondary Education

Since the era of Sputnik, there has never been higher concern with the apparent degradation of secondary science education as there is right now. This includes attacks on traditional pedagogy, the instructional media and the materials, and most importantly, astonishment at the disappointing student outcomes. Studies of students' engagement in normal lectures shows that in some cases, less than 1% of teacher talk elicits any thoughtful response on the part of the students, yet teacher talk fills 70% of many classrooms' time. In addition, 9 out of 10 use textbooks alone 95% of the time. Our high schools, most apparently patterned after the college lecture model, are not working. Unfortunately, as Robert Tinker of TERC has pointed out, "Bad education is very cost-effective."

III. The Power of Hands-On

Hands-on astronomy strives to engage students in activities that speak to the condition of the adolescent learner. Teenagers and children are generally fascinated by the cosmos and issues of the origin and fate of the earth and the universe. They enjoy learning appropriately delivered physical laws. The way hands-on programs enable the students to learn is by letting them take control of their own computer, their own telescope, their own simulation, their own data, or their own model and allowing them to control their environment and manipulate it to discern its inner structure. Chicago educator Vivian Hoette has noted: "Teenagers are concerned about two things: power, and how things look". Hands-on programs addresses both of these needs.

Another key strength of the hands-on philosophy is its departure from memorization of facts. These activities usually strive to teach the processes of genuine science, not demanding rote regurgitation of alien scripts. For example one common activity found in a number of successful hands-on curricula uses the motions of Jupiter's moons to teach projected motion, graphing, Newtons's laws, and other mathematical and physical principles.

Hands-on activities change the role of the teacher from that of "sage on the stage" to "guide on the side". In many cases, students and teachers will not know the answers to questions that arise in examining data. Could any scientist look a child straight in the eye and honestly elaborate on why some galaxies form as spirals and others as ellipticals? Our current wealth of ignorance concerning virtually every topic in astrophysics makes the subject matter alive, vivid, and current. These are truly exciting times in astrophysics. Our knowledge is growing at an incredible rate, but virtually every recently discovered truth is fertilizer for the growth of our ignorance.

IV. Pitfalls of "Hands-On"

Hands-on curriculum by itself is not enough. For example, earlier high-powered curricula were laden with"cook-book" style hands-on activities, where students followed carefully researched and piloted recipes to achieve the "right answer." These activities gave disappointing student outcomes. Students might as well have "dry-labbed" these activities. A key for the student is the sense of newness and learner-centered exploration.

Another pitfall is assuming that simply supplying some number of hands-on activities into a random classroom will yield good results. Many thousands of CD-ROMS and computers are sitting in school warehouses, unused and useless. Why did these attempts fail? We believe that one of the main causes is failure to take into account the ecology of science education: reform consists of understanding the situation of the teachers, the classroom environment, the students' cognitive development, the technology available in the classroom, the teacher training available, the districts and school constraints, etc. CD-ROMs tossed into the classroom as solutions to reform will simply create disappointment, and make teachers and students consider themselves failures. Teachers have to feel very comfortable with the material; they have to have a context where it will work; the kids have to respond to it rapidly – basically a finely-tuned set of parameters has to be adjusted before we will see the kinds of changes we desire.

V. Some Typical Hands-On Activities

A number of successful hands-on programs are thriving in schools across America. Typical activities include (see the list by Fraknoi, in this volume, for a complete list of astronomy programs/activities):

- building and using telescopes and binoculars,
- measuring photographs,
- building and using spectrographs,
- observing sun spots and the sun's rotation and motion,
- building celestial spheres and astrolabes,
- measuring the sun's temperature,
- measuring angles and positions of solar system and other astronomical objects (for example, the moons of Jupiter),
- studying planets from photographs,
- measuring variable stars and quasars
- discovering and measuring supernovae.

Organized hands-on astronomy curricula such as Project STAR (Center for Astrophysics/NSF)and GEMS (Lawrence Hall of Science) allow students to work with a number of media to understand the cosmos and fundamental math and science. Project AASTRA will train teachers to use existing materials, like STAR and GEMS, in their classroom. GEMS has an exciting activity based on Jupiter's moons for grades 4-9. Students working in teams of 4 to 8 students, measure the positions of one moon of Jupiter from slides. Students learn how to estimate the periods of the orbit, discuss trends in their data (such as period and orbit size), and hypothesize about what they see. They then will engage in nighttime activities to compare their ideas and slides with visual observations. One of Project STAR's activities is to measure the height of the rims of craters of the moon. Working from photographs and knowledge of the positions of the sun, the earth, and the moon students measure the length of the shadows. Similar triangles are then used to find these heights. CLEA, from Gettysburg College, enables schools to undertake simulations and explorations in contemporary astrophysics, on the PC computer. Project ASTRO, has developed curriculum materials and a network of amateurs and professionals to work together with schools to teach important astronomy experiments.

VI. Automated Astronomy

The development of a new generation of automated telescopes of various apertures over the next decade is almost a certainty. Inspired and informed by the first automated telescope, built by Stirling Colgate, the Berkeley Automated Supernova search has been regularly churning out up to 500 CCD images from a 30" aperture telescope per night for over a decade. Other telescopes are rapidly joining this growing community of telescopes that can attain these levels of imaging quality and quantity. The Mt. Hopkins Automatic Photometry telescopes have attained many hundreds of thousands of measurements of variable stars, and dozens of papers in scholarly journals co-authored by students. A number of other telescopes are available on line now or soon in the future, including the Bradford (UK) telescope, the Iowa State telescope, the SARA (Florida State 0.9m at KPNO) telescope, the Indiana telescope, the NF/Observatory, the TIE telescope on Mt. Wilson, the UCSB remote Astronomy Project, and eventually the Center for Astrophysics micro-observatories. In addition, the trend in remote or queue observing at major telescopes is accelerating, including remote or queued observations at the NTT (New Technology Telescope of ESO), the ARC 3.5m telescope, the WIYN 3.5m on Kitt Peak, and many other large telescopes that will be remotely operated.

With the advent of computer controlled telescopes and desk-top super-computers (viz., the 486PC) for image processing, it is now possible to bring a new universe of fresh images into classrooms. Kids and teachers usually love this initially, but we have found that it is key to train the teacher, provide appropriate *open-ended* curricula and activities, provide appropriate software, and link astronomers with teachers. We believe that the probability of success in this type of technology-mediated education, when we are dealing with one of the least technologically rich areas of our culture (classrooms), depends critically on the ability to create a complete learning environment. This environment allows empowered students and teachers to get from Alpha Centauri to the Omega Nebula without getting stuck on learning how to double-click rapidly in succession or alternatively not have anything to double-click upon.

VI. The Hands-On Universe Project

One example of a successful hands-on project is the Hands-On Universe (HOU) project of University of California's Space Science's lab and the Lawrence Berkeley Lab. HOU enables high school students to request their own observations from professional observatories. Students download CCD images to their classroom computers and use the HOU image processing software to visualize and analyse their data. HOU also provides a comprehensive curriculum that integrates many of the topics and skills outlined in the national goals for science and math education into open-ended astronomical investigations.

The HOU software includes a variety of image display functions and palettes as well as plotting tools, arithmetic manipulations, and photometry routines. Through the investigation of the solar system, galaxies, variable stars, and supernovae, students develop problem-solving techniques and critical thinking skills. Along the way students discover the need for algebra, geometry, various data representations and interpretations, as well as physical principles such as force and motion, energy transformations, and properties of light. The classroom computer, along with the image processing software and telecommunications, are regarded as research tools that enable students to pursue their investigations.

HOU provides extensive teacher training during summer workshops where previous HOU teachers serve as Teacher Resource Agents (TRAs) to train and support new teachers. Following the workshops, TRA's give ongoing support during the school year. The HOU WWW pages provide a forum for communication among teachers, students, HOU staff and professional astronomers. Through e-mail, classroom visits, and video-conferencing (where available), professional astronomers mentor teachers and students to help guide their HOU investigations.

Through external evaluations during the field-testing of HOU it was found that students were more motivated to learn math and science in the context of astronomical investigations; the classroom computer was used as a research tool and was more integrated into the classroom activities; and the teacher's role was perceived by both teachers and students as a co-investigator. The use of pedagogical and assessment techniques based on group collaborations, computer projects, performance task assessment, and journal keeping increased dramatically by HOU teachers, not only in their HOU classes but in their other classes as well.

Startling discoveries and valuable science can occur when high school students are given access to professional telescopes as witnessed by two Oil City High School students in spring 1994. Melody Spence and Heather Tartara requested observations of M51 (The Whirlpool Galaxy) during their investigation of spiral galaxies. A few days later they received a phone call informing them that they had captured the first light of SN1994I, the ninth supernova of 1994. Ms. Spence and Ms. Tartara will appear as co-authors on the SN1994I photometry paper.

This supernova discovery is perhaps the ultimate example of hands-on astronomy in which students are not only empowered as learners but also as research participants collaborating with the professional astronomy community. Other examples in this hierarchy of hands-on uses go from giving students the tools to work with real-time astronomy images, interactive simulations, canned activities and exercises, and delivery of information such as with a planetarium program. The common ingredient in all these examples is enabling students to work within an interactive context involving real astromomy examples.

Discussion

Mattei.

Two hands-on projects complement each other well - the Hands-On-Astrophysics project of AAVSO provides the tools and the archival for astronomical research in variables, and Hands-On-Universe provides excellent opportunities to obtain more data - remotely.

Textbooks and Electronic Media

Jay M. Pasachoff
Williams College–Hopkins Observatory
Williamstown, MA
USA 01267

E-mail: jay.m.pasachoff@williams.edu

Abstract Diffusion of astronomy in schools and universities is broadening from text only to include electronic media. I summarize trends in American textbook publishing and the current books available. I further discuss educational material available for classroom and other use on videodiscs, computer disks, and CD-ROM.

1. Textbooks

Over the last few years, U.S. textbooks for the survey course have become increasingly sophisticated in the use of color diagrams and photographs and in the availability of supplemental material. This course is by far the largest exposure of astronomy to American students, with enrollments of over 200,000 per year. The increasing cost of preparing the books and auxiliary packages has led to many authors and publishers bringing out both higher and lower level books, with different lengths and different levels of mathematics.

Some 15 authors (or sets of co-authors) have brought out two dozen different textbooks with current copyrights (1993, 1994, or 1995), including (with publishers in parentheses): Jay M. Pasachoff (Saunders), Michael A. Seeds (Wadsworth), William K. Hartmann (Wadsworth), William J. Kaufmann III (Freeman), Michael Zeilik (Wiley), Andy Fraknoi, David Morrison and Sidney Wolff (Saunders; one book includes the name of the late George Abell), Eric Chaisson and Steve McMillan (Prentice-Hall), Thomas T. Arny (Mosby), Dinah L. Moche (Wiley), James B. Kaler (HarperCollins), Sune Engelbrektston (Brown), Theodore P. Snow (West), William H. Jefferys and Robert R. Robbins (Wiley), and Karl F. Kuhn (West). See Table 1.

Free supplements given to adopters of class-sized numbers of copies include special magazine or newspaper supplements; overhead transparencies of artwork and photographs; slide sets of artwork and photographs; videotapes, videodiscs, and computer programs; printed or computerized test banks of exam questions; and instructor's manuals including recommendations for audiovisual materials, laboratories, and answers to questions from the text.

Table 1. Current U.S. Astronomy Textbooks, 1994-95

Jay M. Pasachoff:
Astronomy: From the Earth to the Universe, 4th ed., 1995 Version
Journey Through the Universe, 1994 Version
Saunders College Publishing

Michael A. Seeds
Foundations of Astronomy, 1994 ed.; 4th ed due 1997
Horizons: Exploring the Universe, 1995 ed., 1995
Wadsworth

William K. Hartmann
Astronomy: The Cosmic Journey, 5th ed., 1994 (with Chris Impey)
The Cosmic Voyage: Through Time and Space, 1992 ed.
Wadsworth

William J. Kaufmann III and Neil F. Comins
new editions of Kaufmann's *Universe*, 4th ed., 1994 and CD-ROM
Discovering the Universe, 4th ed., 1996
W. H. Freeman and Co.

Michael Zeilik
Astronomy: The Evolving Universe, 7th ed., 1994
Conceptual Astronomy, 1993
John Wiley & Sons

David Morrison and Sidney Wolff
Frontiers of Astronomy, 2nd ed., 1994
Saunders College Publishing

David Morrison, Sidney Wolff, and Andrew Fraknoi
Abell's *Exploration of the Universe*, 7th ed., 1995
Voyages to the Cosmic Realm, 1996 (superseding *Realm of the Universe*, 5th ed., 1994 Version)
Saunders College Publishing

Andrew Fraknoi (Resource Book)
The Cosmos in the Classroom: A Resource Guide for Teaching Astronomy
Saunders College Publishing, 1995

Eric Chaisson and Steve McMillan
Astronomy Today, 1st ed., 1993
Astronomy: A Beginner's Guide to the Universe, 1995
Prentice Hall

Thomas T. Arny
Explorations: an Introduction to Astronomy, 1st ed., 1994
Mosby

Dinah L. Moche
Astronomy: A Self-Teaching Guide, 4th ed., 1993
John Wiley & Sons

James B. Kaler
Astronomy!, 1994
[short version, as yet unnamed], 1996
HarperCollins College Publishers

Sune Engelbrektson
Astronomy: Through Space and Time, 1st ed., 1994
Wm. C. Brown Publishers

Harold P. Coyle, Bruce Gregory, William M. Luzader, Philip M. Sadler, and Irwin I. Shapiro
Project STAR: The Universe in Your Hands, 1st ed. 1993
Kendall/Hunt

Theodore P. Snow
The Dynamic Universe, 4th ed., 1991
Essentials of The Dynamic Universe, 4th ed., 1993
West Publishing Co.

Robert Robbins, William Jefferys, and Stephen Shawl
Discovering Astronomy, 3rd ed., 1995
Wiley

Karl F. Kuhn
In Quest of the Universe, 2nd ed. 1994
West Publishing

John D. Fix
Journey to the Cosmic Frontier, 1995
Mosby

Addresses of publishers

Saunders College Publishing
800 237 2665/708 647 8822
Suite 1250
150 South Independence Mall West
Philadelphia, PA USA 19106

Wadsworth Publishing Co.
Belmont, CA USA 94002

W. H. Freeman and Co. 800 347 9411
41 Madison Avenue
New York, NY USA 10010

John Wiley & Sons 800 248 5334
science@jwiley.com
605 Third Avenue
New York, NY USA 10158

Prentice Hall
Englewood Cliffs, NJ USA 07632

Mosby-Year Book, Inc.
11830 Westline Industrial Drive
St. Louis, MO USA 63146

HarperCollins College Publishers
10 East 53rd St.
New York, NY USA 10022

Wm. C. Brown Publishers and
Kendall/Hunt
2460 Kerper Blvd.
Dubuque, IA USA 52001

West Publishing Co. 800 328 9352
610 Opperman Drive, P.O. Box 64526
St. Paul, MN USA 55164-0526

2. Computer planetaria

Computer programs are now available for personal computers that allow teachers and students to recreate the sky for any given time or place. In addition to simulating sky views, such astronomical phenomena as eclipses, conjunctions, occultations, daily motion, precession, and so on, can be simulated. Programs such as Voyager II for the Macintosh and Visible Universe and Dance of the Planets for IBM compatibles are examples of this genre that are available on floppy diskettes. Similar programs for the IBM compatibles are Superstar, LodeStar Plus, TheSky, PC-Sky, and StarGaze. The programs are available for less than $100. Some include CD-ROM's with additional stars or images; other such programs are available only on CD-ROM (next section). See Table 2. John Mosley's "Software Roundup 1995" appeared in *Sky & Telescope* for August 1995, pp. 58-61, and reviews continue to appear.

3. CD-ROM's

CD-ROM's are now available that include not only the above planetarium features but also hundreds of photographs, short movies, on-line astronomical dictionaries or encyclopedias, and other features. Further, the number of astronomical objects available is increased to even the 16 million stars in the Hubble Guide Star Catalogue, over 5000 asteroids, and so on. One of the most popular is RedShift, written in Russia by Russian scientists for an English company and marketed widely in the US to general audiences through computer stores and catalogues. It is also available from Apple in an educational bundle of software. I have written instructions and lessons to accompany it for The Learning Team, 10 Long Pond Road, Armonk, NY 10504, 800 793 8326. *Guide* is another CD-ROM planetarium program. Voyager II and Distant Suns have CD-ROM disks beyond the original programs. See Table 2.

Other astronomical CD-ROM's include a disc about the 1991 total solar eclipse, a disk about the Murmurs from Earth set of images and recordings on a record being carried out of the solar system on the Voyager spacecraft, and a Comet Collision with Jupiter disc about the crash of Comet Shoemaker-Levy 9 into Jupiter.

Portable systems add a CD-ROM player weighing only about 1 kg to a laptop computer, and allow the demonstration of imaging and sky positions.

Table 2. Computer Planetariums and CD-ROM's

Computer Planetarium

Voyager II (Carina Software, 12919 Alcosta Blvd #7, San Ramon, CA USA 94583, 510 355 1266). CD-ROM available. Also, Sky Gazer, a new program for novices. For the Mac.

Dance of the Planets (A.R.C. Science Simulations, P.O. Box 1955, Loveland, CO USA 80539, 303 667-1168, 800 SKY 1642). For the IBM PC and compatibles.

Superstar (picoSCIENCE, 41512 Chadbourne Drive, Fremont, CA USA 94539, 415 498-1095). For the IBM PC and compatibles.

LodeStar Plus (Zephyr Services, 1900 Murray Ave., Pittsburgh, PA USA 15217, 412 422-6600). For the IBM PC and compatibles.

TheSky (Software Bisque, 912 12th St., Golden, CO USA 80401, 303 278-4478, 800 843-7599). For the IBM PC and compatibles. CD-ROM available.

PC-Sky (CapellaSoft, P.O. Box 3964, La Mesa, CA USA 91944, 800 827-8265). For the IBM PC and compatibles.

Visible Universe (Parsec Software, 1949 Blair Loop Road, Danville, VA USA 24541, 804 822-1179). For the IBM PC and compatibles. The professional version even shows eclipse tracks.

StarGaze (CEB Metasystems, 1200 Lawrence Dr., Suite 175, Newbury Park, CA USA 91320, 800 232-7830). For the IBM PC and compatibles.

CD-ROM

RedShift (Maris Multimedia, 3871 Piedmont Ave., Oakland CA USA 94611, 510 652-4746. For Macs, Power PCs and PCs.

Distant Suns (Virtual Reality Laboratories, 2341 Ganador Court, San Luis Obispo, Ca USA 93401, 800 829-VRLI). For Macs and PCs

Guide (Project Pluto, Ridge Road, Box 1607, Bowdoinham, ME USA 04008, 207 666-5750). For PCs.

4. Videodiscs

The capabilities of laserdiscs include 54,000 frames on each side, so one can include thousands of still images plus approximately an hour of movies. The widely used videodisc on astronomy from 1984 was superseded in 1993 by a second edition called Beyond Earth. Images include NASA spacecraft movies, supercomputer simulations of black holes, Yohkoh x-ray solar images, and so on. For information, contact Optical Data Corp, 30 Technology Drive, P.O. Box 4919, Warren, NJ USA 07059, fax 908 668 1322. It is for the NTSC television standard.

I have consulted on both the above videodisc and a videodisc for the Saunders astronomy texts: Saunders Videodisc Library #1: The Solar System, released in 1994.

The American Association of Physics Teachers has a Physics Cinema Classics disc with many demonstrations.

5. NASA Space Educators' Handbook

A NASA scientist has put together a free series of Macintosh disks in Hypercard that include a dozen QuickTime very brief sound movies (a few seconds each) of space exploration and various written material. To receive them, send 10 high-density 3.5" formatted Macintosh diskettes to Jerry Woodfill, NASA JSC (IA12), Houston, TX USA 77058.

6. Gems of Hubble

The Space Telescope Science Institute has prepared a series of "Electronic Picturebooks" for Macintosh computers, where they use Hypercard to display a few dozen images and text on each subject. Subjects now available are Gems of Hubble (a set of highlights), Magellan Highlights of Venus, Images of Mars, The Planetary System, Apollo 11 at 25, Scientific Results from the Goddard High Resolution Spectrograph, The Impact Catastrophe that Ended the Mesozoic Era, and some other terrestrial subjects. They are available as computer disks or as a CD-ROM that includes them all. They can also be downloaded over the Internet. Get information from clallo@stsci.edu.

7. Astronomical Society of the Pacific

The Astronomical Society of the Pacific (390 Ashton Avenue, San Francisco, CA USA 94112, fax 415 337 5205, Internet: asp@stars.sfsu.edu) is a supplier of a wide variety of electronic media resources, including computer software, CD-ROM's and videodiscs.

8. Other Suppliers of Materials

Among suppliers of astronomical computing materials are: Sky Publishing Company, the publishers of *Sky & Telescope*. 800 253 0245. Kalmbach Publishing Company, the publishers of *Astronomy*. 800 558 1544. Mac-Connection/ PC-Connection, discount software suppliers, 800 MAC-LISA.

9. World Wide Web

The Internet has brought almost instant access to scientific data to many sites around the world. For example, the morning after observations were taken of the impact of Comet Shoemaker-Levy 9 on Jupiter, many observations from major observatories were posted on the Internet. Use of a graphics interface like Netscape or Mosaic, and calling the Jet Propulsion Laboratory's "News Flashes" on http://www.jpl.nasa.gov, brought a choice of dozens of images via the World Wide Web. Hubble Space Telescope observations are available from the Space Telescope Science Institute on http://marvel.stsci.edu. Many links are included in my own textbook's homepage, http://albert.astro.williams.edu/jay. Recent major articles about the Internet (Stuart J. Goldman, "Astronomy on the Internet," *Sky & Telescope*, vol. 90, no. 2 (August 1995), pp. 20-27; and David Bruning, "Blasting Along the InfoBahn," *Astronomy*, vol. 23, no. 6 (June 1995), pp. 74-83) contain other hyperlink addresses as well as File Transfer Protocol (ftp) addresses. The World Wide Web is growing at a tremendous rate.

Table 3. Astronomical Sites on the World Wide Web

http://www.stsci.edu/resources.html (Space Telescope Electronic Information Service)
http://anarky.stsci.edu/astroweb/net-www.html (WWW Index for Astronomers)
http://marvel.stsci.edu/net-resources.html (Astronomical Resources)
http://blackhole.aas.org/AAS-homepage.html (AAS Home Page)
http://www.digital.com/gnn/wic/astro.toc.html (AstroWeb: Astronomical Resources)
http://marvel.stsci.edu/top.html (Space Telescope Science Institute)
http://www.usno.navy.mil/ (U.S. Naval Observatory)
http://maxwell.sfsu.edu/asp/asp.html (Astronomical Society of the Pacific)
http://info.er.usgs.gov/network/science/astronomy/index.html (General astronomy)
http://www.metrolink.com/seti/SETI.html (SETI Institute)
http://www.esrin.esa.it/ (European Space Agency)
http://jpl.nasa.gov/ (Jet Propulsion Laboratory)
http://skyview.gsfc.nasa.gov/skyview.html (Skyview)
http://crux.astr.ua.edu/gallery2.html (Messier Gallery)
http://sohowww.nascom.nasa.gov/ (SOHO)

Discussion

Bishop.

I wonder if some visual portrayals by computer astronomy programs may lead to or perpetrate misconceptions. For example, the earth rotating in steps rather than smoothly and incorrect scale of diameter and distance. As with *any* model, the observer should understand what is wrong with a model as well as what is right or useful in it.

Hill.

[Re: Sadler's remarks] Studies of Textbooks must study the teacher's style and his/her use of the text.

Kinney.

[Q: to Phil Sadler] Is there any evidence that interactive computer programs do anything besides amuse the astronomers who design them?

Sadler.

There is a community of researchers who assess the effectiveness of educational software. To the best of my knowledge, I have not seen any studies of astronomy software. While one can surmise that such software is popular among astronomers, both amateur and professional, we do not know if it helps or even hurts the understanding of our students.

I do not mean to single out high-tech materials. In a study we are conducting at the CfA to assess the effectiveness of high school practices on success in introductory college physics courses, we have been quite surprised to find that there is very little difference between the students who use different high school physics textbooks. Indeed, students who do not use a textbook in high school physics perform significantly better in college physics than students who use a textbook. There is no data on this effect for astronomy courses, but I wonder if the results would be any different.

How can this be? The use of textbooks may increase the pressure for "coverage" of astronomical concepts. Teachers who reduce coverage, i.e. "post-holing" by covering few topics in great depth (no doubt, carefully selecting materials) may be using the most effective strategy.

Schatz.

New resources provide an overkill of images, especially two-dimensional images, which can disconnect students from three-dimensional images of real-world experiences. How can we overcome this disconnection in writing textbooks or in using computer simulations and other resources?

Pasachoff.

Even in the sky, we tend to see two-dimensional images when we look out at what has long been known, significantly, as the "celestial sphere." Transferring close-up views of Ping-Pong balls made to revolve around basketballs from that three-dimensional close-up to distant objects billions of kilometers away is by no means easy for students.

It may well be that some of the simulations on computer screens that students can carry out will lead some large fraction of students to understand these matters more clearly than they do with in-class demonstrations, or that the computer simulations will at least add to the understanding from the in-class demonstrations. After all, as we often forget, for many of today's students things on computer screens seem more real than things they see directly.

Using a computer planetarium program like RedShift, for example, a student can look at the Earth from the Moon as the Earth resolves, and then can look down from high above the Earth, as well as from intermediate positions. Day and night show clearly and can lead elementary-school students to understand why we have day and night. Similarly, they can see the Moon from many angles and follow the Moon's changing phases. Empowering the students to move their vantage points around with their computer's mouse can allow them to play personally enough and interestingly enough that their comprehension will be greatly increased. We should be figuring out the best ways to guide students in the use of the newly available technologies so that we are not merely saying that "technology is good" but that "technology can help us, if we use it correctly."

Tyson.

When I last visited the Dalton School in Manhattan I noticed that Malcom Thompson (of Jastrow & Thompson textbook fame) designed an astronomy laboratory class around the Mac-based Voyager II program – two students per computer. It appeared to me to be extraordinarily successful.

Wiley.

As a fifth grade teacher I see my role as one who builds the base for a life-long interest in astronomy. I can do this in my classroom best when I have an astronomy textbook supported by high-interest-level hands-on activities. Through outreach in Santa Barbara I have worked with professional astronomers, as well as amateurs, who were willing to spend time with my students. Their input was invaluable because my students, if they were interested, could project themselves into an astronomer's role.

However, my textbook, *Exploring the Universe* (Prentice Hall) was critical in helping my students understand new concepts. I am not sure a strictly "hands-on" approach benefits students. While these activities are fun, I am not sure the comprehension level stays for the long term. We need both text and hands-on.

Many school districts cannot afford up-to-date astronomy texts. In fact, most school districts would not consider the expense.

Is there a possibility that NASA, NSF, or ASP can provide a soft cover astronomy text for elementary schools to use? I think teachers would be more willing to teach astronomy if they had this resource.

Model for Reaching Underserved Audiences: The Cultural Arts Program for Children Living in Temporary Housing

Neil D. Tyson
Acting Director
Department of Astronomy / Hayden Planetarium
American Museum of Natural History
Central Park West at 81st Street
New York, New York
USA 10024-5192

and

Ismael Calderon
Department of Education
American Museum of Natural History

Tel.: (212) 769-5933
Fax: (212) 769-5934
Internet: tyson@amnh.org

Introduction

Many academic and cultural institutions throughout North America are trying to create and implement educational programs that will reach underserved audiences. These institutions either feel a moral obligation or envision a potential financial profit from these new audiences. Traditionally, underserved audiences have been identified as members of minority groups, such as people of color, but it also includes women, individuals with limited financial resources, or a combination of these. This audience, for the most part, has been alienated by academic and cultural institutions. The definition for audiences identified as undeserved, continues to expand as institutions grow and enlarge their activities. These can now include any group with special interest or affiliations such as senior citizens and individuals with specific educational needs. Here we present the framework for a successful educational outreach model, the Cultural Arts Program for Children Living in Temporary Housing, conceived by the New York City Department of Cultural Affairs and Board of Education, and implemented by the American Museum of Natural History in New York City. The total educational partnership includes the New York City Public Schools, the New York City Department of Homeless Services, private foundations, and nearly forty of New York City's cultural institutions. Organizations provide either financial resources or services, whichever best fits their mission. When shared among all partners, the services can be rendered without overburdening any one of them.

The Cultural Arts Program for Children Living in Temporary Housing is a unique science and arts outreach initiative instituted in 1988 for elementary school age students living in temporary housing shelters. This group comprises a growing segment of the city's public school population. Today in New York City there are over 19,000 homeless families, with about 14,000 school-age children, living in shelters. These children and their families are subject to the psychological trauma of moving through the shelter system, the stress of having little or no privacy, and the social humiliation of not having a home of their own. Such living arrangements produce fear and anxiety in children, which adversely affects their personalities and thwarts their social, educational and cultural potential. Such circumstances are precursors to failure in society, which can lead to non-functional citizens who will perpetuate the cycle of poverty prevalent in lower socio-economic communities.

The goal of the Cultural Arts Program for Children Living in Temporary Housing program is to integrate the young people and their families into the life of the city and to provide them with skills and discipline in the arts, sciences and humanities. Through a variety of arts and science activities the program strives to: provide daily cultural programming for homeless children that creates an ambiance of stability; encourage development of each participant's ability to make independent decisions; and foster children's ability to perceive and respond to everyday life challenges through their engagement with the arts and sciences. The program's objectives provide an avenue for children to: become acquainted with the cultural resources of the city; experience a sense of ownership of these facilities; provide learning experiences that stimulate curiosity; and promote self-esteem.

Institutional Responsibilities

The Department of Cultural Affairs provides administrative support and financial support, and is the City-wide coordinator of the program. They also select the participating arts and science organizations. The New York City Public Schools support the project through funding, food services, and technical assistance by social workers and school district personnel. The Department of Homeless Services provides support by assisting with the selection of participating organizations, providing bus transportation, technical assistance for care givers, recruitment of students and orientation of participating groups.

Each cultural institution contributes its unique expertise to develop meaningful and engaging arts and sciences experiences. The activities and resources contributed by the city agencies are extremely important because they relieve the cultural institutions from difficult and time consuming work, which allows them to focus their resources on the development and implementation of programs based on sound educational principles.

Communication between the cultural institutions and social workers of the housing facility is extremely important in helping sensitize the museum professionals to the unique circumstances of the temporary housing population. Moreover, communication between the housing staff and the cultural institution allows the housing staff to organize and prepare students for their contact with the museum culture. During the program, close attention is paid to students' progress and individual needs. Supervision is provided by housing facility workers, the program coordinator and instructors. As necessary, supervision is increased if behavioral or adjustment problems arise. Regardless, every attempt is made to provide a low student to instructor ratio. In the Cultural Arts Program a minimum of two educators per workshop session of typically twenty students was provided.

Organized bus transportation to and from the housing facility site, provided by the Department of Homeless Services, is crucial to the success of the program. The journey itself helps to promote a sense of camaraderie and thus steady student participation.

The New York City Public Schools provide funding for brown-bag meals to be served to the participants at the cultural institutions. This is a significant part of the program, since a nutritious meal guarantees that students will not be hungry and will be able to concentrate on the task at hand. The institutional attention allows the student to quickly perceive they are special.

To further ensure the success of the program it is essential that the partners, particularly the cultural institution, make the children feel accepted by providing them with a nurturing and friendly environment. Personnel of the cultural institution, from security guards to administrators, must be cognizant of the special needs and situation of the students – they must be mindful not to offend. This does not mean that discipline and order are discarded, but that the students should be treated courteously and respectfully.

Program Details

Through a Request for Proposal, the Department of Cultural Affairs (DCA) selects institutions to participate in the program. Interested institutions are required to submit proposals for a project organized around a central arts-science theme, yet allowing each session to stand on its own. This criteria allows for students' maximum benefit, given the transience of the population. DCA provides grant awards up to $10,000 for a ten-session project to participating cultural institutions. The grant covers costs of staff, supplies and equipment necessary to implement the project.

The American Museum of Natural History has participated in the Cultural Arts Program for Children Living in Temporary Housing since 1989. The Museum's Department of Education has developed a project of hands-on participatory science experiences organized around the concept of light as the curriculum theme. Successful outreach programs, such as the aforementioned, should be embedded in sound educational learning theory. But they must break away from traditional classroom instruction to engage the student in meaningful contextualized learning. The curriculum followed by the students in the program is based on constructivist learning theory, utilizes cooperative learning situations, and provides learning experiences for students with different learning modalities based on Gardner's theory of multiple-intelligences. These theories guide the development and implementation of curriculum activities, permitting diverse opportunities for learning.

The curriculum for the Cultural Arts Program is based on the constructivist learning theory because Museum educators are cognizant that not all people have the same interest and abilities – they learn in different ways and possess different particular profiles of intelligence. This learning theory takes into consideration the students' prior learning experiences and provides new situations for meaningful learning. Program instructors are made aware that individuals develop idiosyncratic knowledge structures as they perceive, interact with, and interpret events and objects encountered in their environment. This understanding allows a successful program to be delivered. Every effort is made to put the learners at the center of the learning process for them to make sense of concepts in a dynamic interplay between prior knowledge and new information. Each workshop is designed to keep the student cognitively active in the learning process. Throughout the program, instructors are mindful of misconceptions that the students might possess regarding science phenomena and are prepared to provide situations to challenge students so that they may construct appropriate science models. Through this process, conceptual change (learning) is achieved as the instructor provides a program of activities from which knowledge and skills can be constructed. The constructivist curriculum creates an environment for students to explore novel and unfamiliar ideas. The implementation of constructivist learning theory in the Cultural Arts Program curriculum encourages students' use of knowledge to solve the problems and complete the tasks that they may confront in the wider community.

Throughout the Cultural Arts Program for Children Living in Temporary

Housing, students are provided with opportunities to learn cooperatively, to encourage them to share their knowledge and skills. This cooperative learning approach takes into consideration the students' multiple perspectives, diverse backgrounds, learning styles as well as the aspirations which the students bring into the learning environment, thereby permitting them to explore information from different perspectives. Cooperative learning situations in the program curriculum reveals to the students that learning is a social activity involving mutual exploration and feedback, which leads to better understanding of science concepts. This learning approach involves students more actively in their learning, fosters an understanding for cooperation and teamwork, and encourages students to develop an active voice in shaping their ideas while considering the ideas of others. Students become cognizant that they all possess different knowledge and that it should be shared.

Gardner's research in cognitive psychology (Gardner, 1993) indicates that individuals use different learning modalities such as visual, linguistic, interactive, etc. to process information and learn. Gardner's identification of seven intelligences, language, logical-mathematical analysis, spatial representation, musical, bodily-kinesthetic, interpersonal, and intrapersonal, reflect different learning modalities which individuals employ to learn, remember, perform and understand concepts in different ways. Outreach educators must be mindful of these different learning modalities in order to develop experiences which reach all of their audience. They must understand that not everyone can process information through all modes, but rather each person has at least a few modes of processing information. The implementation of the cognitive phenomena of multiple entry points in learning situations is necessary to promote understanding. The curriculum for the Cultural Arts Program for Children Living in Temporary Shelters employs Gardner's (Gardner, 1993) five entry points which coincides with the seven learning modalities in designing program experiences: narration (story or narrative), logical-quantitative (deductive reasoning), foundational (philosophical and terminological facts of a concept), esthetics, and the experimental approach.

Workshop Activities

Workshop sessions take place after school to allow for maximum participation. The curriculum activities for each session are highly participatory and use the Museum's extensive exhibit base as well as the resources of the Hayden Planetarium. Activities have involved learning about the properties of light, including a study of optical instruments such as the microscope, telescope, camera, and eye, and how these devices are used to explore the Earth and the rest of the universe. During the first and second workshop students are introduced to light as a form of energy, the production of light by luminous sources such as light bulbs, television screens and bioluminescent organisms. Students use light boxes, mirrors, prisms, lenses, and filters to explore light's physical properties. They gain an appreciation for the importance of light as a physical phenomenon and its application in technology. During the third and fourth workshops students gain insights into the amazing faculty of sight and explore how animals have adapted different visual structures to permit them to perceive their environment.

Using the Museum's extensive dioramas, students identify a variety of life forms with different eye structures and hypothesize how they perceive their environment. The instructor guides them through an understanding of and appreciation for organisms and their perception of light. Students are then given an assortment of materials and asked to create their own eye structure and explain how it works. The fifth and sixth workshops apply the student's understanding of the principles of light by engaging them in the construction of a pinhole camera and the development of an image (film processing and developing) produced by light interacting with a metal (silver compounds). Parallels are drawn between the use of light by the human eye and the pinhole camera. Students have an opportunity to enter an oversized pinhole camera made from a refrigerator box to understand what happens to the image as it is projected on the inside of the camera. Participants also make collages on light-sensitive paper, which is then developed and displayed as art.

During the seventh workshop students learn the mechanics and use of the compound microscope. They study the planktonic life in New York City's Central Park pond, learn to prepare slide mounts, develop drawing techniques, and examine video-microscopic images of simple organisms such as protozoa. During the eighth and ninth sessions, students visit the Hayden Planetarium adjacent to the Museum. They learn to identify constellations using the Zeiss projector, make their own sky charts with the assistance of an astronomer and learn to use a star locator.

Program Evaluation

Throughout the ten-session workshops students' learning is assessed by the program coordinator and instructors through observation of their participation in activities, completion of tasks, and questions which they pose. Reports are written daily for each student to document and monitor their progress. Throughout the program students keep a journal of their own experiences, which is reviewed by the program coordinator. They also have opportunities to voice their successes and concerns about the program in daily rap sessions. These are audio recorded and are also analyzed by the program coordinator. Periodically an external evaluator, arranged by the Department of Cultural Affairs, is present at programs to ensure that the curriculum's goals and objectives are clearly articulated to students and successfully implemented. Information gathered through these methods is evaluated and is used to make adjustments to the program. Dialogue is maintained throughout all phases of the program between DCA, the cultural institution and the housing facility to guarantee that the needs of the participants are being met. As the Cultural Arts Program for Children Living in Temporary Housing evolves, components are adjusted to meet the needs and interest of the students.

End-of-Program Events

The culminating activity for the program is a "star party" which includes participants and their families (however fragmented), and all program personnel that have participated in the project. During an awards ceremony, participants are acknowledged for their achievements, given participation certificates and a one year Museum membership. Public acknowledgment of the students' participation in the program reinforces their self esteem and legitimizes their learning. The Museum memberships keeps the parent involved in Museum activities for one year, serving as a bridge to their future involvement with the institution. During the culminating activity everyone is treated to dinner in the Museum's employee dining room. Participants set up and display their projects for invitees to view. During this time the students make presentations of their work on display and explain the science that they have learned. The group then convenes at the Hayden Planetarium where they are treated to a short presentation of the evening sky in the Planetarium's sky theater. Finally, participants move outdoors where members of the New York City Amateur Astronomers Association, in conjunction with a Planetarium astronomer, set up telescopes to view the evening sky and identify planets and constellations.

After the program, additional visits are made to the housing facility to involve the entire housing community. Amateur astronomers and Museum educators engage the community in star gazing with telescopes set up in the city streets. As curious passersby stop to look at planets and stars, news of the activity rapidly spreads throughout the neighborhood. The interest for this activity has been overwhelming as witnessed by the number of individuals of all ages who became involved.

The program's influence is evident in some of the program anecdotes. Some illustrate cognitive and affective learning as well as a growth in self esteem:

Amateur Astronomer's visit to the living facility '94: Neighborhood kids walking by were invited to look at a planet through a telescope and at first declined. They returned and eventually looked with one commenting, "Oh sh@*, that's f*#@ing Jupiter!"

1995 Program - Microscopy Session: When questioned about their aspirations and interest in science prior to the workshop, three students said they did not like science, but after the workshop, they said they'd changed their minds.

1995 Open House: After activities, one student was asked if he'd be attending the program and replied, "If it has anything to do with science, I'll be there".

1995 Program: One girl commented on the increased confidence in her abilities at the end of the first week.

1995 Program: One student commented on being bored during the Phases of the Moon worksheet, but became excited when she received the constellation locator and was told it was hers to keep.

Culminating Event: One parent expressed appreciation for the program and how well the children were received by the Museum.

Culminating Event: Following the program, one parent commented that she would definitely be returning to the Museum with her children.

End-of-Program Dinner: Upon realizing that he was drinking Coca Cola out of a long stem wine glass, one student noted, "We may be homeless, but we have class!"

Before the students are selected to participate in the program, members of the Museum's Education department staff visit the housing facility to introduce students and their families to the Museum and its activities. These open house-style sessions simultaneously spark interest among potential participants, motivate both students and staff, and inform parents about the activities in which their children will be involved. As the Cultural Arts Program has evolved, periodic visits to the housing site have been made during the run of the program (such as the one described above with the amateur astronomers) to maintain students' interest and engage the housing community in science activity. These visits generate excitement, while providing valuable educational and social opportunities. They help reaffirm the cultural institution's commitment to children's science learning.

In conclusion, a significant criterion to ensuring a successful outreach endeavor such as the Cultural Arts Program for Children Living in Temporary Housing is that an honest and sincere approach be taken by the host institution. A complete effort, with a commitment of maximum institutional resources, should be undertaken. Respect for the audience and participating organizations should be exhibited throughout. The audience for whom programs are developed should be appropriately identified and understood for the successful delivery of services.

Reference

Gardner, H. (1993). Multiple Intelligence: The Theory in Practice. USA: Basic Books.

Discussion

Fierro.

I have worked with children that live on the streets in Mexico City. some of them are extremely brilliant. They are true survivors. So it's very gratifying to work with them; they are eager and quick to learn!

Grice.

The Charles Hayden Planetarium in Boston has also been involved in programs for underserved groups. We created a planetarium show called "WSKY: The Radio Station of the Stars," specifically for inner city kids. Several planetariums around the U.S. are now playing this show for their general audiences.

To supplement this program, we developed and taught the "City Stars" workshop for inner city teachers and provided the teachers with instruction and a packet containing hands-on activities (and materials needed) to work with their students back in school.

The Boston Museum of Science, which we are a part of, has created a partnership with an inner city area of Boston known as Egleston. We invited 300 people (made up of families) to the Musuem for a special planetarium show and star party. They were provided with a complimentary dinner and materials to take home. We plan to provide additional collaborations with this group and possibly other inner city groups in Boston.

Mechler.

I want to supplement the speakers' presentation on outreaching to underserved groups by mentioning that one other such group is those men and women in prison. This past spring I was approached by the local humanist group to present a series of presentations to interested inmates at a nearby federal corrections institute (middle-level security). I did so, basing the presentations on the Cosmos programs, preceded and followed by discussions. It was a most interesting and positive experience for all concerned, I think that includes me. For people so physically constrained, what potential there is for a passionate interest in a most unconstrained topic – the universe! Assuming I continue this volunteer effort in the fall, I should give a more complete presentation on this in the future.

The Role of Planetariums in Astronomy Education

James G. Manning
President, International Planetarium Society
Museum of the Rockies
Montana State University
Bozeman, MT
USA 59717-0272

E-mail: ammjm@gemini.oscs.montana.edu

Good afternoon.

In the 1940's, chemical engineer and philanthropist John Motley Morehead approached the eminent Harvard astronomer Harlow Shapley with a question. Morehead wanted to build either an observatory or a planetarium for the children of his native North Carolina, and he asked Shapley which the astronomer would recommend. Shapley suggested that a planetarium would be better – because he felt the people of North Carolina were among the "most astronomically ignorant" people in the United States.

Morehead responded that if Shapley would amend his statement to say that the people of North Carolina were among the "most ignorant in astronomical matters," he would build a planetarium. Shapley did, and Morehead did, and the rest is history – 46 years of history, to be precise, during which more than four million people have been introduced to the wonders of the universe at the Morehead Planetarium – thanks to the hard work of two generations of planetarium staff.

In many ways, the Morehead story is not unique; planetariums for decades have been created to serve the cause of astronomical enlightenment – to offer people knowledge and understanding and a sense of place in a universe far bigger than themselves. It's an important role and one that we continue to play – changing, we hope, as times, technology, education, and our view of the universe change.

To best appreciate the role of planetariums, it's important first to understand the broad reach that they have today. Consider these facts and figures...

The first projection planetarium was demonstrated by the Zeiss Optical Company at the Deutsches Museum in Munich, Germany in 1923. It was permanently installed there in 1925 – 70 years ago. By 1970 – the height of the Apollo moon program – there were an estimated 700 to 800 planetariums in the world, half of them less than six years old, constructed during the peak years of the Space Race. Today, 25 years later, that number has more than doubled to a little over 2,000.

Based on figures compiled in the 1994 IPS Planetarium Directory, we find that slightly more than half of the world's planetariums are located in North America, with large numbers in Asia and Europe, and relatively few in other parts of the world. If we consider distribution by country, we find that half are in the United States, more than 300 are in Japan, and Germany ranks third with nearly 100. Nineteen countries have ten or more planetariums.

Some 33 percent of these planetariums are located in primary or secondary schools; 17 percent are at colleges and universities; 15 percent are part of museums and science centers; 7 percent are associated with observatories or other institutions; the settings of the remaining 27 percent are uncertain, but would probably be distributed among the other categories.

Housed within these planetariums are a variety of star projectors: Spitz models, familiar dumbbell shapes as in the model Zeiss, the more recent single-sphere type such as the Minolta Infinium or Goto Helios projectors, the fisheye lens of the Evans & Sutherland Digistar computer graphics system. The theaters themselves vary greatly as well, ranging from small and cosy to the cavernous tilted-dome theaters, to hundreds of portable theaters like Starlab, operating around the world in addition to the 2,000 fixed theaters, bringing astronomy directly to schools, auditoriums, shopping centers, and other venues.

In all of these theaters with all of these instruments in all parts of the world, it has been estimated that 20 million people visit planetariums each year. Personally, I believe this estimate is low. But if one takes it at face value, assuming an average of 10,000 visitors per planetarium per year, and calculates backward over seventy years, one ends up with more than 450 million people served since the first planetarium opened at the Deutsches Museum. Even for McDonalds' that's a lot of hamburgers!

Clearly, planetariums represent one of the biggest and most visible avenues for presenting astronomy and related subjects to the public – surpassable perhaps only by television. This gives planetariums an enormous potential for supporting both formal and informal astronomy education.

In defining the role that planetariums play, it's useful to review their strengths. We all know that planetariums can reproduce the night sky for any place on earth on any day of the year for many years past and future, creating a view that simulates the three-dimensional "backyard" sky. And that they can accurately reproduce the apparent motions and cycles of the sky – in speeded-up fashion so people can see what happens over time. But planetariums also create environments that encompass the audience, bringing them into the experience in a way that classroom, book, television or computer screen cannot. They can combine and effectively use audiovisual technology to help create these experiences. And they possess tremendous flexibility in how these audiovisual resources can be used.

It is these strengths which allow the planetarium to effectively demonstrate astronomical principles and to present and reinforce concepts and information in ways that other media cannot. And we take advantage of these strengths in the goals that we set for ourselves.

First and foremost, we strive to educate, in ways ranging from curriculum-based school lessons to popular-level programs. We also strive to enlighten, which I think is not quite the same as to educate; we want people not just to know but to understand and to incorporate this understanding in their lives. And yes, many of us also try to entertain – on the principle that you catch more flies with honey, that learning *ought* to be fun, and that people probably learn more when they're enjoying themselves. And not least, we strive to inspire. Our time with people is brief, and it is perhaps less important that someone remembers the diameter of Jupiter than that he or she remembers Jupiter as a neat planet, and buys a book or enrolls in a class or comes to the next star party to learn more – or takes time to look for it on the next clear night.

In setting these goals, planetariums operate in all three realms of learning: in the thought-processing of the cognitive realm; increasingly in the psychomotor area as we offer more interactive experiences involving physical action; and also in the affective realm, the realm of feelings, as we encourage greater appreciation and enjoyment of the sky and try to cultivate a sense of the adventure of science.

We meet these challenges in as many ways as there are planetariums. Perhaps the best way to illustrate this point is to present a small sampling of the things planetariums and planetarians are doing around the world.

The so-called "star show" continues to be our stock-in-trade, offered in variety of forms, from the traditional live-narrated current-sky program to automated multimedia presentations on popular space-related topics. Most of us also present educational programs specifically designed for school grades to meet science curriculum objectives. Where possible and appropriate, we often key on current news and icons to attract visitors and send our messages as we did in a show we presented at the Taylor Planetarium at the Museum of the Rockies in Bozeman, Montana, in conjunction with the Columbus quincentennial. The Hayden Planetarium in New York has developed children's programs featuring Star Wars' C-3PO and R2-D2 and the Muppets, which I understand are very well-attended.

Supporting our efforts are improving technologies. The new planetarium at the Forum Der Technik in Munich has a new Zeiss Model VII projector, which uses fiber optics to create stars that look like true points of light, creating an exceptionally realistic-looking starfield. The Digistar computer graphics system allows the audience to travel through a stellar data base in "Star Trek" fashion, can show the proper motion of the stars over hundreds of thousands of years, and creates many other three-dimensional learning opportunities.

The London Planetarium, just this week, reopened after extensive renovations that retired its aging projector, replacing it with a Digistar and other new technology to carry it forward into the new millennium. The new planetarium at Brevard Community College in Cocoa, Florida is the first in the world to have two planetarium projection systems: a Minolta Infinium to produce a crisp starfield, and a Digistar for 3-D simulations and computer graphics. Wherever and whenever possible, planetariums are updating themselves with new technology to better meet their goals and serve their public.

Computer systems are increasingly in use in planetariums today, both to control the planetarium projector and to automate auxiliary effects in programs. Video projection, pioneered by such planetariums as the Armagh Planetarium in Northern Ireland and at the Munich Planetarium, is becoming an audiovisual staple. Some planetariums such as the Hayden have access to Silicon Graphics workstations for creating animated video sequences for their programs and for distribution to other facilities.

Slide projection is still a mainstay of planetariums, but in a variety of forms: images zoom and slew and create panoramic scenes, and so-called "All-Sky" projections engulf the audience in all-dome images – from a cathedral in the Munich Planetarium to the Eagle Nebula at Hayden. Astronomer David Malin's photographs are distributed as all-sky sets by Sky-Skan, Inc., one of a number of companies offering products and services to the world-wide planetarium community.

At the South African Museum Planetarium in Cape Town, Anthony Fairall also uses a Silicon Graphics workstation to manipulate data into all-sky images – including the radio sky and the distribution of galaxies out to 250 million light years as seen from the southern hemisphere. Galaxies are color-coded white to blue to indicate distance, and Tony is working on a color-coded version that – when used with ChromaDepth glasses available from a laser company called Audio-Visual Imagineering – will produce a three-dimensional effect to demonstrate the clumping of galaxies. Mickey Schmidt at the Air Force Academy Planetarium in Colorado

Springs has created a Digistar data-base of these galaxies that can be flown through.

Planetariums such as the Hansen Planetarium in Salt Lake City, the Buhl in Pittsburgh, and the Munich Planetarium have installed a kind of interactive technology: seat buttons which allow the audience to vote on a choice of space destinations or topics within a program, and respond to questions.

Hands-on experiences are manifesting themselves in more traditional ways as well, especially in school programming. The Holt Planetarium at the Lawrence Hall of Science in Berkeley, California, has been a pioneer in interactive programming; its activity guide series called PASS - Planetarium Activities for Student Success, created by Alan Friedman and Alan Gould and others - is currently being translated into Japanese for use in that country's many planetariums. Among its myriad activities are lessons involving the identification of features on the moon and the use of models to demonstrate the moon's phases.

Sheldon Schafer at the Lakewood Museum Planetarium in Peoria, Illinois, creates mysteries in which the time of the crime and the culprit can be determined by knowing when and where certain constellations appear in the sky. At the Suginami Science Center in Tokyo, Shoichi Itoh engages students in discovery through lessons in which they photograph the planetarium starfield, create planetariums of their own, and find the constellations on their own using star maps.

Jeanne Bishop at the Westlake Schools Planetarium in Westlake, Ohio, reports that several Cleveland area planetariums are outfitting a Mobile Observatory with telescopes and computers for use by students; Jeanne plans to have some of her astronomy students prepare and conduct interactive lessons for elementary students using this equipment. And the staff at the Air Force Academy Planetarium uses its facility extensively for hands-on lessons of a special kind: Air Force cadet training in topics ranging from aeronautics to survival skills using a compass and the planetarium sky.

In all of these educational efforts, planetarians work to meet the local and regional science curriculum objectives to be relevant to schools. Planetarium educators have thus been interested reviewers of the U.S. National Academy of Sciences' draft proposal for national science education standards, which advocate, among other things, universal curriculum goals, activity-based teaching, emphasis on scientific inquiry, and improved teacher preparation and resources. A similar movement is currently afoot in Europe, with key participation by planetarians.

While our primary focus remains astronomy education, we've also come to realize that our planetariums have evolved into multimedia environments capable of performances in the arts and humanities as well. Musical concerts, poetry under the stars, and actors in performance are periodic additions to many planetarium's schedules. And the Hansen Planetarium staff can tell some hair-raising tales about Van de Graaff generators. At the Taylor Planetarium, we've had storyteller Lynn Moroney in performance under the stars, telling Native American sky tales. And at Hayden, Bill Gutsch and staff have produced live performances of African songs, dances, and sky stories.

The most ubiquitous of these special programs is the laser light show featured at many planetariums, drawing in non-traditional audiences to be introduced to the planetarium and encouraged to attend other offerings. But even the laser is being diversified in its use: facilities such as the Adler Planetarium in Chicago use them regularly in their astronomy programs. Jack Dunn at the Mueller Planetarium in

Lincoln, Nebraska, has developed a program for people with visual impairments such as retinitis pigmentosa, using the intense light of lasers to create starfields and visual effects that give them back a night sky they thought was lost to them forever.

And the Hayden Planetarium in New York employs the Omniscan system developed by Audio-Visual Imagineering, in which laser light is projected though a wide-angle lens to create all-dome laser images for use in star shows ranging from grids to astronomical effects.

But the efforts of planetariums extend to more than just the star theater; astronomy classes, seminars, and workshops regularly combine classroom, planetarium, and outdoor learning at many of our facilities. Teacher workshops offer in-service training and resource materials to teachers of all grades; one took place in a tipi at the Museum of the Rockies' palaeontology field camp, and included work in a portable Starlab and observations under a pristine Montana sky.

At places like the Museum of the Rockies, outreach efforts are vitally important. We live in a big state with a sparse and scattered population, much of which lies too far away to come for a visit. So we go to them – shipping four Starlab planetariums around Montana with lessons, for which we offer periodic training sessions, and which have served well over 100,000 people – mostly students – in the last eight years. We also ship small travelling telescope trunks to schools; these trunks contain Edmund Astroscans with instructions and exercises for learning about telescopes and observing the real sky.

Starlabs and other portable planetariums are often used by facilities that have no fixed planetarium facility, widening the reach of astronomy educators. Sue Reynolds in New York heads an IPS committee which regularly collects and distributes tips and educational exercises to assist portable planetarium users in their efforts.

Having already referred several times to telescopes, I should point out that planetarians recognize that one of our primary objectives is to teach people with our simulated skies so that they can better understand and enjoy the real one outside. Many planetariums have telescopes associated with their facilities, and even more have regular programs of real-sky observation. During the day, we show people the sun using safe methods, and at night we have star parties, helping them find their way around the sky and showing them cosmic sights for real.

And at the Buhl Planetarium at the Carnegie Science Center in Pittsburgh, Martin Ratcliffe uses the Internet to link students with the Mount Wilson Observatory in California, letting them control one of the telescopes there and engage in research projects.

And of course, when special events come around, we generally pull out all the stops; we were there with large audiences watching the remarkable discoveries of the Voyagers as they happened ... we were a big part of Halley-mania a decade ago and got lots of people looking at the sky ... and managed to impress quite a few with views of Comet Shoemaker-Levy Nine's revenge on Jupiter last summer.

Planetarium efforts also extend to exhibits. For example, last summer our staff curated an exhibit called "Pioneering Space" at our Museum, built around NASA scale models and chronicling the history of the U.S. manned space program. A few steps away, people could see a complementary program called "The Final Frontier" in the planetarium. Many planetariums have extensive exhibitions to solidify the astronomy experience – just two that leap instantly to mind are the

Griffith Observatory in Los Angeles and the Adler with its marvellous collection of antique instruments.

The Lakeview Museum Planetarium in Peoria has developed a scale model of the solar system that has made the Guinness Book of World Records. The planetarium dome represents the sun, and scale models of the planets are placed at locations throughout the city and beyond – with Pluto 40 miles away! This July, the museum is sponsoring a bicycle ride from the sun to the planets, certain to give participants a unique perspective on the solar system!

And while all of this is going on, we regularly check our sources for the latest-breaking information from the universe to pass on to the public. The Space Telescope Science Institute in Baltimore has been particularly helpful in supplying planetarians with the latest materials and information on the remarkable discoveries of Hubble, and has assisted prominently in the development of several planetarium shows on the subject. More and more planetarians are blasting in cyberspace as well, sharing information and snaring resources from NASA and other places, all of which help us to do our jobs.

We also help each other through organizations such as the International Planetarium Society and its eighteen affiliated planetarium associations world-wide. The affiliates meet annually and IPS biennially, bringing together planetarium professionals and astronomy educators to discuss and debate the issues which concern us, to share ideas and experiences, to examine the latest products and technologies, to see what others are doing in their planetariums, and to support and encourage each other in our work.

Recent meetings have been held in the mountains overlooking Salt Lake City and at Cocoa Beach, Florida. We meet next year in Japan, hosted by the Science Museum of Osaka, and in 1998 we travel to London where we will be hosted by the London Planetarium.

The exchange continues between meetings through affiliate newsletters and *The Planetarian*, the IPS Journal edited by John Mosley. IPS also engages in special initiatives and special publications – one of the most prominent of which is the biennial directory of world planetariums, which facilitates communication among planetarians around the globe.

Well – I hope this small sampling from just a few planetariums helps to give some dimension to the role of planetariums in supporting formal and informal astronomy education. Our challenge is how to use this resource most effectively to help solve some of the problems outlined here today. But if this is where we are – where will we be going in the future?

Prediction is always tricky, but if I had to, I would summarize future trends in eight words – which I think not only reflect trends in education, technology, and society, but which will also be survival strategies for at least the near term.

Diversification: it will become increasingly important for planetariums to diversify their activities and offerings appropriately to serve and maintain the widest possible constituency and to support their operations.

Technology: we can't compete with an $80 million Spielberg move, nor should we really try, but we can employ new technologies to enhance our strengths and improve the quality of both our education and our productions. Updating our facilities where and when we can will be important.

Basics: nothing can currently beat the planetarium in its ability to demonstrate basic astronomical principles and simulate the backyard sky. This will be of vital importance especially to school planetariums and planetariums with a large student clientele. People will still need to know the basics.

Interactive: hands-on, activity-based programs will become increasingly important in meeting science curriculum objectives. People learn better by doing, and planetariums will be getting on board that bandwagon in greater numbers.

Multidisciplinary: schools and governments are working increasingly toward the integration of math and science disciplines to better prepare students for real-world experiences. The advantage of the planetarium is that it can synthesize these disciplines, relating astronomy not only mathematics and the other sciences, but even to history, the arts, and language. This will be a strength.

Multicultural: as our societies become more diverse and the modern world shrinks even further, it will continue to be important to learn about other cultures. Everybody's got sky stories and most have astronomical histories, and the planetarium will continue to serve a role here.

Outreach: the ability to serve constituencies outside of the star theater will become increasingly important as a way to diversify, to broaden our reach, and to maintain visibility. But it may work in the other direction, too, as research grants increasingly require an education or outreach component which planetariums may be able to help fulfil.

Financing: many planetariums will continue to have to scramble for the dollar or peso or ruble that keeps them open and operating. We will probably be looking increasingly to grants, endowment opportunities, and other fundraising methods to maintain our operations.

And finally, I might add one more word – perhaps less a prediction than a hope. And that word is **partnership**. I hope that the community of planetariums and the communities of scientists, educators, and astronomy groups will continue to forge closer ties and open new avenues of dialogue – as represented by this symposium. I think it would be of mutual benefit, and would help to advance our common goals.

The Englishman Thomas Carlyle once lamented: "Why did not somebody teach me the constellations, and make me at home in the starry heavens, which are always overhead, and which I don't know half to this day?" Well, planetariums have been doing just that for the last 70 years.

Not only do we teach people the constellations and their age-old stories; we teach them about the stars that comprise them ... how they came to be, how they live and die ... that some of them, like the sun, may have planets ... that they are part of a vast family called the Milky Way ... just one galaxy among billions wheeling through the universe.

We teach them the cycles of the sky, and about the wanderers we find there ... and why the earth is a special place and worth taking care of.

We explain the tools and methods which help us to think what we think and know what we know ... we tell them where we've been ... and where we may go in the future ... we remind people that they, too, can explore the space frontier – as future scientists, or astronauts, or one day as the voting public that will decide how far and fast we will go. Our task is to help make those decisions enlightened ones.

In the planetarium profession, ours is not so much the astronomy of research and discovery as it is the astronomy of interpretation and enlightenment. We help to bridge the distance between the scientist and the public, between the individual and the universe. We strive to channel human reaction to the pretty lights in the sky into a deeper understanding of the universe from which we spring.

This is the role we play, every day, each in our own way, in two thousand planetariums across the planet. More than 450 million served and counting ... one show ticket, one class enrollment, one peep through a telescope, one person - one glimmer of understanding - at a time.

Thank you.

Addendum

John Mosley, Griffith Observatory, wished to add two points of emphasis: that in large urban settings, the planetarium often provides the only opportunity for people to really see the sky, and that planetariums today are the public centers for astronomy education.

Discussion

Crawford.
I show many people dark skies in IDA activities, and often I hear "It's just like in the planetarium." Shows in planetaria are doing a great job, but it's sad that so few ever see (or will see) a "live" dark sky. It would be sadder yet if in only a few generations they would be the only place anyone could see the universe.

Duncan.
Why is the partnering of astronomers interested in education and planetariums so rare?

Manning.
We're starting to change that.

The AAS Education Initiative in Astronomy

S. Edwards[1] and S.E. Strom[2]
Co-chairs, AAS Astronomy Education Policy Board
[1]Smith College
Astronomy Department
Northampton, MA
USA 01063

E-mail: sedwards@smith.smith.edu

[2]Astronomy Program
University of Massachusetts
Amherst, MA
USA 01003

E-mail: sstrom@donald.phast.umass.edu

Abstract: The American Astronomical Society is preparing an Education Initiative aimed at mobilizing its 6000 members to define effective programs which can contribute to the fundamental goal of strengthening science education at all levels in the United States. To make the best use of the relatively limited human resources in the Society, the Initiative is designed to take advantage of the unique characteristics of astronomy in attracting students to science and to ensure that investment is concentrated in areas where the membership has its greatest strengths and skills.

1. Introduction

It is most appropriate that the first presentation of the Education Initiative of the American Astronomical Society (AAS) be made at a meeting of the Astronomical Society of the Pacific (ASP), a society that has been devoted since its inception to astronomy education and to forming links between the science education and research communities. The Education Initiative of the AAS is designed to increase the emphasis on education within the primary society of professional astronomers in the US and to stimulate the growth of programs that define the most effective ways for the AAS membership to respond to the dual problem in science education facing our nation:

- the vast and alarming science-literacy gap between the US scientific elite and the majority of our citizens who lack the most basic scientific knowledge and quantitative reasoning skills
- the growing perception that the scientifically talented students leaving our undergraduate and graduate institutions are trained as relatively narrow specialists rather than as problem solvers armed with the broad mix of skills and experience needed to meet the challenges of the next century

The Executive Branch, Congress, and academic and industrial leaders have called on the nation's scientists to assume a leadership role in helping to solve these problems (Science in the National Interest, 1994). Astronomers, though small in number, have the potential to make a significant contribution to both of these issues. Our field is distinctive in its drawing power, in its long tradition of informing and inspiring the mind, and in its ability to resonate with some of the most basic questions at the root of human curiosity. These characteristics make astronomy a powerful vehicle to develop public awareness and to advance public understanding of science. Creative presentation of scientific and mathematical concepts in an astronomical context offers the potential of first capturing the imagination of our nation's children, and then leading them toward a deeper understanding of the role of science in the world they will inherit.

At the same time, astronomy provides a natural vehicle for developing greater breadth among our scientifically talented students as well, because of its multidisciplinary nature, incorporating methods from scientific disciplines as diverse as meteorology, geology, molecular chemistry and particle physics, and its use of advanced instrumentation based on sophisticated state-of-the-art technology and engineering. In order to respond to the needs facing our nation and to fulfill the educational promise of our discipline, we as astronomers must find a way to match our dedication to excellence in research with an equally strong commitment to advancing science literacy and to educating a workforce well equipped to meet the challenges of the next century.

2. Goals

A successful approach by the AAS requires that we focus our efforts in areas where we as astronomers have both the clearest understanding of the issues, and a reasonable expectation of being able to shape outcomes. We thus have chosen to direct our initial efforts on examining critically the curricula and culture in the colleges and universities in which AAS members involved in education teach and carry out research. The first goal of the AAS Education Initiative will be to conduct a series of policy discussions aimed at identifying specific changes in curricular goals for undergraduate and graduate astronomy education in the 21st century that will ensure that our teaching efforts in colleges and universities:

- will continue to succeed in producing new generations of first-rate research scientists
- will produce science students who are broadly trained and able to move into a variety of careers requiring a strong technical and scientific background
- will make significant advances in increasing science literacy among college students, especially pre-service K-12 teachers
- will attract outstanding minds to careers in science education

A second goal of the Education Initiative is to develop a mechanism within the AAS to inform and coordinate science education efforts at all levels (public outreach, planetaria, museums, pre-college students, college students, graduate students) through (1) the creation of an accurate, accessible, and regularly updated electronic *Astronomy Education Database and Resource Network*; (2) the formation of a network of sub-committees within the AAS to coordinate ongoing education efforts at each of these levels; and (3) the establishment of partnerships with other professional scientific societies carrying out science education reform programs. The Astronomy Education Database and Resource Network would build on extant databases, such as those compiled by the ASP, but would include a broader range of topics, such as:

- Education Policy: This section would document the strategic planning and policy actions of the AAS, and include up-to-date policy discussions emanating from the NAS, the AAAS, funding agencies, the executive branch, and congressional subcommittees.
- Curriculum Issues: This section would describe innovative courses and programs being implemented in graduate and undergraduate astronomy programs, and

outline national/state efforts to define curriculum standards for pre-college science education.

- Curricular Resources: This section would identify curricular resources developed by AAS members and other members of the astronomical community, including textbooks, laboratory resources, educational software and audio-visual materials, organized according to its appropriateness for graduate, undergraduate, high school or elementary school audiences.
- Professional Development Opportunities: This section would identify resources available to assist in expanding the skills of instructors in undergraduate and graduate astronomy courses.
- Teacher Preparation and Enhancement Programs: This section would list and briefly describe programs related to astronomy designed to improve the math and science skills of our nation's schoolteachers.
- Public Outreach Programs: This section would include descriptions of programs for and by Planetarium/Museum Directors, outreach programs at National Centers, programs working with the media, etc.
- Funding Opportunities: A regularly updated list of funding opportunites in science education from the NSF, NASA, Dept. of Defense, Dept. of Education and other relevant funding agencies.
- Related Activities in Other Professional Scientific Societies: A list of key individuals or committees involved in education reform in the ASP, in other professional scientific societies, education societies such as the AAPT and the NSTA, and astronomical organizations, and a short overview of their educational activities.

By gathering and synthesizing this information, the AAS will be in a position to coordinate and partipate in substantive dialogs and collaborative efforts with other astronomical groups, professional societies and organizations involved in science education reform. Armed with this information and a developing understanding of what constitutes successful programs in science education for pre-college and public audiences, the AAS can then begin to identify the most effective mechanisms and programs in which astronomers can contribute significantly to fundamental science education reform -- information which can both marshal human resources within the society, and influence financial investments from the funding agencies.

3. Implementation and Outcomes

To achieve the goals of its Education Initiative, the AAS hopes to appoint a skilled and energetic Education Coordinator who will work with the AAS Executive Officer and the newly appointed Astronomy Education Policy Board to plan and carry out both the policy discussions on graduate and undergraduate curricular reform and to develop the plans to inform and coordinate education efforts within the astronomical community.

The policy discussions with members of the astronomical community will be conducted at AAS meetings and in regional workshops for department chairs. The discussions will culminate in a strategic plan for astronomy education that will identify changes in curricular goals in graduate and undergraduate programs and suggest potential investment strategies and programmatic initiatives that, were they adopted by the NSF and/or by NASA, might be particularly effective in advancing science education reform.

Although specific recommendations await the outcome of the policy discussions, alternative approaches to teaching science students at both graduate and undergraduate levels might include: (1) new mechanisms to encourage talented science students to pursue careers outside of narrow research disciplines and prepare them for such paths; (2) identifying ways for astronomy departments to take maximum advantage of plans under development within the NSF to decouple funding for the education of graduate students from research objectives: and (3) establishing a broader reward structure and recognition of achievement in the profession. Similarly, discussion of new approaches to undergraduate teaching of non-science students might (1) identify the most effective ways of teaching science through astronomy to non-scientists; (2) establish whether a new genre of introductory class for pre-service teachers would be the best means of preparing tomorrow's teachers to meet the standards being developed by the National Research Council; and (3) identify how best to forge ties between astronomy and education departments at our colleges and universities.

In addition to specific directives aimed at curricular reform, we anticipate that implementation of the AAS Education Initiative will elevate the level of discussion of education policy issues within the AAS, and provide the momentum for a continuing dialog involving all segments of the Society--both those already engaged actively in science education, and those whose primary interest is focussed on research. Furthermore, by establishing an informed membership, the AAS will be capable of participating effectively in national efforts to mobilize the talents of scientists to address pre-college education reform, and to promote public awareness of science and science literacy.

Acknowledgements. The plan for the AAS Education Initiative was formulated by the AAS Astronomy Education Policy Board, with members Alexei Filippenko, Andrew Fraknoi, Catherine Garmany, Eugene Levy, Lawrence Rudnick, and Michal Simon, appointed by AAS President Frank Shu. Seminal ideas for the mechanism to inform and coordinate science education efforts within the AAS were provided by an ad hoc AAS Education Committee with a one year term during 1993/94, chaired by Suzan Edwards, with members Bruce Balick, Robert A. Brown, Doug Duncan, Andrew Fraknoi, William Herbst, and Julie Lutz. The dedicated commitment of the volunteer AAS Education Officer, Mary Kay Hemenway, and the growth of the AAS Working Group in Astronomy Education led by Steve Shawl were essential motivations to this effort.

Discussion

Asbell-Clarke.

I do not think the requirements for applicants for the Education Coordinator of the AAS to have a Ph.D. in astronomy is wise. The Coordinator will have to interact and coordinate the education community, including knowing the needs of teachers and students. I am not convinced that Ph.D. astronomers (with the exception of everyone in this room) are the best suited for this position.

Smith.

1) The Education Coordinator need not be a Ph.D. astronomer. (I am one of them.) While astronomers may feel more comfortable with a Ph.D. person in this position, a professional educator may be able to work more effectively with other educators.

2) It is unlikely that most departments teaching astronomy would be able to add a separate course for pre-service K-12 teachers. Since many of us teaching

college/university astronomy know the people in our science education departments, it might be more feasible for us to team-teach with them in their courses. For the smaller number of us with planetariums on campus, the science ed. classes can come to the planetarium. On my campus this occurs now on a regular basis, and we cover both content and methods.

Zeilik.

I become very leary when I see astronomy education for teachers split off from those of the general non-science majors. Conceptually, both start at the same level. The strategies to incur conceptual change are basically the same, I don't think it would be good policy to treat these populations differently – teachers with a special small class, the others in a standard survey.

Edwards.

I agree completely that both groups of students start at the same level, and that all would benefit from strategies for conceptual change. However, teaching future K-12 teachers is one of the most valuable investments in the future we can make, one where we have the potential to have the greatest leverage in improving science education over a decade-long timescale. Something extra for this group of students that will help them to understand how to use scientific thinking in their own curricular planning is worth striving for. One of the goals of the Education Initiative is to understand how astronomy departments can work together with Education Departments at schools where both exist, to define effective ways to do this. No doubt different solutions will exist at different institutions, but if we have limited resources and time, I would like to see future teachers as one of our prime beneficiaries.

Buckingham.

I strongly encourage AAS to partner with the 1,000 U.S. planetariums, thousands of science museums, science centers, natural history interpretive sites, etc. These institutions reach more citizens per year than all universities (combined) in five years. Such institutions can provide tremendous multiplier effects for your reform efforts (including outreach to in-service teachers).

DeVore.

The AAS is commended on beginning to plan formally for educational reform. However, I am dismayed that this appears to be a top-down solution that has (so far) precluded the ultimate public, etc. The strategic plan needs the contributions of these consumers. (If Ph.D. astronomers alone knew the solutions to these problems, there would be no need for this group.) This committee needs to work in partnership with working professionals – not just university professors.

Thus, the undergraduate education session which in part aims at preservice education should be held at the January meeting of AAS in conjunction with the AAS Astronomer-For-A-Day program. Doing this session in June during the last days of the K-12 school year virtually precludes the participation of teachers.

Dukes.

While I agree that a restructured introductory astronomy course along the lines suggested by Zeilik and Hoff would be effective for preservice K-12 teachers these teachers need something more. In particular they need training in specific activites (geared to their state and local frameworks as well as national standards) that they can use in the classroom with their students. These activities are not and should not be identical to activities used in the introductory astronomy class. A separate class section might not be necessary but a separate lab or recitation section is definetly required.

Hill.

Currently, an astronomer who wishes to address a specifically astronomical education problem cannot publish his/her results in a journal for astronomers. About the best he/she can do is publish the results in the American Journal of Physics, which is read by a small, but significant number of astronomers. I would suggest that the flagship journals of both the American Astronomical Society and the Astronomical Society of the Pacific should <u>at the very least</u> carry occasional reviews of results in astronomy education and historical astronomy.

Shawl.

The *Journal of College Science Teaching* is an excellent place for astronomy education papers. Astronomy educators should read this journal. It is broadly based so that readers can be exposed to a larger set of ideas than would occur in an astronomy-education-only publication.

Hoff.

Changes in undergraduate general education classes can be made to reflect better teaching practices (i.e. cooperative learning). The use of a participatory lecture can accomplish a great deal to model good teaching. To suggest separate courses or sections for pre-service teachers is unrealistic and will not happen, hence we should concentrate on that which is doable.

Schatz.

I first want to applaud AAS for putting this education initiative forward and giving it the prominence that it appears it is getting. The focus so far is on graduate and undergraduate education. What is the discussion about initiatives for the K-12 level?

Wasiluk.

Why doesn't the AAS quit their seemingly covered and hidden hatred toward educators? Instead of this meaningless effort towards "educating" educators, work on leaning more of their true needs. The "Astronomer for a Day" is a great start, but how about expanding it.....offering AAS memberships to K-12 teachers and teacher trainers in the university.

Being a teacher myself, we don't need to be talked down to, or get a program that is already happening in other venues, SPICA, STAR, POPS. If you are a small number of people, don't waste money, talent, time on programs already done elsewhere. Better, use your money to send us continuing information.

Sakimoto.

In juxtaposing an emphasis on alternate careers for students trained in astronomy with an absence of an emphasis on outreach to underrepresented minorities, your policy statement gives the impression of trying to create a closed society – of trying to preserve careers in professional astronomy to those already in careers. Given that probably less than 0.5% of the 6000 AAS members are under-represented minorities, what steps are you taking to build diversity into the ranks of professional astronomers?

Edwards.

You are correct that this is an important issue for the future, and the Education Policy Board has not yet begun to address explicitly programs to recruit and retain minorities in its deliberations. This is certainly not due to lack of commitment to these principles, but I think that this issue is sufficently important that it needs to be dealt with more directly than as an "add-on" to our examination of curricular goals in astronomy education.

Getting Astronomers Involved in the Teaching of Astronomy

Julieta Fierro
Instituto de Astronomia UNAM
Apdo Postal 70-264
04510 Mexico DF
Mexico

E-mail: julieta@astroscu.unam.mx

One of the most difficult problems of education is to convey scientific updated knowledge to all levels of formal education: enough comprehension of science at the convenient time. Astronomy is such an interesting subject that it can be a good vehicle to transmit part of this knowledge. A way to approach the problem is to get astronomers to contribute periodically in non-formal education and to have them review curricula and school books when their field of expertise and related topics are discussed. If astronomers get a chance to write, participate in radio and television programs, collaborate with planetaria and multimedia projects, give public lectures, workshops and conduct astronomical observations, as well as contribute to science center projects, the quality of education will increase. And if they can supervise the national curricula and books, their influence will be better and longer lasting. The reason is astronomy is especially appealing and if it is conveyed by someone who understands the subject it can be made simple and interesting. That is why it is extremely important for research evaluation to consider formal and non-formal education as a priority.

In Mexico this has been possible to a certain extent, simply because astronomers were given a chace to get involved with education at all levels.

Education problems in developing countries are so great that an isolated effort to improve it cannot make a great difference, even if it is very costly and time consuming. In Mexico, the average education is only three years of elementary school, so it is not sufficient to improve curricula, or write better books, have more influence of scientists on education, have better parental care, improve support for teachers; all efforts must be carried out simultaneously, during a considerable amount of time, in order to have the desired influence on mass education. Many isolated efforts have been done with little success in spite of many hours dedicated by parents, teachers and children.

We shall discuss some of the projects Mexican astronomers have been involved with and what success they have acheived. One of the first great involvements of astronomers, and scientists in general, with public education was the series of government sponsored books: "La Ciencia desde Mexico". Presently more than 130 titles have been edited on many subjects, 15 of which are on astronomy. The books are cheap and widely available. The main problem with this kind of project is the heterogeneous quality of the books: the level of clearness of the books varies greatly; so does the level at which different topics are covered. Several authors did not finish their texts, so that certain fundamental topics were not covered. Nevertheless, this experience encouraged editors to have Mexican professionals write books, with better quality paper and sometimes with color illustrations. Now, fortunately, many scientific books written in Mexico are used in schools and are available in the 300 public libraries throughout the country (prevouisly only books in English and a few translations were available). I feel such projects are excellent for developing countries, mainly because it encourages youngsters to study science because they do not feel it is only cultivated abroad. The language used in such books is easier to understand and includes experiences that are not alein to them, so it is easier for students, and public in general, to relate to the subjects.

Mexican astronomers have also participated actively in radio programs which are relatively easy to produce and reach enormous audiences. Some of the radio programs are specially recorded for rural teachers, in order to give them encouragement and updated information. For countries where access to small communities is difficult and where many native languages are spoken such projects are extremely useful. Astronomy programs for motor vehicle drivers, specially for those having to cope with rush hour traffic, can be attractive as long as they are kept interesting and clear, and conducted informally.

Another interesting experience has been scientific exhibits, specially sponsored by local state governments and the National University. In particular the latter has opened a great science museum and is constructing a new one specially dedicated to light. It has held temporary exhibits in all sorts of places including subway stations, schools and even the Palace of Fine Arts in Mexico City, accompanied by concerts performed by the National Orchestra. Science centers are an ideal way to get the population to contact astronomy knowledge, especially if they have follow up programs for teachers and clubs for adults. Summer courses held at science centers are very popular and an ideal way for youngsters to get excited about astronomy.

The National Academy of Sciences has sponsored "The Wagon of Science" that is literally an old train wagon that has been converted to carry out science workshops for children. There is almost one of these per state and they are becoming more popular because of their relatively low cost and great success. All kinds of experiments are carried out there. Astronomy projects include: properties of light, telescope viewing, constellations and sun dials.

Recently our National University has begun a program that gives high school pupils a very small scholarship and to "adopt a researcher during the summer". Although most of the students enrolled in such programs get truly motivated to study science, unfortunately the number of people involved is very low.

Mexican astronomers' long term goal is to participate more actively in basic education. The problem in Mexico is so great that our efforts have not had the impact they would like. Presently they have nothing to say concerning curricula. Little by little we are being invited to write a few chapters of the books or in some cases to check what others have done.

What astronomers need to be able to promote science is for them to have a chance to do it!

Discussion

Hawkins.

How is your education program integrated with efforts of the research astronomers?

Fierro.

Very well integrated. In particular, astronomers who work in education are not ostracized, and are viewed under the same "prestige" standards as research-only astronomers.

Education Programs At The National Science Foundation[1]

Gerhard L. Salinger
National Science Foundation
Room 885
4201 Wilson Boulevard
Arlington, VA
USA 22230

Tel.: (703) 306-1620
E-mail: gsalinge@nsf.gov

The astronomy research community has been exemplary in its efforts to communicate how astronomy is done to the general public. Almost no other discipline, for example, has a *Sky and Telescope*. Increasing the contact of research astronomers with the schools is a most welcome goal.

Several professional societies have divisions that channel the interests of the scientists: In physics and in mathematics, there are separate professional societies concentrating on educational matters; but recently there is the Forum for Education of the American Physical Society and special sessions at meetings of the American Mathematical Society - both of which are primarily research societies. In chemistry, chemical education is a division of the American Chemical Society. Before implementing such activities in the astronomy community, a review of successful practices in other professional societies should pay dividends. In preparation for making contributions to science, mathematics and technology education, particularly for the K-12 system, it is important to be familiar with the new materials and with the developing Standards.

The main purpose of this talk is to provide some information about educational reform, particularly in elementary and secondary education, that may guide efforts of astronomers in changing the teaching of science, mathematics and technology education in schools. Of particular interest to astronomers are the programs of the Directorate for Education and Human Resources at the National Science Foundation that fund the development of instructional materials and the in- and pre-service education of teachers and faculty to implement educational reform from kindergarten through undergraduate school. A significant part of the success of the projects funded by these programs is the interest and contributions of practicing scientists to all of the activities.

The reform of science, mathematics and technology education has been enhanced by the development of national, not federal, Standards. Led by the National Council of Teachers of Mathematics, the mathematics community developed the Curriculum and Evaluation Standards for School Mathematics (1). The Standards provide guidance about what mathematics students should learn and how they should learn it. They are not "hurdles" specifying what a student should learn in each grade, but rather goals that can be used to determine if good mathematics is being taught and learned. Students are expected to reason and communicate mathematically; they should become mathematical problem solvers and see connections between mathematics and other disciplines. Mathematics should be taught from rich engaging contexts. The mathematics research and education

[1]Opinions, findings and conclusions or recommendations expressed in this article are those of the author and do not necessarily represent the views of the National Science Foundation.

communities support the mathematics standards.

There are two sets of Standards for science which, fortunately do not differ significantly. Historically the American Association for the Advancement of Science (AAAS) developed Project 2061 – a comprehensive restructuring of all of science-based education. The initial document, *Science for All Americans* (2), describes what the scientifically literate person should know upon leaving high school. Chapters in *Science for All Americans* describe discipline knowledge in the sciences and "habits of mind" which scientifically literate people should have. Understanding of technology and design are also emphasized. Project 2061 has provided *Benchmarks for Science Literacy* (3) which provides guidance for what students should know at grades 2, 5, 8 and 12 in each area. Project 2061 is producing a series of documents for implementing the Benchmarks and educational reform.

In the early stages of the development described above, leading science education organizations joined together with the National Academy of Sciences to develop the *National Science Education Standards* (4). There has been much opportunity for comment and the final document is to be released in December, 1995. The Standards cover content, teaching, professional development and assessment. Although there may be some slight differences in emphasis, either the Standards or the Benchmarks provide a good basis for judging whether good science is being taught in schools.

More recently, the National Science Foundation and NASA have funded the International Technology Education Association (5) to develop a rationale and structure document for technology education. Technology education has replaced industrial arts. It is the study of the built world using design, systems analysis, and three-dimensional modeling as ways of knowing. In schools, technology education serves to provide technological literacy for all students and as a background for students preparing either for the workplace or for further education in engineering.

Implementing the Standards requires exemplary instructional materials and teachers who understand how to guide students' learning. The instructional materials should help the teachers in their transition to new ways of teaching to maximize student learning of concepts and processes. Basing materials on the Standards ensures that the content is important, that concepts are emphasized over memorization of facts, and that the pedagogical strategies are appropriate. Scientists are needed in the development of materials to ensure that the content is accurate and that science is taught the way science is done. The Instructional Materials Development program is particularly interested in materials that emphasize the learning of design, that have authentic research experiences for students, that motivate the study of science through actual workplace examples and that contain innovative and appropriate uses of educational technology. The materials are carefully piloted and field tested and have plans for national dissemination.

The Teacher and Faculty Enhancement programs provide opportunities for continued professional growth. Emphasis is on both content knowledge and instructional strategies. College faculty must model good instructional practices both for preservice teachers and for all students. In Teacher Enhancement, more of the projects involve all of the teachers in a school district to systemically change the way in which science, mathematics or technology instruction is delivered. Teachers must be involved in establishing the need for the project and its form. Faculty Enhancement projects should emphasize serious astronomy for faculty who teach astronomy but are not astronomers. This can include serious research projects on small telescopes.

There are several programs in the Directorate for Education and Human Resources that should be of interest to astronomers seeking to implement changes in content and instruction. The Division of Elementary, Secondary and Informal Education (ESIE) has an annual budget of close to $200M. About half is for Teacher Enhancement; about 20% is in Instructional Materials Development and slightly less in Informal Science Education. The Young Scholars Program is funded at about $10M. The Division of Undergraduate Education has a budget of about $84M. Collaboratives for Preservice Education of Teachers and Faculty Enhancement has a budget of about $25M; Course and Curriculum Development receives about $22M and the Instrumentation and Laboratory Improvement Program funds about $21M as matching grants for equipment. This program is particularly interested in the use of modern equipment to do astronomy in which student have to analyze data. The Advanced Technological Education Program for technician education at the two-year college level and preparation for it at the secondary level is jointly managed in ESIE and DUE. Current Funding is about $23M.

(1) NCTM, 1940 Association Drive, Reston, VA, 22090,1989
(2) AAAS, Oxford University Press, New York, 1990
(3) AAAS, Oxford University Press, New York, 1993
(4) National Research Council, Washington DC, 1995
(5) ITEA, 1914 Association Drive, Reston, VA 22090

Discussion

Fraknoi.

Many underserved groups and two-year college faculty or small nonprofits are daunted by the complexity of the application process for NSF grants. Has any thought been given to experimenting with simplified application and proposal procedures for smaller grants? Would it be possible to have some experiments in this direction in each education division?

Salinger.

All the programs of the Division of Elementary, Secondary and Informal Science and the Advanced Technological Education Program require preliminary proposals. The preliminary proposal is a six page letter to the Program describing what is to be proposed, who is doing it, how it will be evaluated and disseminated. These are reviewed inhouse and advice given.

The NSF tends to want national impact. There is very little funding for equipment (except for colleges) and local workshops. These projects should be proposed to state and local funders.

The Role of Education in NASA's Space Science Research Programs and Missions

Jeffrey D. Rosendhal
Office of Space Science
NASA Headquarters
Washington, DC
USA 20546

E-mail: jrosendhal@smtpgmgw.ossa.hq.nasa.gov

1. Introduction

This paper briefly describes the approach being adopted by NASA's Office of Space Science (OSS) for making education at all levels and the enhancement of the public understanding of science integral parts of space science research activities. A full discussion of these new policies is contained in "Partners in Education: A Strategy for Integrating Education and Public Outreach into NASA's Space Science Programs" which was released in March 1995. This strategy is one component of NASA's overall contribution to a major national initiative to dramatically improve science, technology, and mathematics education and scientific and technological literacy in the United States.

While the goal of OSS remains first and foremost to plan and carry out a world class program of scientific research, it is also now clear that both the public and the political system expect benefits broader than purely scientific ones to be derived from NASA research programs and missions. Global economic competitiveness and sustained leadership in science and technology depend on greater public literacy in science, mathematics, and technology, and on the appropriate production of scientists, technologists and engineers to meet the future needs of the workforce. Producing the finest scientists and engineers for the twenty-first century, and raising the scientific and technological literacy of all Americans are specifically identified as national goals in the White House report "Science in the National Interest" released in August 1994. The NASA Strategic Plan also singles out education as one of the fundamental operating principles to be embodied in the conduct of every NASA activity.

The OSS Education/Public Outreach Strategy was developed as a direct response to these imperatives. It articulates a long-term vision for OSS involvement in education and public outreach and outlines the initial steps to be taken to begin to realize this vision. In so doing, it acknowledges the fact that the responsibilities of scientists must include explaining scientific results to the public and communicating the role and importance of science and technology in contemporary society.

2. Goals of the OSS Education/Public Outreach Strategy

- OSS will use space science missions and research programs and the talents of its research and development communities to make significant and measurable contributions to meeting national goals for the reform of science, mathematics, and technology education, particularly at the precollege level, and the general elevation of scientific and technological literacy throughout the country.
- OSS will continue to support the education and training of graduate students and postdoctoral fellows in space science in order to create the talented scientific workforce needed for the 21st century. OSS will expand its efforts in support of the education and training of undergraduate students.

- OSS will promote the involvement of women, underrepresented minorities, and students with disabilities in its educational programs and their participation in space science research and development activities.
- OSS will facilitate and cultivate strong and lasting partnerships on local, regional, and national scales between the space science research and development communities and the professional communities in science, mathematics, and technology education.

3. OSS Education Strategy: Environment/Boundary Conditions

Neither NASA nor OSS are primarily education organizations. At the present time, very large investments in educational reform, curriculum development, teacher training, and the development of science, mathematics, and technology standards are being made by individual states and school boards, and other agencies within the US government. The space science community will have to meet three challenges in order to be effective in enhancing these larger ongoing efforts. Space researchers must understand what is happening nationally, become aware of and sensitive to the needs of the education community, and identify those areas where space scientists can make significant contributions. A wide variety of small-and large-scale contributions are possible. The key to making a useful contribution is to identify and participate in those activities that draw on the unique character of the space science program and the special talents of the space science community, and that can be highly leveraged to have large impact.

Meeting these new responsibilities will require some redirection of resources. Budgets are not likely to grow over the next several years and a new emphasis on education and public outreach will require the application of some resources to these activities. But major changes should not be necessary. Incorporating education and public outreach components into the planning and implementation of missions and research programs should be relatively easy provided that it is understood from the beginning that the inclusion of such activities is important. Such efforts must also be focused, well coordinated, and carried out in collaboration with individuals who can supply critical expertise.

4. The Role of Space Scientists in Education/Public Outreach

The OSS research community can make significant contributions to education and the public understanding of science because it is a source of continuing new knowledge and remarkable new discoveries that can inform teachers at all levels and excite both students and the public about science. Space scientists can provide information, ideas, and materials in formats useful to educators and understandable to the public. They can write articles for the science sections of newspapers, trade books intended for public consumption, or introductory textbooks intended for both the high school and college level. They can provide information and materials directly to students through lectures in classrooms and to teachers through presentations at workshops. They can develop educational materials and technology or consult on the implementation of science standards and policies. The ultimate effectiveness of such activities will, however, depend on how well scientists understand the knowledge, skills, and needs of teachers, students, and the public. Achieving such an understanding can best be done by collaborating with individuals who can supply critical expertise.

The development of such collaborations and the formation of lasting partnerships between the space science and education communities lies at the very heart of the OSS strategy. It is through partnerships that activities undertaken by space science researchers can be highly leveraged and significantly affect national efforts directed towards education reform. It is through working with educators that the results from the space science program can be incorporated into museum displays, planetarium shows, science textbooks, and curriculum units throughout the country. It is through working with partners that the results from NASA's publicly supported space science programs can be taught to teachers, displayed on the walls of classrooms, placed in the hands of students, and appreciated by their parents. It is by partnering with the education community that space scientists can best share what they have learned with the people who have supported their work and, in so doing, contribute to increasing the public understanding and appreciation of science.

In assessing the contributions that scientists can make to education/public outreach, there are also roles that go well beyond supplying purely technical expertise. In particular, scientists:

- Have an understanding of how science really works;
- Have access to the tools and technology for the manipulation and distribution of information;
- Possess an enthusiasm for science and a way of looking at the world around them to try to understand why things work the way they do;
- Are a living demonstration that science is a human endeavor carried out by people of all kinds who have many different types of skills and who approach their science in many different ways;
- Have a presence in colleges and universities, research laboratories, and industries and businesses of all types across the country.

Any effort directed towards raising the public understanding of science and technology must communicate the fact that the products of science have a central role in modern society and are created by people who are an integral part of that society.

5. First Steps/Implementation

OSS is now taking the first steps in the process of implementing its Education/Public Outreach Strategy. OSS will develop the mechanisms to oversee and orchestrate a coherent program of education and public outreach activities. Emphasis will be on the infusion of education and public outreach activities into all programs and missions, making education and public outreach an explicit part of ground based and space flight research programs and projects conducted by NASA and NASA-supported space scientists throughout the country.

OSS will provide incentives and legitimize educational and public outreach activities across all disciplines, programs and projects. It will support education and public outreach supplements to research funding and grants for education and public outreach activities. It will arrange for the training of scientists in education reform and how to develop and present information and materials in forms that are useful to teachers and effective in the classroom. It will facilitate access to educational and public outreach materials in space science. It will work with the space science community to increase the participation of underrepresented groups in space science programs. It will carry out an ongoing evaluation of its education and public outreach programs to assess their educational impact and effectiveness in contributing to the broad public understanding of science and technology.

6. Concluding Remarks

The strategy sketched here represents an important new direction for OSS. It is a crucial element in OSS's response to the need to make broader contributions to attaining the country's larger scientific and technical goals. While many factors will contribute to the ultimate effectiveness of the OSS Strategy, there are a number of considerations which seem to be absolutely required for success:

- A recognition by individual scientists that they have a responsibility to do more than their own scientific research in return for public funding;
- An acknowledgement by the scientific community of the importance of this new role;
- Recognition by scientists' home institutions of the critical national importance of education/public outreach activities;
- Removal of institutional barriers particularly between science departments and departments of education at colleges and universities;
- A willingness by scientists to adapt to the needs of the educational system rather than expect that the system will adapt to them;
- Respect by scientists for professionals in other areas if there are to be genuine partnerships and collaborations;
- Respect by scientists for their intended audiences. The point of communicating science to the public is not to convince the public of how smart or how clever the scientists are but to convince the public that what scientists do is important and contributes something to the lives of the people who are the ultimate supporters of the research;
- Realistic expectations;
- A long-term commitment;
- Patience on the part of everyone involved.

References

National Aeronautics and Space Administration 1995. Strategic Plan.

National Aeronautics and Space Administration 1995. *Partners in Education*: A Strategy for Integrating Education and Public Outreach into NASA's Space Science Programs.

Office of Science and Technology Policy/Executive Office of the President 1994. *Science in the National Interest*.

Discussion

Tucker.

One key to successful astronomy education is student motivation. NASA has a key card to play here. When students actually touch a piece of the moon, it has a profound effect on many of them. NASA has made the process of obtaining the loan of a lunar sample for classroom use extremely difficult. Instructors must travel to a regional site (often hundreds of miles) on one day out of the year to learn how to display the specimen. In many cases, including my own, this is not feasible. This instruction could be done over the Internet! Also, NASA should provide lunar touchstones to some public facility in all urban areas with populations over 100,000.

The Role of National Observatories in Science Education: From Moonlighting to Mainstream

Laura Danly
Space Telescope Science Institute
3700 San Martin Drive
Balitmore, MD
USA 21218

Tel.: (410) 338-4422
E-mail: danly@stsci.edu

I. Introduction

There can be no doubt that science education is emerging as a central theme in the national debate about the role of science in the future of our society. Numerous studies showing poor performance by American children in science, and the alarming rate of scientific illiteracy among the American public (as was vividly illustrated in Andy Fraknoi's talk this morning; see contribution, this volume, p. 9), have inspired policy makers to reorder priorities, emphasizing the importance of science education as part of the scientific enterprise of the country.

Success in improving science education and scientific literacy nationwide will be achieved only through a concerted and coordinated effort by several different institutions and individuals. For astronomy education in particular, some of the players include:

- astronomers
- educators (teachers, professors, administrators)
- amateur astronomers
- colleges, universities, and two-year colleges
- industry and contractors
- funding agencies such as NSF and NASA
- scientific professional societies
- education professional societies
- national observatories and labs
- youth organizations
- curriculum developers
- education specialists
- publishers
- news and entertainment media
- planetaria
- science centers

Each has a unique role to play in the process of obtaining, translating, and distributing the process and results of astronomy to a wide and diverse audience.

Of course, many of the individuals and organizations listed above (and I suspect most of us in attendance here) have been engaged in science education long before being directed by national policy. Motivated by their love of astronomy and a desire to serve their communities, astronomers, students, and staff from university departments and national observatories have initiated educational outreach activities throughout the country. Frequently, however, their efforts have been conducted without institutional support, usually on volunteer time, and, in the worst cases, to the detriment of research careers. But that is changing. Recent policy initiatives suggest that we now have the opportunity to conduct educational outreach within the institutional framework and, in many cases, with budgetary support.

Our challenge is to create an effective national coalition which builds on the strengths and unique assets of individual participants and promotes responsiveness to local needs. Any national effort should be oriented to serve the individual needs of local communities throughout the country, providing resources to the institutions that have direct contact with those communities. The ideal network would permit all participants to have access to infinitely more resources, learn from the experiences of others, and share the fruits of their labors. Therefore the title of the talk, from "moonlighting to mainstream": how can we incorporate the talent, experience, and individuality of the educational outreach programs being developed and conducted at the national observatories with the unique assets those instituions possess, to build an effective national program?

First, I will review the educational activities of the national observatories to date. I then address the potential role to be played by the national observatories in the context of a coordination national effort. Finally, I raise some issues that are likely to arise in the transition to a national effort.

2.0 Educational Programs at National Observatories

2.1 National Optical Astronomical Observatories (NOAO)

Since 1964, NOAO has maintained a Visitor Center which serves upwards of 70,000 to 100,000 visitors per year. The Center includes exhibits and a gift shop which sells crafts from the Tohono O'odham tribe on whose reservation the Center sits. Current developments and future plans call for interactive exhibits and even a 16" visitor telescope. NOAO provides tours and public evenings, and maintains the NOAO photo collection which provides astronomical images upon request. NOAO is also an REU site (funded by the NSF), providing research experiences for undergraduates at NOAO observatories. More information on these activities and other news and information services can be obtained from Yvette Estok, Manager of the Public Information Office (yestok@noao.edu).

This past year, NOAO created the new position of Education Officer (currently held by Suzanne Jacoby; sjacoby@noao.edu), to develop a K-12 educational outreach program. Her poster (see contribution in this volume, p. 225) summarizes their activities, which include:

- increased and coordinated involvement in local classrooms with teams of Tucson-based NOAO staff making multiple visits to elementary classrooms, and
- use of the extensive computer facilities of NOAO to electronically disseminate educational materials, seeking out those programs without resources to do it for themselves.

Future plans are underway for a more expanded program, and funding is being sought. Up-to-date information can be found on their web page at www.noao.edu.

2.2 National Radio Astronomical Observatories

NRAO offers daily public tours at their observatory sites in both Green Bank, WV and Socorro, NM. Special tours can be arranged by request (contact Sue Ann Heatherly in Green Bank, and Dave Finley in Socorro). Roughly 20,000 visitors go through each observatory per year. In addition, the Green Bank Observatory has for several years run summer teacher training workshops where teachers conduct observations using the 40' telescope on site. All 40' data is available on line via anonymous ftp. Further information can be found at www.nrao.edu, or www.info.gb.nrao.edu for Green Bank in particular.

2.3 Arecibo Observatory (National Astronomy and Ionospheric Center - NAIC)

Arecibo Observatory is the world's largest single-dish radio telescope. With funds contributed by local Puerto Rican government and private organizations, they are constructing the Arecibo Observatory Visitor and Educational Facility. The poster by Daniel Altschuler and Jo Ann Eder summarizes their activity (see contribution, this volume, p. 156).

Located in Arecibo, Puerto Rico, they are uniquely situated to serve the large minority population of the island. Their programs are all bilingual, and develop awareness of the "invisible sky" -- the universe which is not accessible by direct sensory experience. Emphasis is also placed on atmospheric science, and on the daily life of the observatory. In these ways, the Observatory's program highlights those characteristics unique to Arecibo.

Arecibo is also an REU site (NSF funding) for summer undergraduate research experiences. More information about these programs can be found by emailing jeder@NAIC.edu or daltschuler@NAIC.edu, or on their web site at www.naic.edu.

2.4 Space Telescope Science Institute

From its inception, STScI has supported extensive news and information services, including the provision of images, explanatory text, newsletters, and science writers' workshops, particularly near the time of the HST-related shuttle missions and the Jupiter-SL9 encounter. STScI does not support regular public tours or maintain a visitor's center (the official visitor's center is located at the Maryland Science Center on Baltimore's Inner Harbor), although it does provide monthly public lectures followed by public observing.

During his tenure as head of the Public Affairs Office, Eric Chaisson changed the office name to Education and Public Affairs to reflect the numerous educational activities in which he engaged. Most visible were the national teachers workshops held during the summers and the Star Finder broadcast and video program, a series of 30 15-minute television programs on astronomy which was co-produced with Maryland Public TV for classroom use. During that time, STScI became one of the founding partners in the Johns Hopkins Space Grant Consortium, now the Maryland Space Grant Consortium, which runs a wide range of outreach and teacher training programs.

In November, 1993, STScI created a new position, the Project Scientist for Education (currently held by Laura Danly; danly@stsci.edu), dedicated solely to developing educational outreach. Over the past year, NASA provided funding to develop a significant outreach program at STScI, now headed up by Carol Christian (carolc@stsci.edu). The Education Program Manager is Anne Kinney (kinney@stsci.edu). General questions can be directed to outreach@stsci.edu. Information can also be found on the STScI web site at www.stsci.edu.

At present, STScI runs three outreach programs to local students. Each was started by the initiative of an individual staff member outside his/her official job duties; each was funded by the Maryland Space Grant Consortium, and each uses STScI staff time plus volunteers from other local institutions. Posters by Laura Danly and Flavio Mendez, and Jon Saken (see pp. 188 and 272 this volume) give more information. Programs include:

- The Women's Science Forum, an outreach program for high school girls interested in science
- Space Telescope Elementary Outreach Program, provides classroom visits to 3rd and 4th graders, while offering training for pre-service teachers
- Student Hands-On Physics, a series of hands-on activities presented in eight repeated visits to four classes at West Baltimore Middle School

In addition, STScI maintains a speakers bureau which serves local and national functions, and provides the program for several teacher training programs, (including one geared specifically to introducing teachers to the Internet and the World Wide Web) for the Maryland Space Grant Consortium. STScI also provides, on average, 6-8 internships for high school students, and 3-4 internships for teachers per year. It also runs a summer undergraduate research program, similar to an NSF REU, but funded by the Director's Discretionary research fund.

STScI also runs the Initiative for Developing Education through Astronomy (IDEA) for NASA. IDEA is a grants program for small initiatives in astronomy education. All astronomers are eligible to apply for grants either in the $6K or $20K category. For more information, see the poster with Carol Rest (p. 227), or e-mail idea@stsci.edu.

In the spring of 1994, STScI and the Buhl Planetarium of the Carnegie Science Center co-produced a planetarium show, "Through the Eyes of Hubble". The show premiered in March, 1995, and in the first six months following its premiere, was sold to 48 planetaria worldwide, having been translated into six languages (German, Japanese, Dutch, Danish, Finnish, and Spanish). More information on this program and other planetarium support activities can be found in the poster by Rob Landis on p. 230.

STScI has also developed and distributed several educational posters on HST-related topics such as the first servicing mission, the planets as viewed by HST, the Jupiter-SL9 encounter, and stellar evolution. The Institute supports numerous professional meetings of the scientific, education, and the museum and planetarium communities (e.g. AAS, AAAS, ASP, NSTA, AAPT, ASTC, etc.), providing booths and speakers on a regular basis. Upcoming activities include a joint ST ScI - Smithsonian traveling museum exhibit emphasizing the second HST servicing mission and a presentation of results on cosmology. It is scheduled to open in the spring of 1997 in Washington, DC. Finally, a live television broadcast, "Live From.... Hubble" will be broadcast in the spring of 1996 to educate students and other viewers on how astronomical research is conducted with the Hubble Space Telescope.

3.0 The Potential Role of the National Observatories

The national observatories are a national resource. Through significant tax-payer investment, the national observatories provide the means by which advancement in astronomy occurs by providing equal access to large, state-of-the-art telescopes. The central, coordinating role played by the national observatories in conducting the research enterprise in the country can and should be extended to astronomy education: the concentration of unique resources and direct access to both the astronomical community and federal funding mechanisms make the national observatories well suited to coordinate national efforts in astronomy education.

Unique resources of the national observatories which can be used to enhance national education efforts include:

- the data. The results of astronomical research lie at the heart of what the scientific community has to offer the science education communities. As the data are produced and provided by the national observatories, those institutions are uniquely situated to support individual researchers in making new results and their associated images available to science communicators and the public at large. In addition, most observatories also have an archive or some archiving capability. While a complete archive is unrealistic for some facilities, the production of a uniform archive of data for student hands-on research projects would be desirable, as well as an archive of images which are electronically available.
- computing facilities. The hardware, software, and expertise concentrated at national observatories are well suited to address the national need for digital information on astronomy. As mentioned above, most observatories have experience with processing, displaying, and archiving astronomical images. In addition, the national observatories can serve as central repositories for on-line information.
- expertise. The on-site staff of the national observatories provides a pool of knowledge and experience which can enhance all aspects of a national education initiative.
- the observatory facilities. The impressive technology associated with the observatory facilities attract school groups and visitors by the hundreds of thousands per year. Providing information and services to visitors can be an effective outreach mechanism. In particular, the daily life of an observatory can illuminate the process by which astronomical research is conducted. In addition, the on-site facilities can be ideal settings for hosting educational workshops and other activities which gather national audiences together.

The national observatories must play a national role. However, outreach into the local community is a must. First-hand experience in education is essential if the national observatories are to be effective in coordinating national activities. In addition, most national observatories are situated in the midst of underserved communities (e.g. the Native American population adjacent to Kitt Peak and the VLA, the urban African-American population adjacent to STScI, etc.), and can develop direct access to students and teachers from under-represented groups. Finally, the national observatories can demonstrate leadership in developing programs which engage their own in-house staff and which are effective in the local community, as other observatories, institutions, and universities are being asked to do.

3.0 Issues to Consider

The new policy directives on the role of scientists in the science education enterprise provide a great opportunity for the national observatories' participation. However, engaging in education must not compromise the primary mission of the national observatories in supporting and conducting the astronomical research in this country. Careful consideration must be given in developing a policy which addresses the allocation of resources, and in particular, staff time. Too often, staff are expected to simply add new responsibilities for education to their current activities. Federal funding agencies and policy makers must realize that new funding must be made available to support new activities, or that previous activities will have to be halted to allow new initiatives. In the long run, science education in this country will not be served by starving the research infrastructure.

Entry into a new field necessitates some retraining. Science education and reform efforts have been in existence for a long time, and we scientists must educate ourselves and become familiar with the experience base of those who have come before us. This is essential for both those who directly participate in science education initiatives, and for institution administrators whose decisions set the course and allocate resources for these activities.

Efforts must be coordinated among all national observatories. Astronomy should be taught as a whole, including data from all wavelengths, the way it is practiced professionally. The impact of any individual institution can be multiplied by coordinating efforts, sharing resources and experiences, and avoiding duplication of effort: the whole is greater than the sum of the parts. The various funding agencies should coordinate their efforts in kind. Our success will be measured by the degree to which we improve the public understanding of astronomy as a whole. The astronomical enterprise will not be served if individual projects or observatories compete for the limelight, at the expense of other wavelengths or projects.

If the national observatories are to have national impact, they must seek national dissemination mechanisms. Many such mechanisms already exist which can be strengthened by the participation of the national observatories. It is not necessary to start from scratch in developing a national voice, and would probably not be the best use of the limited resources available.

It is ironic, and perhaps not unrelated that the call for increased participation by national observatories (and indeed the entire scientific community) comes at a time when federal resources are shrinking. In the years to come, we will likely see an increasing role played by the private sector in providing educational materials and distribution services. In particular, we will likely see alliances with industries with whom we have had no prior experience (i.e. non-aerospace industries), such as publishing and other communications industries. Mechanisms must be developed by which public and private institutions can cooperate.

We are witnessing new growth in the field of science education, and with it, new challenges and new opportunities. It is essential to maintain flexibility in order to meet changing conditions. In the end, it is still individual initiative which lies at the root of these activities, and that must be nurtured and supported.

Acknowledgement: This work is supported by NASA contract NAS5-26555.

Discussion

Altschuler.

At the NAIC, one of the three (NSF supported) national centers for research in astronomy, we have decided that we did wish to enter the mainstream and stop "moonlighting". We have pursued this goal for several years. In July 1995, we begin construction of the Arecibo Observatory Visitor and Educational Facility, a $2,500,000 investment. Our poster provides details.

Duncan.

National observatories can provide leadership in good part because they have astronomers with professional training and national reputations. You have no educators with equivalent professional training. How can you play a leading role in education?

Danly.

I agree that there should be no <u>inherent</u> compromise in theory. In practice, however, there are important management decisions to be made when resources are limited and the goals are many. Being aware of the trade-offs and priorities is important, I think, to a successful program.

Fraknoi.

While I strongly applaud the efforts of the national observatories to get more actively involved in astronomy, I have two concerns about what is happening:

1) The various observatories are not yet talking to each other. In their enthusiasm to get going with <u>something</u>, several may duplicate efforts. For example, several groups want to put a "definitive" list of educational resources on the Internet. We hope such groups can cooperate and interface.

2) Some of the observatories are "re-inventing the wheel." They are not really becoming familiar with what has already been done by others over the years before they get their programs started. I would recommend that each observatory at least try to do what STScI is now doing: form an outside Advisory Board of people with long experience in astronomy education.

Hill.

I don't think there is any inherent "compromise" between STScI's mandate to (1) produce superb science with the HST and (2) make available public supported research to the public. To pose the question this way seems to imply divergent basic objectives as opposed to different, but complementary emphasis of the HST endeavor.

Danly.

The astronomical research community has many assets it can "bring to the table" for the national discourse on astronomy education – the data, the facilities, the expertise. Because of the unique resources of the national observatories I described, and in particular, their mechanisms, they are in a position to play a coordinationg role for the research community. I would be indeed foolish for any single entity to attempt to direct a national program with only part of the experience required. It would likewise be foolish to assume that by hiring one or two or even five professional educators that a single organization is in the position to set the national agenda. It is hoped that leadership would be demonstrated by all communities seeking a successful partnership and a successful outcome.

Lebofsky.

STScI is doing a great job in educational outreach, but left out one critical need: support of the home institute. Many of us face lack of support to out-and-out discouragement.

Danly.

I agree completely; I think most of us have felt that discouragement. We have heard discussion here that the funding agencies are emphasizing education in all their grant programs. I suspect change will occur by some combination of urging from the top-down and the bottom-up. My discussion today assumes, perhaps optimistically, that change is underway.

Maciolek.

I compliment the national observatories on their openness to working with teachers. Focus on your unique aspect, which is the interaction of teachers with professional astronomers, doing real science – this is your most important resource. Teachers taking back slides of "behind-the-scenes" tours or sharing details of a conversation with a professional astronomer is of prime importance and has a wide impact in terms of how many people are informed and influenced.

What's Happening in Physics Education?

Ramon E. Lopez
The American Physical Society
One Physics Ellipse
College Park MD
USA 20740

E-mail: lopez@avl.umd.edu

[The author was not able to submit a written version of his paper. This summary was prepared by John Percy from the audiotape of the symposium, and approved by the author.]

Recent developments in physics education have been motivated by the realization that traditional methods of instruction do not work. Students can solve "type problems", but do not understand the underlying concepts, and therefore cannot apply them to new situations. The needs of physics graduates are also changing: there is presently a mismatch between their traditional specialized training, and the needs of the workplace. Many of the developments at the post-secondary level are related to what is happening at the K-12 level.

1. Undergraduate education

The emphasis is on active, collaborative, and problem-based learning. Microcomputer-based labs are increasingly common; students plan experiments, manipulate variables, take and analyze data, and interpret the results. Truly national curriculum materials are being developed, based on research on teaching and learning, and supported by NSF and other agencies. These materials are being properly field-tested – an expensive but necessary process. New and appropriate methods of assessment are gradually replacing the traditional "type problems" and examinations. Graduate teaching assistants are being educated in teaching techniques, (as well as in gender and ethnocultural issues – important topics when so many graduate teaching assistants are from abroad), instead of being thrown into the classroom "cold", as was traditionally the case. Professional development is increasingly being provided for other undergraduate instructors, as it would be for K-12 teachers, and for most other professionals. Necessary administrative support is being obtained, both for the academic changes, and the physical changes – microcomputer labs, networks etc.

2. Graduate education

Changes have been motivated, at least in part, by the job situation for young physicists: there are twice as many being graduated as there are jobs, so many are in PDF "holding patterns". A recent meeting of physics department chairs strongly endorsed a "professional MSc degree" which would not be a "consolation prize" but a valid, industry-oriented certification. Graduates would have more flexibility if they had access to courses in computer science, scientific communication, economics and business. There could be research grants awarded to students (instead of advisors), which would also allow more flexibility. Teaching training and experience also help.

Physics education is in a state of flux, and the number of departments engaging in significant reforms is still small, but the number is growing, and there seems to be real momentum and progress to reform.

Discussion

Bishop.

It is my observation that some science TA's for undergraduate courses do not speak English well enough to communicate. I think it is essential that any person identified as a teacher in a U.S. college/university speak English well enough that all students can understand.

Hemenway.

The University of Texas at Austin has recognized this problem. All non-native English speakers are required to have passed a test in oral English communications before they are permitted student contact (including office hours as a TA or classroom instruction). All TA's are required to take a graduate course in education methodology in the astronomy department.

Kernohan.

I applaud the effort to improve the graduate school teaching of TA's. To teach well you must know how to teach and you must know the material. We've all had teachers and professors who knew the material but couldn't teach. We've also been talking about middle and high school teachers who don't know astronomy and physics. We need both.

Fraknoi.

Physics boasts a separate organization for physics teachers (AAPT) (which also includes high school teachers). Would you recommend such a path for the astronomical community?

Lopez.

I would recommend that a separate organization devoted to scholarship in astronomy education not be formed; rather it should be viewed as an integral part of astronomy as a whole and that section or division within the AAS be formed to accomodate a community whose research is in the field of astronomy education (K-12).

Safko.

The suggestion that astronomy education efforts be concentrated within the AAS is dangerous. Many of 2 & 4 year college astronomy teachers consider themselves to be mainly in physics. They would probably not pay the dues – the K-12 teachers would clearly not pay AAS dues to the AAS. The line between physics and astronomy instruction is often blurred, that is why the AAPT has a strong program in astronomy education. Any program in astronomy education must keep contact with those who don't just teach astronomy and who often do not have a research background in astronomy. There are several present and past chairs of AAPT committees present at this meeting.

Astronomy Education in the Two-Year Colleges

George F. Tucker
Sage Junior College of Albany
140 New Scotland Avenue
Albany, NY
USA 12208

Tel.: (518) 445-1754
E-mail: tuckeg@zeus.sage.edu

Let me begin by introducing myself and my institution. I am the only faculty member teaching physics and astronomy at Sage Junior College of Albany (SJCA). I have been teaching at SJCA for four years and before that was a visiting professor at a four-year institution. SJCA is a private two-year college with an enrollment of approximately 700 full-time students. It is located in Albany, New York, half a block from the original site of Dudley Observatory. Unfortunately that beautiful facility was torn down in the 1960's, and is not available to our students. The reason I was asked to speak to you is because I am serving this year as Chair of the American Association of Physics Teachers' Area Committee on Physics in Two-Year Colleges.

Before I begin to discuss some differences between astronomy education at two-year and four-year colleges, I want to describe the makeup of the two-year college community. The majority of two-year institutions are public community colleges. The remainder includes other public institutions and private junior colleges. Many of these colleges are experiencing declining student enrollment and budget tightening. While no recent studies have been conducted, it is believed many, perhaps a majority, of all college students take introductory astronomy at two-year institutions.

Two-year college instructors face a series of challenges to providing astronomy education. While these challenges are not unique to two-year institutions, they occur far more often than at four-year colleges. While some two-year colleges have their own observatories and planetariums, they are the exceptions. The norm would be for the college to have available a few older portable telescopes that would have to be setup for each observing session. Most two-year colleges are in urban areas where light pollution is often a serious problem. For example, at my college about 15 stars are visible on a clear night. Telescopes with CCD cameras, which can be useful in light polluted areas, are seldom available. Most two-year colleges do not have dormitories and students usually leave the campus before nightfall. Most students have either evening jobs or family responsibilities that make their return for observing sessions difficult. These factors make it difficult for two-year colleges to provide the hands-on experience that many students expect from an astronomy course.

Budgets tend to be smaller at two-year colleges and there is generally less instructional technology available. Many institutions do not have computer labs available for use by astronomy students and Internet access is often unavailable. Computer projection panels for use in classroom computer demonstrations are rarely available.

Two-year college astronomy instructors usually have received their highest degree, not in astronomy, but in physics or chemistry. They generally have heavy teaching loads and are not given much time or support for faculty development.

Qualified student workers who could assist with telescope setup and act as tutors are not often available. This places the burden for all aspects of the course on the instructor. Little emphasis is given to faculty or student research at the two-year colleges level.

Two-year college students vary widely in background and preparation. Women, minorities and older students are strongly represented and teachers should be sensitive to the instructional needs of these groups. In introductory astronomy courses, students who have previously had physics and calculus may be sitting beside returning students who have never had physics and whose last mathematics course was high school algebra taken ten years before. This disparity in preparation can really challenge the instructor to prepare a course that interests and challenges all the students in the class. Finally, for those students who do become interested in astronomy, there are usually no advanced astronomy courses available at two-year colleges so they may be unable to immediately continue their education in the field.

On a personal note one of the most difficult aspects of two-year teaching is the fact that students remain for at most another year at the college. We do not get to see them develop intellectually for four years, and seldom see the effects of our teaching. Occasionally a student will return to campus and stop by to discuss some development in astronomy. This is very rewarding, but I envy my four-year colleagues opportunities for further interaction while the students are still attending their institutions.

Balancing the negative aspects of two-year astronomy education are many positives. Urban settings bring with them the opportunity for instructors to avail themselves of the many resources found in our cities. Planetariums and science museums can be visited. The observatories of other educational institutions and research organizations can often be used. Most cities have amateur astronomy clubs that can provide observing opportunities and guest speakers.

Control over how astronomy instructors at two-year colleges spend their budgets usually resides with one or two people, the instructor and possibly a division chair. This affords flexibility in selecting new equipment and materials without having to convince a large group of a need. Similarly, two-year college faculty have more control over course content and instructional methods. This allows two-year colleges to rapidly adopt innovative teaching methods and incorporate new material.

With less emphasis on research comes less pressure on the instructor to spend large amounts of time on research simply to "publish or perish." The instructor can concentrate on teaching and doing research that really interests them. Since most prospective two-year colleges instructors know this is the case, two-year colleges tend to recruit faculty whose first love is teaching. If support is available, the instructor also has more time available for faculty development.

Class sizes at two-year colleges tend to be smaller than at four-year institutions. If there is a laboratory with the course, the course instructor is usually the laboratory instructor. These circumstances foster more contact between instructor and student and can contribute to a better learning environment for students.

Students at two-year colleges do work to maintain a high grade point average, but grade pressure is usually less than at four-year colleges. This allows students to view astronomy courses as a learning experience not just an opportunity to pad their average. They are more willing to accept the fact that not all aspects of astronomy are completely understood and not every question asked has a pat answer. The diversity of the student population offers an opportunity for the astronomy instructor to provide a wide range of students with an appreciation not only of astronomy but also of science as a discipline. Often the course may be the only science course students are planning to take. Included in this group are students Sheila Tobias has called the "second tier," those are students who are capable of having successful careers in science but whose lack of preparation has left them convinced they cannot succeed. If properly designed, an introductory astronomy course can provide these students with the motivation to attempt to pursue a scientific career and the self-confidence necessary to believe they can succeed.

To summarize, two-year colleges serve an important function in astronomy education and can become even more important in the future. While the two-year college instructor faces many challenges, two-year colleges offer tremendous opportunities for promoting astronomy education. If the astronomical community wants to introduce innovative teaching methods and materials into introductory astronomy courses, look to the two-year colleges. If the astronomical community wants to reach under-represented groups, look to the two-year colleges. And finally, if the astronomical community wants to assist in luring the "second tier" into science careers, look to the two-year colleges.

Discussion

French.

As someone who teaches at a small, 4-year tuition-driven college, I'm impressed by the overlap of positives and negatives from your list. I'm convinced that we in 4-year colleges can learn much from our 2-year colleagues – thanks for sharing.

Hoff.

Most astronomy teaching in terms of numbers of students is happening at two-year and four-year middle scope universities. We must not forget that you are central to the teaching of astronomy.

Wasiluk.

As Adjunct Professor of Astronomy at Lord Fairfax Community College in Middletown, VA we are the home base to the Shenandoah Astronomical Society an amateur group that has recently donated time and talents to having an observatory built on our campus.

Also, I'd like to know any information as to how many astronomy instructors at two year colleges are part time faculty like myself.

Tucker.

The American Institute of Physics Statistics Division has received National Science Foundation funding to survey physics instruction at two-year colleges during the coming year. I have been in contact with the principal investigators and they have agreed to include some questions about astronomy education. So we may have some updated data in the near future to answer your question.

SECTION II. REPORTS OF SMALL-GROUP DISCUSSIONS

In order to provide for focussed discussion on key topics, the symposium participants divided into eight groups, each led by a chair and a recorder/reporter. The groups convened late Saturday afternoon for a preliminary meeting, then reconvened on Sunday morning to prepare a report. In some groups, there were brief invited presentations, some of which are included in the third section of this volume. The reports were presented and discussed in the plenary session on Sunday afternoon. The reports and discussions are contained in the following pages. The writer of the report is the individual whose address is given in full at the beginning of the report; the other leader, and all of the other members of the group, also made substantial contributions to the report. We thank them all.

The evaluations of the symposium pointed out the obvious: much more time should have been allotted to the small-group discussions. It might even have been advisable to divide into sub-groups, to enable more participants to take an active role in the discussion, and to enable more topics to be covered. The session on K-12 astronomy could have been divided into elementary, junior high school, and high school astronomy, for instance. Nevertheless, the groups worked well, and we encourage future symposium organizers to include them.

Post-Secondary Astronomy

Chair: George S. Mumford
Department of Physics and Astronomy
Tufts University
Medford, MA
USA 02155

E-mail: gmumford@pearl.tufts.edu

Recorder/reporter: Neil Comins

An extensive bibliography dealing with many of the topics that our discussions touched on (and more) has been provided by A. Fraknoi (Fraknoi, 1995) and is available from him in care of the Astronomical Society of the Pacific. Among his categories the following are pertinent to post-secondary astronomy: books on teaching resources; articles on teaching resources; articles on computer-based resources; books of lab activities for college level; books of computer-based activities; selected articles on classroom or lab activities; selected articles on sky observing activities; books on astronomy education; articles on astronomy education in general; articles on specific projects to improve astronomy education.

Introduction

About two dozen persons were present throughout most of the discussions on the topic of post secondary astronomy courses. These were about evenly divided between public and private institutions; the majority were from departments of physics and astronomy at four-year colleges and universities. A list of participants is appended.

The session opened with a ten-minute presentation by M. Zeilik, "Conceptual Astronomy: A Cognitive Approach". The use of four-student discussion groups in a large lecture class to promote cooperative learning is a primary feature as described elsewhere in this volume.

Thereafter, there was a wide-ranging discussion that focused on five conceptual elements of a post-secondary, introductory, survey course: audience, goals/objectives, content, tools/methods, assessment/evaluation. In each of these areas various problems were identified and solutions hinted at. We feel it important that these conversations continue and end with some suggestions in this regard.

Audience

In the limited time available, we restricted ourselves to talking about the introductory survey course. There was passing reference to certain concerns related to programs for those who major in astronomy: what backgrounds should these students have? what are some of the best, intermediate-level textbooks? what are suitable topics for advanced laboratories? is adequate preparation provided to a junior-college transfer? Obviously there is sufficient material here for a separate discussion group. But the majority agreed with spending the time on the course or courses designed for those the admissions office throw our way – namely a group of often scientifically illiterate individuals from diverse backgrounds and representing the entire spectrum of academic interests. We should never forget, however, that these are future parents, school committee members, teachers, legislators, and taxpayers whose attitudes towards science we have a penultimate or even the final chance to shape.

Many of these come with a variety of phobias ranging from fear of mathematics in general, to fear of logarithms specifically, to an all-encompassing science anxiety. In the latter case a suggestion was made that one step on the way to helping cure the student of anxiety was to have him or her write a short essay on the experience that proved to be so bad and then discuss the circumstances of this situation. Others have used a "math survival kit" to aid in the transition. Such a kit may include examples of the use of exponents, especially powers of ten and scientific notation, of how order of magnitude estimates are made, of deciding on an appropriate scale for a graph, rounding, ratio and proportion, and other necessary tools.

In addition to the phobias, many come with a variety of misconceptions (one participant has identified some 140 or more) among them that the earth is closer to the sun in the summer; that lunar phases are caused by shadows; that the sun goes around the earth in a day; and more. For a further discussion see the selection in this volume by P. Sadler.

One group of students that we should be particularly aware of, and maybe gear special activities to, are prospective school teachers. If we wish to have an influence at the elementary or secondary level, we have to provide these people not only with appropriate tools but also far better attitudes towards science. If they can be identified, and sometimes they will show on the roster as graduate degree candidates, they might be provided with additional activities and considerably more personal attention than we are likely to pay the normal student.

While institution of prerequisites might yield classes with students of more uniform backgrounds, and might restrict class-size, these are generally frowned on at this level, although the descriptions of some courses do reiterate the college admissions criteria: e.g. a knowledge of high school algebra is required.

To summarize, most of us are faced with designing and/or presenting a course for up to 300 or more students, most having little background or interest in science. One saving factor, however, is that some may have a latent interest in astronomy, which at the elementary school level ranks next to dinosaurs in popularity.

Goals/Objectives

In the course of our discourses we reached a consensus to consider **goals** as lofty ideals and **objectives** as more down-to-earth, prosaic elements. A primary goal, for example, is to increase science literacy while an objective would be to have students understand the cause of the seasons.

Most of the discussion emphasized goals. Besides that mentioned above other goals suggested: to introduce astronomy as a physical science – as a human activity observationally rather that experimentally based; to inculcate students with the concept of science as a process rather than a series of facts; in general, to attempt to change a student's attitude about science. An additional goal is somehow to transmit the excitement of science – the ultimate high of finding out something that no one has known before – as an evolving, continually changing activity carried on by real people.

Somewhat different in nature from the preceding goals but equally important is to reconcile student expectations from the course with those of the instructor.

How do you do this?

One suggestion from G. Mechler:

I have four goals in my lecture courses which form the beginning of my 30-page "Introductory Notes". I start my lecturing by going over these goals in class:

- to increase interest in the universe/astronomy;
- to increase scientific literacy;
- to instill a "cosmic perspective";
- to contribute to their personal growth.

At the end of the course I review these in class and challenge the students to ask themselves how well we attained them. Sometimes I hand out a brief assessment sheet asking the same questions.

A flip through some of the chapters in current texts provides additional examples of goals and objectives. The latter would include understanding how the relative motion of the sun, Earth, and moon lead to eclipses; or to understand Kepler's laws of planetary motion; while a goal might be to develop an understanding of the uniqueness of astronomy among other fields of scientific endeavor. Where else do you find so many hobbyists? In what other realm of science do so many nonprofessionals make major contributions? Just imagine the excitement of the school girl (see C. Pennypacker's article) who discovered a supernova! And consider the use of astronomical terms commercially: pulsar (watches), comet (cleanser), Mars (candy bar), zodiac (boat).... Challenge your students: how many songs can they name that have moon in the title? how many have DNA or lepton or hydrogen peroxide? There's something special about astronomy.

Content

The survey course is seldom a prerequisite for another course. In some cases the individuals enrolled in the second semester course are totally different from those in the first semester and *vice versa*. Today, a typical science distribution requirement can be completed by two courses in disparate fields– one in astronomy the other in, say, geology. The implication is that often the content becomes a personal matter, left to the discretion of the instructor within the confines of a general course description.

There was some general agreement that misconceptions were addressed. But there was virtually no agreement on any other aspect especially on the order of the topics – Earth to universe, or universe to Earth, for instance. Similarly there was no agreement on the depth of topics or the balance between historical/cultural aspects and the more pedantic ones. Where some of us may be intrigued by a study of Isaac Newton's genius and its related pathology to help explain why he may have achieved his unifying theory, others are more prone to simply present the theory. Would Stephen Hawking have had as many insights if he were not wheel-chair bound? To some this is an intriguing questions which to others really has no place.

What about newspaper headlines or the morning telecast that trumpets a new discovery with the Hubble Space Telescope? Do you change the day's agenda to try and provide additional background to your class so that they may gain a greater understanding of the event, or do you ignore it?

A number of issues related to laboratories, observing, and homework were raised. Many of us assign work to students outside the classroom, but are we sure they learn anything? One can always ask a suitable question on an examination to try to find out.

Questions were brought up as to suitable sources for information on hands-on projects including classroom activities and indoor laboratories. A place to begin is with the paper distributed by Fraknoi that was mentioned earlier. We also note that one of the most recent compilations of laboratory exercises for elementary astronomy was a so-called resource letter published almost two decades ago (Kruglak, 1976) in the *American Journal of Physics*. These resource letters are detailed, annotated bibliographies that survey current literature. With the introduction of computer labs and telescope simulations, it would seem that the time is probably ripe for another.

The question of the mix between indoor labs and outside observational activities was addressed. Obviously there is no clear-cut answer since so much depends on local conditions and the types of equipment available. In regard to the need for equipment, as J. Percy pointed out in a later discussion "you don't need telescopes to observe some variable stars". Many instructors have had considerable success with activities of this type. (Some of us are anxiously awaiting the materials being developed in the American Association of Variable Star Observer's "Hands-On Astrophysics" project as described elsewhere by J. Mattei). As instructors become more involved with hands-on and cooperative learning, one anticipates considerable growth in, for instance, the use of student journals and relatively simple observing projects. Often students are required to keep track of the location along the horizon of the setting sun throughout a semester, or the motion of the moon in the sky for a month. Beware of a student being asked by a parent what the bright object visible in the western sky just after sunset is and not being able to give an answer (let alone the correct one) – this will reflect on you and your course!

In connection with projects for large classes, an intriguing suggestion was that contact might be made between students at two schools separated by a considerable north-south distance. On a day decided upon, students at either location might determine the noon altitude of the sun from the length of the shadow cast by a gnomon. The north-south distance between the schools could be established from an atlas or other source and the circumference of the earth found *a la* Eratosthenes.

How much of the observing should be required? How much optional? No ready answers appeared for either of these questions. Part of the answer concerns the number of individuals available to aid in the instruction and the size of the course. Some institutions rely on undergraduates, who have previously taken the course; others use graduate students or other members of the department.

With a growing emphasis on learning by doing there are many of us who are frustrated by not having the time, or space, or help necessary to run discussion or laboratory sessions outside of the lecture hall effectively. This seems to be particularly true for some from departments of Physics and Astronomy, where Physics rates a "P" but astronomy only an "a".

Tools/Methods

The discussion here centered on tools to use in the classroom with the large class and not on the laboratory. Some of us have tried to teach a survey course using a variety of different sources for readings – three or four paperbacks – but have found that most students prefer a crutch – the textbook – to fall back on. There were the usual comments regarding too many topics being covered in the standard texts, but each referee has her or his own prejudices regarding what must be included. With content pretty much the same and with all textbooks having multi-colored illustrations, up-to-date coverage and being well-written, what are the criteria used to select an appropriate one? Maybe the ancillaries are the answer. Most of these include a teacher's guide, sets of slides and colored transparencies. Many provide video renditions of astronomical events. Some dispense software using either the PC or Macintosh format. And for each student purchasing a given volume there is a magazine collection or a newspaper supplement to read. What uses are made of these materials? If one changed a textbook on an annual basis quite a collection of slides, transparencies, video tapes and the like would result. How extensively is this material used? Is it best to display in class a diagram precisely like the one in the text, or does it make more sense to show one with a slight variation?

A question was raised regarding the use of outside reading – that is in books beyond the above-mentioned supplements – but it was not apparently a subject of general interest. There are some interesting possibilities among volumes currently available, but should these be required?

Some instructors assigned projects, or term papers that cater to a student's main academic interest. References to astronomy are found in numerous places in literature, art, and music. The Space Age has opened issues related to law and international relations and to concerns relevant to the psychology and health of persons who spend considerable time in space. Even debunking astrology is a possibility (though the student who wrote a computer program to cast horoscopes is still trying to find the bugs that kept it from "working" with real persons). The point is that for virtually any academic interest some astronomically related material may be found if one searches hard enough and this may get the recalcitrant student involved.

Among the greatest changes in the current classroom are those brought by the computer and VCR. Many instructors have the ability to project an image raised on a computer monitor so that extensive use of various software is possible. We suspect that the use of so-called planetarium programs that have the ability of displaying the sky from many different locations and from long in the past to well into the future is becoming more common. There has been an integration into the classroom here, but has it been assessed?

CD ROM technology makes available on a single computer disc the ability to call up images of astronomical objects; to watch film clips of astronauts walking on the moon; to tour the planets; to demonstrate concepts such as the seasons, day or night, eclipses, and so on. Video discs provide similar capabilities, but few in our group seem to have used them. All of these certainly can make the class more entertaining but do they lead to a greater understanding of the material?

A problem is the lag time between the time you begin to set a particular view and the time the view appears. Sometimes, for no reason, a computer freezes, leaving the instructor to *ad lib* while vainly clicking the mouse or banging console keys. It is well to have an alternative plan in the event of a midclass foul-up.

In his paper in this symposium volume, J. Pasachoff mentions many sources for software. Reviews of these materials appear in *Astronomy*, *Sky & Telescope* and other sources. Some feel that these reviews are really inadequate to assess what a particular piece of software can or cannot do in a particular situation; what its strengths and weaknesses truly are. Textbook publishers provide copies of their products for evaluation. It's too bad software manufacturers can't do something similar, or at least provide more opportunity for more people to play with software at meetings. A day-long poster paper at a recent summer meeting of the AAPT permitted the interested person to stop and run a variety of different software items for as long as he or she was interested. Those who participated seemed to appreciate the opportunity.

Given that another session was putting major emphasis on technology (see the report of the group chaired by L. Marschall) we mentioned only in passing the classroom use of the Internet and satellite transmissions including such items as daily displays of sunspots.

Some wondered about techniques other than lecturing that were effective in large classes. A number of years ago Safko (1972) described a large, self-paced class. *Discovering Astronomy* by R. Robbins, W. Jeffreys, and S. Shawl (John Wiley & Sons) uses a unique discovery- or inquiry-oriented approach. In connection with cooperative learning, Zeilik mentioned that since his class met for two 90-minute periods per week, it was possible to have the 30-minute duration, small-group conversations. He felt that something would be lost in the more common 50-minute periods.

While not addressed as a major issue, of interest would be a survey of the effectiveness of various lecture techniques. There are those who talk and use the blackboard plus slides with the feeling that if students have to concentrate on writing down notes this aids the learning process. The use of transparencies gives students something to copy from that may be superior to writing on a blackboard but in either case, is the writing legible? is it visible from the back of the lecture hall? Others provide their students with detailed outlines or skeletons of lectures with minor gaps to be filled in during the class. Does this increase a student's understanding of a topic? One can see some use of this technique especially in connection with complicated diagrams such as renditions of retrograde motion that may not work so well when being done free-hand at the blackboard. And, finally, there are those who provide detailed notes so that virtually all a student has to do is sit back and listen in class. If such notes are reviewed ahead of the lecture by a student, the lecture could be used to aid in clarification of more obscure points.

The large lecture class is cost effective, but research has shown that lecturing doesn't work to promote higher-level thinking particularly in physics. For example, in a series of tests given students to assess their understanding of and problem solving ability in mechanics, the results of traditional instruction - lecture, demonstration - were uniformly low for all teachers which suggests that methodology is a more serious problem than teacher competence (Wells, Hestenes, and Swachhamn 1995). Maybe the future will see more of the methods being developed by Zeilik, who has commented in regard to lecture techniques "none really make a difference - not lecturing does".

Probably most, if not all, would agree that certainly lecturing is not the best way to teach, nor is it the best way for students to learn. What lies ahead that is cost effective as well as educationally sound? Possibly it is a more problem-oriented approach, with student and instructor interacting over the campus computer system. Undoubtedly there will be a greater use of computer technology.

Questions arose concerning classroom demonstrations: are they effective? If so, what are the most effective ones? Some of us have seen K. Brecher of Boston University or J. Kernohan, Milton Academy, perform their favorites at meetings of the American Astronomical Society (AAS), and AAPT, respectively, but we anxiously await an overdue update of Tattersfield's *Projects and Demonstrations in Astronomy* published by John Wiley in 1979.

Some concerns were expressed about the matters of attendance and retention of students. The latter seems to be dependent upon a school's policy of when the last chance of withdrawal from a course with no record of enrollment is allowed. In connection with the former, some base a course grade on attendance or on class participation, which obviously requires attendance.

In connection with work for extra credit, some associate extra work with honor grades. One individual reported on the practice of requiring that all prospective A students take an oral final examination.

A partial listing of places that one may find articles related to astronomy education or in which one may publish such articles includes:

American Journal of Physics
Astronomy
Journal of College Science Teaching
International Amateur-Professional Photoelectric Photometry
(I.A.P.P.P.) *Communications*
Journal of the Royal Astronomical Society of Canada
Mercury
Sky & Telescope
The Physics Teacher

Assessment/Evaluation

In our discussions the term **assessment** was taken to mean a measure of goals while **evaluations** measure objectives. Thus, assessments provide feedback as to how well goals are being met. They are for the teacher and for course improvement. If they are to be of use they should be done promptly and as one individual suggested midway in courses so that students can provide instructors with feedback and there remains time for the latter to make adjustments before the end of the semester. An alternative to using some standardized form or questionnaire is peer review that may become more common in the future.

Self-assessments help students judge their understanding of material outside of a grading context. These also aid the instructor in establishing whether or not the student really understands what is being said. Beware of over assessment, however.

Considerable discussion took place regarding types of tests used for evaluations. Some took the view that writing for homework and other assignments was preferable to using essay questions extensively on final examinations. As was pointed out, there are both effective and ineffective multiple choice questions. A pool of effective ones might be of interest.

One participant had made a study of the correlations between three types of questions: multiple choice, those requiring short answers, and those that called for long ones. There was no correlation between how students did on the multiple choice and on either of the other sets. However the short answer results were strongly correlated with the long answer.

If you really want to find out how well students understand a topic, commented another, have them write an essay describing the particular phenomenon to a child.

Outlook

Two of the list servers on the Internet that carry items of interest to astronomy teachers are PHYS-L and ASTRO-L. A problem with the former is the number of items daily that one must wade through before finding something with a primarily astronomical bent. On the other hand, the content of the latter is often devoted to relatively esoteric concerns in the historical domain. Apparently no one wanted to undertake the administrative tasks associated with, say, ASTRO-L, so that suggestion fell by the wayside.

The AAS's Working Group on Astronomy Education is served electronically by a newsletter that originates with S. Shawl at the University of Kansas, who said that he would be willing to aid in the further dissemination of information on educational topics through the Internet. If you are not connected, but wish to be, contact him: S-SHAWL@UKANS.edu

We plan to share this conceptual frame with colleagues and invite them to help flesh it out from their own experiences and expertise. Many of us have never had the opportunity to deal in depth with educational research and teaching methodologies. We need the results of research and the products of those who have.

Available forums begin with the expanded educational office of the AAS and education sessions at annual meetings. The AAPT has interest and activities in this area (Mumford, 1995); the American Physical Society (APS) has constituted a Forum on Education (two years ago there was a joint session on astronomical education put on by the Astrophysics Division of APS and the AAPT's Astronomy Education Committee); the Astronomical Society of the Pacific has a clear involvement as does the International Astronomical Union. How does one coordinate and unify these different venues? That was a question addressed at the conclusion of this conference, but one thing is clear there must be a continuing series of these sessions bringing together the diverse groups involved in astronomy education.

References

Fraknoi, A. 1995, *Astronomy Education in the U. S.: A Bibliography,* Version 1.30
Hoff, D. 1982, *Phys. Teacher,* **20**, 175.
Kruglak, H. 1976, *Amer. J. of Physics,* **44**, 828.
Mumford, G.S. 1995, this volume.
Safko, J. 1976, *Amer. J. of Physics,* **44**, 68.
Wells, M., Hestenes, D., and Swachhamn, G. 1995, *Amer. J. of Physics,* **63**, 606.

Participants

Bisard, W.: Central Michigan University
Chaisson, E.: Tufts University
Comins, N.: University of Maine
Donahue, M.: SpaceTelescope Science Inst.
Eckroth, C.: St. Cloud (MN) State University
Edwards, S.: Smith College
Friend, D.: University of Montana
Garrett, T.: Tidewater Community College
Hall, D. E.: CA State University, Sacramento
Levine, J.: Orange Coast College
Lightner, S.: Westminster (PA) College
Linton, D.: Parkland College
Mechler, G.: Pima College
Mumford, G.S.: Tufts University
Poole, W.: Addison-Wesley
Procter, A.: Yale University
Ratcliff, S. J.: Middlebury College
Rudolph, A.: Harvey Mudd College
Osell, F.: Leeward (HI) Community College
Saken, J.: Space Telescope Science Inst.
Seacord, II, A.W.:N. Virginia Community College
Shawl, S.: University of Kansas
Vort, M.: Johns Hopkins University
Zeilik, M.: University of New Mexico

Discussion

Bisard.

A summary of assessment of course goals and ways of evaluating objectives of topical concepts would be a good starting point for dialogue. I feel an electronic network would be useful!

Eastwood.

I am looking for ideas to use in large lecture classes to make them more participatory. I would like to somehow network with other people who want to share ideas.

Garrett.

I teach what I believe is the standard two semester lab course (survey course) in astronomy (assuming a two semester course – after my comment it was suggested to me). I am interested in reducing some of the factual content of the course in order to include more of the historical and cultural connections (such as were mentioned in the early Saturday morning talks) and relating astronomy to my students' majors as personal interests. At present I do this through the semester project (required). Are there texts or supplements available that do something along this line? Do most college teachers who do this use their own (self-generated) materials? [Speaker said he will talk to the publishers – possibly they could use their college reps to collect data – possibly as a publishable consequence.]

Osell.

A problem in teaching a survey course is that each teacher may tend to spend most of the course on a single aspect of astronomy (i.e. star evolution) at the expense of other topics. The problem is in balancing the subject matter in order to present a balanced treatment of general astronomy.

K-12 Astronomy Education

Chair: Jeff Lockwood

Recorder/reporter: Jodi Asbell-Clarke
6208 Regina Terrace
Halifax, NS
Canada B3H 1N5

E-mail: asbellcl@ap.stmarys.ca

I. What Should We Teach?

We have heard many anecdotes in this meeting about the difficulties students have in learning about seasons on Earth and the phases of the Moon. Should we even teach these concepts if they are creating such problems? The consensus of the group is yes, but only with an appreciation of appropriate cognitive development and the recognition that other more contemporary astronomy has valuable lessons for students of all ages.

II. In What Order Should We Teach Topics?

We must be aware of the cognitive development that is necessary for the learning of specific content. For instance, to fully understand the cause of seasons or phases of the moon students must be able to view a two- or three-body system from various perspectives. Students are generally unable to do this before junior high. If we push topics on students before they are ready, not only will they not learn the topic, but they may also be frustrated, turned-off and then subsequently bored when the topic is revisited at a later point in their education.

III. How Do We Teach?

Most of us find this question much more important than the previous question of what do we teach. We have found that students have many different learning styles (i.e. visual, kinesthetic) and material must be presented in a variety of fashions to reach all students. The challenge is to keep the content exciting and engaging while approaching it over and over again using several methods. There are many students that will benefit from seeing the same material several times throughout their education, since each time the topic becomes more clear and it may gradually or suddenly "click".

IV. Standards and Reform

Some of us are very concerned at the lack of astronomy outlined in the recent draft of the NRC standards for science education. Although much of the material that is included in the study of astronomy is covered in other topics (i.e. energy transformations, forces and motion, systems, order and change) we fear that the standards will drive the content that is included by textbook publishers, curriculum developers, and designers of state and local science standards. Others in our group feel that teachers will still be free to teach whatever they please behind closed doors and that there is room in the standards for astronomy, since the standards only outline a minimal curriculum content for all students. Several people in our group have tales of reform within their schools that is occurring far too quickly and without the proper infrastructure in place. Although we agree with the long-term intent of many of the changes, the lack of adequate planning, staffing, and training is resulting in chaos rather than improvements.

V. Reactions to this ASP Educational Symposium

The effort that ASP has undertaken here is a wonderful first step towards establishing a working relationship between research astronomers and teachers. One oversight, it seems, is the lack of any talks given by K-12 teachers during the main sessions. We would like to see a teacher from each level (elementary, middle school, and high school) give talks at the next meeting. They would be able to provide valuable insight regarding the day to day efforts of teaching astronomy to children as well as pedagogical techniques and methodology.

VI. Teacher Training

Members of this group are encouraged by the initiative of AAS to reform science teacher education. Science professors in colleges and universities are rarely prepared to teach pedagogical techniques and the process of uncovering the scientist in each child. Teaching the scientific content to future science teachers is not enough.

VII. Communication

We believe that the science education community should be developed with adequate representation of the following three groups: A) Research Scientists; B) Teachers and Curriculum Developers; C) Science Education Researchers.

The third group consists of those people studying the cognitive development of children and understanding the way they learn and process information. It is only with that information that the teachers and curriculum developers can effectively create and utilize materials to convey the scientific content provided by the research scientists.

We suggest that current means of communication ranging from newsletters (such as those from the AAE and ASP), journals, mass media, and the WWW all include more emphasis on the relationship among these three groups. Any of these publications should be a conduit of information among these three sectors, and also serve as a forum for results stemming from model collaboration among the three sectors.

VIII. Funding

We feel that teachers need to play a key role in the development of any teacher enhancement or educational reform project. This includes working right from the start on the design and proposal of the project. Often times teachers are asked to review a nearly complete proposal to "rubber stamp" it and then the proposers claim to have teacher input. Projects are more apt to target the needs of teachers and students when teachers are an integral part of the development of the project.

IX. Cooperation among ASP, AAS, NSTA and Others

In order for this collaboration to work, it must be an equal partnership among educators and scientists. There can be no hierarchical structure that presumes that the content is "handed down" from researchers to teachers. We applaud the interest that the scientific community is taking in reaching out to children and those that teach them, but that must include taking the time and effort to learn how the educational process occurs. This involves working closely with teachers, curriculum developers, and educational researchers to understand the learning styles and cognitive development of children.

To this means, it is advised that people that fulfill key positions such as the proposed AAS Education Coordinator, have classroom experience at the K-12 level. In addition, research scientists should seek advice and mentoring from K-12 teachers to assist in their outreach pursuits. This can occur informally, in one-on-one situations, or in a more established setting such as an "Educator for a Day" session at regional NSTA meetings (following the model of the "Astronomer for a Day" sessions at the AAS meetings).

Discussion

Duncan.

Practical experience in many fields demonstrates that there needs to be input by professional scientists, professional educators, and professional teachers to create the best science curricula. This shows up in so many cities, so many instances. I now call it a "tripod" rule of thumb. Remove any leg and you're unlikely to have a great program.

Hemenway.

Clarify that National Science Education Standards did receive input from teachers. Over half of the 1700 responses received by NRC were from teachers; 150 were from focus groups (including the AAS). Standards are not just content, but an integration of content, teaching and assessment. If standards are to make a difference in our nation's schools, scientists must support them.

Hawkins.

I suggest that centers of informal science education/planetaria host workshops for professional scientists on how to do hands-on and inquiry-based teaching. This type of exposure and training can help all professional astronomers improve their teaching at any level and to better understand needs of K-12 and general public community.

Osell.

Astronomy courses at the sophomore and junior level should be developed for students majoring in education.

Shawl.

Perhaps we need something similar to the Shapley lectureships that are for teaching, in which innovative teachers in astronomy spend two days visiting departments to help/aid/abet improving astronomy teaching. This could be done for both college teaching and K-12. Finally, innovative teachers should involve themselves in presenting Chautauqua-type short courses.

Wasiluk.

It seems people left out a major problem. In many K-12 education settings astronomy education is not seen as important. You still need to overcome that.

Use of the term "pre-service" teacher is seen as non-complimentary term by many. Wouldn't "student-educator" be a much more complimentary term? "Pre-service" teacher is often seen as an obsolete term in many education circles; try to avoid using it.

Wesney.

1) In December of 1892 the "Committee of Ten" (University Presidents) met at the University of Chicago, and decided that astronomy should not be included in

pre-college science education. Astronomy has come a long way (~1 light century, at least) since then. Since 1982 I have been teaching a high school astronomy course, and since 1989 a college astronomy course. I have found it to be a very rich way of engaging in many, if not most, areas of science. Therefore, astronomy fits the new "national standards" perhaps better than any other single area of science. Astronomy could be considered as an "umbrella" over most of the science curriculum K-12, and at the college level!

2) The AAS has invested a great deal in the "AAS Teaching Resource Agent" program as very forward-looking effort. During the summer of 1994, about 75 K-12 teachers were trained to focus on improving the teaching of astronomy at all levels through teacher workshops. Another ~75 teachers are just starting their training as this summer's cadre. Consequently: A) These AASTRA teachers are available, and they should be used to support astronomers and college/university astronomy faculty in their efforts toward astronomy education involvement/reform. B) Astronomers and astronomy educators should help recruit participants in the AASTRA Programs for subsequent summers. AASTRA training programs were and are being held at 3 sites: University of Maryland; Loyola University (Chicago); University of Northern Arizona (Flagstaff). Contact: Dr. Mary Kay Hemenway, AAS Education Office, University of Texas, Department of Astronomy, Austin, Texas.

Teacher Preparation

Chair: Linda M. French
Wheelock College/Harvard-Smithsonian
Center For Astrophysics
Boston, MA
USA 02215

E-mail: linda@annie.wellesley.edu

Recorder/reporter: Dennis Schatz

Some concepts from astronomy are taught at every grade level in the United States, from kindergarten through graduate school. Yet few teachers of astronomy at the K-12 levels have formal training in the subject matter. Rather, they must rely on extra reading to further their understanding of new developments. At the undergraduate and graduate levels, few scientists have any formal training in effective teaching techniques. Good teachers are viewed as being "born, not made", and teaching is usually not counted towards professional advancement. The focus group on teacher preparation discussed this state of affairs and ventured some tentative solutions.

At the College Level

The survey course in astronomy is the lifeblood of most astronomy programs, whether they are independent departments or not. Most graduate students are teaching assistants for at least one year, and in this capacity they handle the majority of the grading and student help sessions in addition to teaching labs and recitation sections. Yet most graduate students receive no preparation for this work; they are thrown into what may seem a sea of bored, frightened underclassmen to sink or swim. Some graduate teaching assistants (TAs) thrive on the work; a few discover that teaching is an important part of their calling as an astronomer and go on to emphasize teaching throughout their careers. More often, however, the underprepared TAs merely survive their teaching experience. Faculty members whose tenure and funding decisions are not based on successful teaching have little incentive to improve their own teaching or to help graduate students learn about more effective teaching. The group agreed that graduate students should have some preparation before they walk into a classroom or laboratory. This preparation could be done by a team of astronomy faculty, college level education faculty, and graduate TA's experienced in the culture of a particular institution.

At a more fundamental level, the entire culture of most astronomy programs needs to reflect the amount of effort and creative energy that good, innovative science teaching demands from faculty members and teaching assistants. Teaching evaluations should be an important part of promotion and tenure decisions, and release time should be granted to faculty who wish to learn about implementing science education reform in their own college-level classrooms. In this way, future teachers can be taught in a way that models better the way they will teach their own students.

The group also discussed, in a general way, the nature of a survey course specifically designed to enhance science literacy for those who might not plan to major in the sciences. Ideally, it was felt that the course enrollment should be small enough to enable students to interact directly with faculty, and to participate actively in hands on learning and discussion. The topics chosen should be fundamenatally important and ideally range across a wide variety of scientific disciplines. Some sample topics might include light and color, cratering, and size and scale in the universe. Whether such courses should be developed for the entire student population was the subject of a spirited debate among the discussion group, as well as in the plenary session.

At the K-12 level

The discussion group was composed of approximately 50% each astronomers and teachers. The astronomers in the group were particularly interested in learning what teachers think are important contributions from scientists. Opinions were far-ranging and varied. Some teachers value well-written, accessible textbooks that are up-to-date and appealing. Some would welcome a partnership with an astronomer, in which the astronomer would come to the same classroom several times during the school year. All the teachers stressed that any efforts by scientists to work at the K-12 levels much be consistent and coordinated with the new National Science Standards, as well as with statewide and local standards projects.

Some astronomers expressed an interest in inservice programs, which reach out to teachers already in the classroom. A wide variety of such programs is funded by the National Science Foundation, and astronomers who are interested in beginning such a program should seek guidance from other scientists, from master teachers who could collaborate, and from NSF. In particular, scientists need to do market research about the wishes and needs of teachers in their target area. Do teachers need graduate credit? Stipends? Is science an area that is likely to be emphasized in the schools systems with which the program will work? No scientist would venture into a new research area without some library research, and the same care and professional respect should be brought to developing educational outreach programs.

Discussion

Duncan.

University of California Santa Cruz graduate students have done a survey of 30 years of Ph.D. recipients from their department asking: 1. What are they doing now? 2. Did graduate school prepare them for it well? A prioritized list of items which should be added to graduate school is part of the survey result. Better training in education/communications skills headed the list.

Procter.

1) Graduate student courses on teaching are definitely necessary. 2) Where can I find information on reducing science anxiety? What have people done?

Hemenway.

In response to the question of teachers becoming AAS members: The AASTRA program, under the sponsorship of AAS and NSF, provides two years of AAS membership to all participants. Thus, the 1994 summer participants (74 teachers) became associate members in January 1995.

Hawkins.

In restructuring undergraduate courses to be more inquiry-based and hands-on, we need to include technology in a way that fosters access and analysis of data (otherwise we are limited to studying issues that involve visual observations of nearby objects). Perhaps projects such as CLEA and Hands-on Universe, etc. should be made more widely accessible.

Osell.

Survey astronomy courses should not be altered to accomodate special groups, such as teaching majors. Sophomore level astronomy courses should be developed to address needs of teaching majors.

Role of Informal Science Centers in Formal Education.

Chair: Janet A. Mattei
AAVSO
25 Birch Street
Cambridge, MA
USA 02138-1205

E-mail: jmattei@aavso.org

Recorder: Isabel Hawkins

Issues Discussed:

1. Communication

a) between planetarium and science museums/centers among themselves.

b) between planetarium and amateur groups & research institutions/colleges/ universities.

- Consensus that there is a lack of communication between centers of informal science education/planetarium and professional astronomers.
- Consensus that planetarium and science centers communicate and share resources/ideas among themselves quite effectively.

2. Different "Cultures"

Different "cultures" and attitudes in the informal and professional science worlds appears to be a very important barrier that is hard to overcome – issue of attitudes of each "culture" toward public and K-12 and their right to know what professionals do and why.

Two "cultures" need to seek commonality and respect each other's accomplishments and unique strengths:

- Planetariums are generally the forums where students are turned on toward astronomy.
- Researchers provide latest and very exciting cutting-edge results.
- At times planetarium staff is undertrained and lacks science background which partnership with researchers can alleviate.
- Education needs to be validated by researchers and their institutions.
- Need to collaborate in real projects, not just have forced meetings for the two cultures to work things out.
- Researchers can help make their research relevant in the context of peoples' lives, in collaboration with educators.
- Amateurs can be middle ground between public and researchers, but amateurs and researchers in general do not work together very often. Professional scientists can help if they themselves work with both communities. Excellent examples of effective and successful collaboration is seen in organizations like the AAVSO and IAPPP.

- Few "good/active" amateurs and professional can address public and K-12, but they can get overburdened with huge need from the audience for their time and effort.
- Undergraduate, graduate students can help, as well as professionals if their institutions have policies that validate education from the top. Create a mechanism for addressing public's "right to know."
- Role of professionals can be good as role models for students if they are trained in K-12 education and are sensitive to multicultural perspectives.

At meetings such as this we need more representation from astronomy educators, informal science museum personnel, teachers and amateur astronomers.

3. Connection to Formal Education

How do informal science centers connect with K-12 formal education?

- Can serve as a link between professional astronomers and educators.
- Can help bring latest research to classroom (mediators for researchers, K-12)
- Technology can bridge the gap between research data and classroom as long as an appropriate interface is developed, and informal science centers can help define such interfaces.
- We need to support and engage teachers, and informal science centers are very involved in teacher professional development.
- We need to support and engage parents, and informal science centers are magnets for families.

4. Proposed Solutions

Communication:

- Connect researchers with planetaria by: sending press releases to planetarium directly via Internet; sending press releases via surface mail using International Planetarium Society mailing list.
- Connect Professional community, Planetarium, Public and K-12
- Identify master teachers through informal science centers' own lists, or through school districts; relay information through master teachers.
- Establish special database of master science/astronomy teachers (done by ASP, AAS, ?)

Connection with Formal Education

Primary mechanism is teacher in service training

- Before starting program: assess needs of teachers in the context of: research, outreach program, curriculum materials
- In activities, do not reinvent wheel, find out what other planetariums have done and have students (undergraduates) disseminate in K-12 after training.

5. Conclusions/Recommendations for Solutions

- In every activity that is developed include master teachers in planning and development and implementation.
- Keep up with master teachers in follow up support particularly financial.
- Support needs to be long term and sustained, so that the process of assimilation by other teachers can take place and formalize.

Every partnership should be on equal ground.

Discussion

Asbell-Clarke.

Adler Planetarium is exemplary in this area. They have at least one teacher on staff full-time; they have created a wonderful teacher training center in cooperation with Argonne National Lab., the Teaching Academy of Math & Science, and local industry. They are training teachers throughout Chicago on Hands-on Universe and other programs and have excellent minority outreach programs. They should be a role model for science centers around the world.

Grice.

Astronomy Day and Spaceweek are excellent opportunities to bring professional astronomers, teachers, planetariums, amateur astronomers, students and the public together. It's a great opportunity for people to learn about resources in their back yards. I encourage everyone to participate in these events.

Public Education in Astronomy

Chair: Andrew Fraknoi
Chair, Astronomy Department
Foothill College
12345 El Monte Road
Los Altos, CA
USA 94022

Recorder: John Mosley

Other Participants

Daniel Altschuler (Arecibo Observatory, Puerto Rico)
David Bruning (*Astronomy* Magazine)
Stuart Chapman (Harford County (MD) Public Schools)
David Crawford (National Optical Astronomy Observatories)
Julieta Fierro (Instituto de Astronomia, UNAM, Mexico)
John Ginder (Toronto Center, Royal Astronomical Society of Canada)
Stuart Goldman (*Sky & Telescope* Magazine)
Tony Heinzman (Apple Valley (CA) Public Schools)
James Manning (Taylor Planetarium, Museum of the Rockies)
Sharmi Roy (Davis & Elkins College)
Marc Wetzel (McDonald Observatory, University of Texas)
plus a few others who dropped by for quantum-mechanically determined intervals.

Overview

The world of *public* education in astronomy (defined as those areas of education that take place outside of formal classroom experiences) is perhaps the broadest segment of the field and impacts the largest number of people. It was, we agreed, also the one least well represented at this conference. In part this was because many of the practitioners (authors, reporters, editor, producers, planetarium and museum staff, tour guides, etc.) would not normally come to the annual meeting of an astronomical society, unless specifically invited (and financially underwritten). Nonetheless, our group had a good representation of astronomers and educators from a variety of settings and we discussed a broad range of issues relating to problems and potential solutions in the area of our charge.

We first listed some of the different settings in and through which astronomy education outside the classroom takes place. On our list were:

- observatory visitor centers
- planetaria
- science centers and science-oriented museums
- amateur astronomy groups (and their newsletters)
- magazines (for astronomy buffs and for the general public)
- newspapers
- radio and television
- popular astronomy books for adults

- astronomy books for children
- astronomy computer software
- outreach activities of scientific societies
- newsgroups, list-serve's, discussions, and web sites on the Internet
- telephone hot-lines.

Problems and Challenges

We then brainstormed for a while about some of the problems and challenges in this area of astronomy education. Among those we discussed were:

1. General Problems

- the field of public education consists of many communities, which are not in good communication
- the practitioners come from a wide range of backgrounds and do not always have shared values or vocabulary
- in some areas, the need to educate has to compete with the need to entertain
- the colleges and universities where most astronomers and astronomy educators are trained often do not demonstrate to their students that they value work in education
- public outreach is not a high priority at many scientific institutions
- there are not enough funding opportunities for work in this field, and many people are not aware of the few that do exist
- small grants for education, such as NASA's IDEA program, or the National Academy's Slipher Fund, only allow for part-time staff for educational projects; thus efforts are often disjointed or intermittent

2. Problems Relating to Planetaria, Museums, and Visitor Centers

- staff at such institutions often feels isolated from the rest of the astronomical community
- staff at such institutions does not have sufficient opportunities for upgrading their knowledge of astronomy (especially current developments)
- budget problems often limit the amount of outreach such institutions can do
- there is no central database of projects, exhibits, programs at such institutions
- many people feel such institutions are more for children and not adults, and so adults are not attending programs in sufficient numbers

3. Problems Relating to the Media

- education is not always a high priority for the media
- science often does not get a fair hearing in the general media (especially compared to pseudo-science)

- scientists are often not very good at treating the media well and communicating effectively with them
- there are few opportunities for training the media to work better with scientists or training scientist to work better with the media

4. Problems Relating to the Amateur Community

- amateurs are under-utilized resource for astronomy education (there are many more amateurs than professionals in the U.S.)
- amateurs doing educational projects often work in isolation and are not aware of what others are doing
- professional astronomers often show disdain for amateurs, even when it comes to their participation in educational projects

5. Problems Relating to Publishers of Books

- under pressure to make a profit, publishers often do not distinguish between science and pseudoscience
- publishing is a world with most scientists do not understand well and do not know how to influence
- editors and publishers change companies frequently and unpredictably
- advertising and promotion are reserved for the books judged to be the most successful; many superb science books are not promoted

Some Potential Solutions

Next we discussed some of the possible solutions to the problems that had been raised. In the short time we had, we could not really elaborate on these, but list them in the hope that others may want to pursue them further:

1. General Solutions

- It would be useful to have another meeting like this one, focusing particularly on public education, and specifically inviting the kinds of groups who did not come to this meeting.
- We need to change the culture of universities and research institutions to value educational work (for example, tenure and promotion decision should include work in education as an important criterion)
- We need to make information on projects in public education (and funding opportunities) more widely available
- The National Science Foundation (or other funding agencies) should fund public education and outreach positions at more research institutions
- We need many more partnerships for education among different scientific and educational institutions
- It would be useful for all the major astronomy education institutions to fund an 800 number for basic astronomy information: with modern technology, you could have an information bank from which callers could select topics of interest to them

- At institutions which have several separate small grants for educational projects, it would be useful to find ways to pooling such grants and developing full-time staff with educational expertise (science centers do this regularly, but research institutions in astronomy do not)
- Revive the programs we had some decades ago to assist rangers in national and state parks in covering more astronomy as part of their nature education programs

2. Solutions Relating to Planetaria, Museums, and Visitor Centers

- Need more open houses and public nights at astronomical institutions
- Some of these institutions could establish a telephone hot-line for information on astronomy
- It would be very useful to get scientists more involved with their local science center or planetarium
- We need a database of projects, programs, exhibits that new people could tap into
- We need other meetings to bring workers in these institutions together

3. Solutions Relating to the Media

- Need programs to train astronomers to work with the media
- Need programs to help train the gate-keepers and reporters of the media to be able to work better with scientists, and to distinguish between science and pseudoscience
- Schools of journalism need to offer more courses on science and covering science (and institute graduation requirements that include a substantial amount of science)
- Need clearer, more effective news releases from astronomical institutions
- Like any other group, the media need patient cultivation
- More planetaria and colleges need to make contact with and assist local news media in covering astronomical stories, including what is in the sky (bearing in mind that the media need good graphics)
- The A.A.S. and A.S.P. should offer workshops on dealing with the media at some of their meetings
- Astronomers and educators with a talent for explaining astronomical developments well should make more of an effort to be heard on the media
- Astronomers need a good database of media contacts interested in working with astronomers (need one nationally, and one in each local area). The World Wide Web may be a good place to store such information, although not everyone working in astronomy education has access.

4. Solutions Relating to the Amateur Community

- Establish closer ties between local amateurs and science centers and planetaria
- Perhaps amateurs can be used as volunteer agents for public outreach at astronomical centers and institutions

- Amateur groups around the country sponsor a National Astronomy Day each year; it would be useful to get more professional astronomy institutions involved
- The magazines and newsletters read by amateurs could include more information on astronomy education projects and programs

5. Solutions Relating to Publishers

- Scientists should make their opinions about both good and bad books known to publishers
- Scientists and educators can offer to get more involved in the process of reviewing books in the news media
- Astronomers and educators need more information about good books on astronomy that have been published but insufficiently advertised. A database of such books may be useful.

Conclusion

We agreed that these few thoughts merely skimmed the surface of the many issues and possibilities in the area of public education in astronomy. We were pleased that this was a major topic for discussion at this meeting, and wanted to encourage future meetings that focused on this area more specifically (or at least continue to include an examination of this area.)

We felt that communication was the key and bemoaned the fact that there was no journal, magazine, or newsletter specifically designed to keep workers in the field of astronomy education in touch with each other. We wanted to encourage the American Astronomical Society and the Astronomical Society of the Pacific to pursue ways to facilitate such communication in the future.

Discussion

Bishop.

An excellent opportunity for connections to media is information about any current astronomical or space event. Accurate satellite data in the form of well-written sound bites or paragraphs often will be used by TV stations, radio, and local papers, such as: "Mir" will be seen by us in the Baltimore area at 11:23 pm rising in the northwest to the right of the bowl of the Big Dipper, then disappearing in Earth shadow midway up in the north a few minutes later." (Actual prediction)

Kernohan.

One thing I think you forgot to add was the public's desire to know, see and learn. People always stay later (at the observatory open houses I run at the Milton Academy) than I plan. Also it's very rewarding when someone sees Jupiter or Saturn through the telescope for the first time.

Mechler.

1) In addition to astronomers going on radio – take advantage of community-access cable TV. 2) Take advantage of known upcoming pseudoscience events, e.g. Jupiter Effect in 1982 and May 5 2000 "doomsday."

Fraknoi.

Fraknoi replied with encouragement for local organizations to do local news releases.

Pasachoff.

Eclipses of the sun and of the moon provide excellent opportunities to take advantage of public interest in astronomy. John Percy, as President of the IAU Commission on the Teaching of Astronomy, has put together a committee consisting of Julieta Fierro of Mexico (Vice-President of the Commission), Ralph Chou (a Canadian amateur astonomer and optometry professor), and me (as Chair of the IAU Working Group on Eclipses). We will collect information suitable for distribution to the public not only in North America but also, as the occasions arise, around the world. The information will include descriptions of the science behind the events and discussions of how to observe the events advantageously and safely.

Wasiluk.

Andy Fraknoi made an excellent comment about inviting local media professionals to ASP meetings, etc. At the last IPS (International Planetarium Society) meeting in Cocoa Beach, FL, then president, Bill Gutsch organized a non-expensive meeting of editors from local papers, TV, radio, magazines etc. in the area where we held the meeting and they gave us great tips about how to foster liaisons, write press releases, get better coverage. It was extremely informative, helped to alleviate misconceptions I had about the press and a direct result was that I've been getting more and better press for my planetarium.

The comment that Jeanne Bishop made was excellent about where to look for MIR, shuttle, local sky stuff. If you don't know this information yourself, call the local amateurs; they might have the information you need to "hook" the media by participating in what's going on and what the public can do to be part of the event, sometimes even in their own backyard.

Reaching Under-Served Groups

Chair: Paul Knappenberger

Recorder/reporter: Kathy DeGioia-Eastwood
Dept. of Physics and Astronomy
Northern Arizona University
Flagstaff, AZ
USA 86011-6010

E-mail: kathy.eastwood@nau.edu

This break-out group was chaired by Paul Knappenberger of the Adler Planetarium, and recorded by Kathy DeGioia Eastwood of Northern Arizona University. The need in this area is so great that it became clear after the first half-hour that in two sessions we could barely scratch the surface.

Existing Problems

The session started with short presentations by Ismael Calderon of the American Museum of Natural History, and Meg Urry of the Space Telescope Science Institute. Calderon, who has run the successful "Cultural Arts Program for Students in Temporary Housing" in New York City, noted that as technology evolves, the gap between the technical "haves" and "have-nots" widens. He also pointed out that we have a very diverse population with equally diverse problems. On a more optimistic note, Calderon stated that some programs have been very successful in interesting members of under-served groups to go on and study science in college, and that we should disseminate information about these successful programs and reproduce them.

Urry maintained that women's issues were the same as those of other under-served groups, and that one real problem is a cultural gap between those who are "in-charge" and those who are "under-served". She pointed out that even after ten years of affirmative action, both women and minorities are still rare in the community of professional astronomers. Urry argued that science will be better off if all the available talent is tapped. She then discussed several myths held by scientists which tend to hold women back. These include elitism on the part of main-stream scientists, such that they believe that they are already as good as possible and don't need any new kinds of scientists; the idea that you have to be born a scientist; the myth that science is a calling and that scientists must sacrifice such things as a family in order to be a scientist; and the belief that in order to be a successful scientist you have to be just like an existing one. Urry summarized by saying that there is still a serious problem, and that we need to encourage young women and improve their experience at all levels.

The discussion which followed the presentations brought more problems of various under-served groups to light. One of the common problems is students dropping out of school at an early age, or dropping out of college once they get there. Several participants felt that with many under-served groups, drastic measures are needed at an early age. Another common problem is lack of resources or funds for teaching materials in poor areas. Some under-served groups, such as the disabled, are often forgotten as a group and can't participate in activities we take for granted. For those groups who come from a different cultural background, common problems can be a language barrier or a lack of cultural integration when presenting scientific principles.

The discussion then focused on recognizing the various under-served groups. The groups considered by the group included women, various ethnic groups including American blacks, Hispanics, Native Americans, and Pacific Islanders, the disabled, and the economically disadvantaged. In all cases the problems are different for rural, urban, and suburban members of each group. Obviously it is possible to belong to more than one group simultaneously.

Aspects of Successful Programs

The group next worked on defining essential ingredients for a successful outreach program. Everyone agreed that the most basic tenet is to not re-invent the wheel. Programs which have been proven to be effective should be duplicated. Another very important aspect is to know the needs and characteristics of your target audience. Funding is obviously a necessity. Less obvious is the idea of forming partnerships with other organizations. Astronomers can supply content, but they need other professionals to supply knowledge of social problems and educational techniques. Forming partnerships can also help with the problem of funding, since the funding can now be shared across several organizations. It was also pointed out that partnerships look attractive to funding agencies.

Another basic idea is that it is necessary to work with the under-served group itself to determine its needs; don't presume that you know what is best for them. Some of the most successful programs have involved families or community adults as well as children. Mentoring is virtually always an important aspect of successful programs. Not surprisingly, it is also important to honor cultural traditions and world-views of the group being served.

Several long-term needs were pointed out as essential. One is a long-term evaluation process which addresses the real outcomes rather than simply bean-counting. Another need is continued, long-term contact with the group being served, along with continued commitment on the part of the institutions involved.

One interesting aspect was that the staff involved in programs are sometimes afraid of contact with the group being served. The staff needs to be educated so that they can not only go into a community, but make the group feel welcome in their own community.

The group discussion then turned to identifying examples of existing programs. This list is certainly not meant to be exhaustive, but simply to provide some examples. The programs mentioned included Calderon's Cultural Arts Program for Students in Temporary Housing, the program for disabled visitors at the Boston Museum of Science, Expanding Your Horizons, Operation Smart by Girls, Inc., trio programs such as Upward Bound, and the Howard Hughes Partnerships. Many other programs are administered by NASA, the National Science Foundation, and the American Association for the Advancement of Science (AAAS).

Follow-Up Actions

The final item of discussion concerned follow-up actions which the group could pursue. One suggestion is the creation of a working group, perhaps by the Astronomical Society of the Pacific, to continue this discussion. The group agreed that sometime in the future a two-day meeting similar to the present one should be held on the specific topic of reaching under-served groups. The discussion group felt that lobbying for expanded funding in this area is important. Internet suggestions included a list of existing sources of funding and a list of people working on similar programs, perhaps managed by the American Astronomical Society, as well as an E-mail list (or list server) of the discussion participants plus others interested in continuing work in this area.

Discussion

Unknown

How do centers of informal science education/planetaria handle the important issue of training for your own staff in issues pertaining to underserved groups?

Knappenberger.

Through special programs that bring in experts to train staff through partnerships.

DeVore.

An excellent reference for scientists working with schools and communities is *Science Education Partnerships*, ed. Art Sussman, San Francisco State University Press. Art works at Far West Labs in San Francisco in science and math education programs.

Fraknoi.

I want to commend two issues of the ASP's *Mercury* magazine related to these concerns to everyone's attention: The special issue on Women in Astronomy (Jan/Feb 1992); The special issue on Minorities in Astronomy (May/June 1995).

Hoff.

(After a remark concerning the need to serve underserved populations – the need to serve those in prison) I added that a recent article in a national newspaper pointed out that there are now an estimated 1.2 million people incarcerated in the US. I had also seen an article earlier that showed that (excluding the medical profession) the number of professional scientists in the U.S. is only about a million. This is an important fact for K-12 teachers to keep in mind. Those who believe that their goal is to prepare future scientists have an equally good chance of preparing a future criminal!

Roettger.

Long term commitment, mentoring, other things you mentioned take a lot of time. Is your group recommending that we put our time into reaching a few people deeply rather than a large number of people broadly? This will affect how programs are measured and funded – it's easy to measure numbers.

Sakimoto.

If you are constrained by resources, I recommend building a comprehensive program for a small group of people. The other alternative – one-shot programs for larger groups – tends not to have lasting effects. You can also partner with community groups or minority universities who already have programs but lack astronomy expertise.

Creating Networks and Coalitions

Chair: Mary Kay Hemenway
AAS Education Officer
Department of Astronomy
University of Texas
Austin, TX
USA 78712-1083

E-mail: marykay@astro.as.utexas.edu

Recorder: John Percy

Reporter: Elizabeth Roettger
The Adler Planetarium
1300 S. Lake Shore Drive
Chicago, IL
USA 60605

E-mail: eroettge@midway.uchicago.edu

"Networks", "coalitions," and "partnerships" are key words, but what do they mean? Why do we need them? How can we use them for mutual benefits? How can we best choose, form, and maintain connections? These questions, and some possible answers to them, were the main issues our group identified. After some discussion of computer networks, it became clear that the networks we, the participants, had actually used were networks of people - although computer networks figure prominently as a means of communication and dissemination.

Counting individuals and groups as "nodes," we defined a network as interacting nodes with issues in common. Although networks connect many organizations and communities, they tend to work through connections between individuals. We believe this is because good networks often rely on mutual respect and trust, which are easier to achieve among individuals. We need networks because an individual can't do everything alone; we need to pool our skills and share the work. No single human being can keep track of the enormous amount of information being generated; we need to filter the information and share the important things with each other. We waste an enormous amount of time and resources re-inventing projects that others have tried and perfected or discarded; we need to share our experiences and learn from each other.

We want astronomy education networks to be used to connect and empower the separate nodes. We hope these networks will:

- draw on the best ideas of other disciplines
- provide astronomer contacts (to others who need them)
- provide astronomy clearinghouses
- enable scientist/educator coalitions to develop materials (preferably those which involve field testing the materials)
- communicate results and lessons learned
- inform all educators of astronomy opportunities
- avoid duplication of effort
- disseminate good projects

- include representatives from the groups that are the ultimate recipients of the projects
- help us allocate resources more effectively
- inform us of what's going on locally and elsewhere
- teach us about alliances and other mechanisms that make networks and coalitions work in other places

Two main themes, with a lot of overlap, kept recurring. We found ourselves trying to connect communities and trying to share resources.

Communities

Who are the communities that have something we need for astronomy education? We started a list of network subsets, and invite you to add to the list depending upon your own interactions and needs:

- amateur astronomers
- American Association of Physics Teachers (AAPT)
- American Astronomical Society (AAS)
- American Geophysical Union
- American Institute of Physics (AIP)
- American Physical Society
- Astronomical Society of the Pacific (ASP)
- astronomy departments in colleges and universities
- Coalition for Earth Science Education
- education departments in colleges and universities
- evaluators
- individual teachers
- International Planetarium Society
- National Aeronautics and Space Administration (NASA)
- regional planetarium associations and teachers with planetaria
- schools
- Space Telescope Science Institute (STScI)
- state Departments of Education
- students
- teacher representatives
- teachers' associations, such as National Science Teachers Association (NSTA)
- youth groups
- You!

How do we choose, reach, and work with these diverse groups? Most of us join a network because we have a need that the network can fill. Recognizing that others have something to offer us makes it easier to respect them and their skills. Astronomers can learn from teachers, planetarium staff, and others as well as vice versa (consider that teachers study how to design good bulletin boards, and then compare that to your experience as you walk through a poster session at an astronomy conference). It's too easy to focus on our own expertise and fail to value the expertise of others; mutual respect is one key to make a network connection work. Another key to successful networks and coalitions is communication – it is more effective to reach someone or some group by using their modes of communication. This may be a choice of vocabulary or a choice between e-mail and phone calls. Although e-mail is a convenient form of communication, we noted that many people and organizations do not have e-mail, and felt it was important to maintain multiple modes of communication to avoid separating the haves from the have-nots. E-mail can break some of an astronomer's or teacher's sense of isolation, but there are isolated astronomy teachers that must be reached through other modes of communication.

Communities

- need mutual respect, two-way participation
- in the astronomy community, we need to know who's doing what
- how do we coordinate efforts?
- how do we share "lessons learned"?

Other Resources (sharing the stuff and information)

- how do we find out what's out there?
- how can we tell what's good?
- how should we allocate our resources (including our time)?
- how do we avoid duplication of effort?
- how do we use networks to disseminate resources?

We need networks for surveying, gathering information, finding out what's been successful, and disseminating information and resources. We wondered who or what would be the best "node" for astronomy education. Directories of who has what resources, projects, and programs would be useful. We particularly want to find ways to make resources available to teachers, and discussed a central clearing-house, or just trying to link teachers to the local "nodes." We wondered how such efforts could be funded: through professional societies, private funding, or funding agencies (we hope to see more cooperation between National Science Foundation, NASA, and other agencies, and between the education and research branches).

Challenges and Lessons Learned

- breaking into new networks takes time and perseverance (lesson)
- LISTENING and learning from others' expertise is vital! (lesson)
- overcome bias (ours and others') (challenge)
- overcome institutional barriers (challenge)
- find effective modes of communication (challenge)
- reach diverse groups (challenge)
- have REALISTIC expectations (challenge and lesson)
- empower without controlling (challenge)
- don't re-invent the wheel (lesson)
- meet the funding challenge (challenge)

We discussed the possibilities of an astronomy education journal or newsletter, moderated discussion groups or information services, Internet "questions and answers," and WWW homepages as a source of interpreted information. We talked about the need for more respect for teaching, particularly in academic culture. To meet this need, we discussed the possibility of an AAS Division for Education. Such a division would promote higher visibility and better recognition of the value of astronomy education, but the benefit of such a division should be compared to the current AAS Working Group on Astronomy Education (WGAE). Although AAS members are restricted to presenting only one paper per meeting as first author, members of an AAS working group may present an addition paper in a session sponsored by a working group. The WGAE permits only AAS members to be members of the working group; AAS divisions may have affiliate members who are not AAS members participate in the meetings and functions of the division. Divisions involve a more formal structure, including dues and officers, than the currently informal WGAE. However, the affiliate members of a division pay lower dues than full AAS members in order to join the division, and don't have to meet the formal AAS membership guidelines in order to join the division (they are expected to belong to another appropriate professional society). Divisions commonly have meetings both in conjunction with AAS meetings and independent of AAS meetings.

Recommended actions:

1. Use existing structures and modes of communication. We discussed STScI as a possible national institute for science education. We listed some existing resources: Physics Education News (PEN is a electronic newsletter produced by AIP), the AAS WGAE (with its electronic newsletter), the AIP monthly magazine "Physics Today", the AAS "Newsletter", ASP's magazine "Mercury", the proposed AAS Initiative in Astronomy Education, and NASA's Space Grant Colleges & Coalitions and teachers' resource centers. We also discussed possible new publications, organizations, or divisions of existing organizations, and are forced to admit that it's quite tempting to reinvent the wheel.
2. Individuals:
   - become your own network node
   - attach to several major nodes
3. Major nodes:
   - make access easy
   - keep continuity
   - make interactions easy
4. Include quality control in the information you communicate. There are a lot of good materials and programs around, but there are some that aren't so good. Many of us don't want to criticize, but it's possible to recommend quality items or describe how something meets your needs. (The worth of a resource lies in the "hands of the user"; some items are excellent for certain circumstances or require special equipment and are useless for others. Therefore, just a list of "what's available" doesn't provide much value.) Andrew Fraknoi (ASP's Project ASTRO) and several others have compiled lists of recommended resources; although a single combined list might be more convenient, multiple annotated lists may serve better as a review and filtering system.
5. We need to provide information on how to help each other connect to others with which they may wish to network. For science teachers, NSTA (with a membership of over 52,000) might be an appropriate place to start.
6. We need to
   - use the AAS Working Group on Astronomy Education more effectively
   - consider an AAS Division on Astronomy Education (we did not reach consensus; strongest argument for a Division was clout and recognition; the strongest argument against a forming a division was recommendation #1)

- find publishers willing to publish regular articles on astronomy education since forming a journal devoted to the topic seems unlikely at present
- promote astronomy education as a research-based endeavor

Discussion

Dukes.

As the founding chair of the AAPT Working Group on Astronomy Education (formed at a time when AAS was not interested in education) I am pleased to see AAS becoming interested in education. AAPT split from APS when APS was not interested in education. In forming two societies some political "clout" was lost. I now feel that combining AAS with ASP education efforts is a better move than forming a new group.

Pasachoff.

Since Bob mentions the American Physical Society, let me mention that the APS's Forum on Education, wanting to broaden its coverage into astronomy, asked me to run for election to its Executive Committee, and I was elected APS/AAPT Member-at-large.

Another organizaiton in the alphabet soup is the AAAS (the American Association for the Advancement of Science). Its Astronomy Division sponsors at least one or two talks at each of its annual meetings. You are encouraged to attend its meeting, especially if it occurs in a city near you, and you are welcome to make suggestions for future symposia.

Eastwood.

Since the AAS is finally showing some interest in education, I suggest that rather than creating yet another new group, those of us who belong to the AAS work from within to make a sub-group. That sub-group could argue for more emphasis on education within the AAS. I do like Andy Fraknoi's idea of a jointAAS/ASP working group which can involve teachers as well as professional astronomers.

Hawkins.

I would like to suggest that AAS and ASP facilitate having teachers become members of AAS and ASP and also that they have the opportunity of participating in K-12 policy subcommittees on education within our societies.

Sakimoto.

I strongly favor the concept of a joint ASP/AAS working group on education. The number of astronomy educators in the country is small, so pooling resources is the best way to maximize output. Also, it would make little sense for an individual to split time between two organizations working on the same issues.

Shawl.

The Working Group on Astronomy Education of the AAS is a very loose group. If you have ideas of what we might do, let us know. Also, your contributions to the e-mail newsletter are always needed. Contact me at S-Shawl@Ukans.edu.

Technology and Astronomy Education

Chair: Laurence A. Marschall
Department of Physics
Gettysburg College
Gettysburg, PA
USA 17325

Fax: (717) 337-6666, Internet: marschal@gettysburg.edu

Recorder: John Safko

Introduction: Identifying Useful Technology

Technological advances in the last decade, without a doubt, have made many new modes of teaching and learning possible in astronomy. At the session on Technology and Astronomy Education we began by identifying those aspects of recent technology which were applicable to astronomy education.

Among the technological advances the members of the group were aware of were:

- Computers: which are used to run telescopes, process digital data, run simulations, and provide demonstrations for lectures and labs.
- Software: including programs for data analysis and simulations. Images can now be stored and manipulated digitally. Planetarium programs like *The Sky*, *Dance of the Planets*, and *Voyager* have become popular in introductory astronomy labs, and simulation programs like those of Project CLEA are making advanced techniques available for hands-on exercises.
- CCD Cameras: make it possible for students to make their own images of deep-sky objects much more easily than they could using photographic techniques.
- Automatic Photometric Telescopes and other robotic and/or remote-controlled telescopes: Provide data for instructional purposes with a relatively low amount of student labor.
- Videotape: There is a growing library of films and supercomputer simulations available for lectures.
- Videodisk: Provides a wealth of still image and motion picture data for use in the classroom and the lab.
- The World-Wide-Web: Makes a wide variety of data, animations, software, etc. available to students and teachers at the click of a mouse. Facilitates communication between people of shared interests. Provides a medium (html documents) for generating hypertext documents that can be used in classes.

Problems Raised by Technology in Astronomy Education:

Participants were in general agreement that there were great opportunities for increasing the effect of astronomy education through the new technologies. But they voiced a number of concerns:

One needs to know what materials are available, what they do, and how to get them: the proliferation of technologies, especially software and web sites, is occurring so rapidly that it is an effort to keep up. Teachers and students need an efficient way of matching their needs and interests to the technology.

It takes time and effort to learn to use the new technology in old courses. Change in classrooms takes place much more slowly than change in technology, especially when a new technology opens up a whole new mode of teaching. Video, for instance, can easily be incorporated into existing lecture courses. Using hypertext on the web, however, implies an entirely new structure to the course, perhaps including a strong self-paced element. Both high school and college teachers indicated that they needed free blocks of time and outside instruction to learn particular new techniques.

Students need to be computer-literate to learn the new technology. Participants seemed to feel that this was less of a problem than in the past. Depending on their earlier educational background, of course, students in the last few years seem to take computers for granted. They are less intimidated by them than they used to be, and they are familiar with the fundamental grammar of the most popular operating systems.

Technology should not be used just because it is new. It should be used with an eye to:

- Balance: Technology cannot solve all the problems of astronomy teaching.
- Appropriateness: Technology need not, and probably should not, be used if a more direct and simpler method with do. For instance, it is better to learn the constellations outdoors, if possible, than to memorize their appearance on the screen of a computer monitor.
- Applicability: Software that runs on a workstation is hardly applicable for use in 5th grade classes. Nor do we want to teach spectral synthesis to students just because the software is available. Teachers should be cognizant of learning objectives in applying technology. Substance should take precendence over glitter.
- Effectiveness: Does the technology work in meeting learning objectives? There is very little systematic evidence one way or the other

Technology costs money.

- Is it worth it? Equipping a classroom or a lab with telescopes, CCD'/s, or computers and network connections costs much less than it used to, but it is still more expensive than using printed photographs and naked eye observations. What is a reasonable cost for a particular situation?
- Who will pay the long term costs? Some of the costs of technology, such as the long-term maintenance and replacement of computers, or the cost of network connections, are not evident in the beginning. It is possible to become dependent on a technology that becomes overly expensive in the future.
- How will we deal with equity between the haves and the have-nots? Some schools have large numbers of computers and Internet connections, others have none. It is possible that we may be creating two classes of students (in the nation and internationally) learning astronomy in quite disparate ways.

The effectiveness of specific technologies in teaching and learning needs to be evaluated systematically. Much of the technology is so new that it is difficult to make informed choices about its wise and economic use.

It is difficult to know what lies ahead. Technology is changing so rapidly, and in so many unforeseen ways, that one is either (1) far behind and perhaps not doing things as well as one could, or (2) riding blindly on the crest of the wave, unsure of whether one is being truly innovative or just trendy.

Suggested Solutions

The following suggestions were put forward, not as all-purpose solutions, but as measures which might prove useful in aiding teachers and fostering wise use of the new technology. Existing institutions and publications may be able to provide the means for meeting some of the challenges we have noted.

We should encourage the creation and maintenance of online resource listings for astronomy education, such as that proposed by the AAS Education Committee.

Popular journals, such as *Mercury*, *Sky and Telescope*, *CCD Astronomy*, and *Astronomy* are appropriate places for news and information on educational resources.

There may be call for the creation of an online journal of astronomy education, and/or the creation of astronomy education news groups or mailing lists.

High school teachers and college teachers do not read the same journals. News of astronomy technology useful to specific audiences should be made available through journals read by those audiences. For instance *The Science Teacher* and other publications of the National Science Teachers' Association are the appropriate media for reaching teachers of K-12 students, while the *American Journal of Physics* is best for college physics teachers.

Professional organizations (through their small grants programs) and governmental agencies should support grants for research into the effectiveness of technology in astronomy education. Innovators in the use of technology should be encouraged to carry out such studies.

Professional organizations, at national meetings perhaps, should support training sessions and workshops on the use of technology for teachers. The American Association of Physics Teachers, for instance, organizes several days of training workshops at each of its semiannual meetings.

Astronomers at "well connected" institutions should encourage their schools to foster Internet access at all high schools and colleges. Colleges should, at minimal cost, offer Internet access to local K-12 teachers and students if they do not already have it. Users of technological innovations should mentor local teachers in useful items and techniques.

A national meeting such as this, on astronomy education, should be held every few years.

Summary

The above suggestions seem to point to a need for better communication between users and potential users of the new technologies, for active outreach to train potential users, and for the fostering of evaluation projects to help teachers decide the most fruitful way to use the new resources. Professional societies and governmental funding agencies can play a role in outreach, training, and evaluation. Professional societies can also play an important role in maintaining central web-sites where resources can be listed along with evaluative information. Technological resources for astronomy will continue to grow in an anarchic, organic fashion, but organizations can do more to foster effective application of the latest technology.

Discussion

Garrett.

What is involved in setting up a mailing list (electronic)? Could anyone here volunteer to do it?

Marschall.

It is easy to do but can turn into a time-consuming ("full-time") job.

General Discussion

Dukes.

We need to remember that astronomy is basically an observational and not an experimental science. Much astronomy in this country is taught in physics departments by physicists. Physicists are trained as experimental scientists and are not used to dealing with an observational science. As such they are not aware of the problems with dealing with a body of data on individual objects. One thing which could be done would be to devise ways to make physicists aware of this problem. A suggestion is to attempt to involve physicists in observational astronomy research. A related problem occurs when dealing with the K-12 community. Teachers are trained by science educators in the scientific method. This involves certain things that are difficult, if not impossible to accomplish in astronomical research. An example is the control of variables. Astronomers usually cannot do this. We should make teachers and science educators aware of this problem.

Reynolds.

Astronomers (scientists) need to learn something to be educators and teachers need to be enlightened about observational methods to expand their narrow view of scientific research methods ("scientific method") where in physics ("not thinking about individual data") or astronomy ("you can't control the variables") things may be really different.

French.

As the DPS (Division of Planetary Sciences, AAS) Education Officer, I will be organizing a workshop for DPS members after our fall meeting in Kona, Hawaii. Please contact me for more information.

Fraknoi.

I would very much like to see an ongoing organization and newsletter, jointly sponsored by the AAS and ASP, for astronomy educators at all levels. It would be very nice if it could have a low membership fee. It could be a joint division of the AAS and ASP, and could subsume the small and often struggling Association of Astronomy Educators.

Hill.

Until the reward structure for professional astronomers (in terms of prestige and salary) rewards teaching as much as research, it will be difficult for us to expect that most professional astronomers will be committed to genuine educational reform in education.

Hoff.

Let's not wait 23 years until the next ast-ed symposium, and even if we can't reach agreement on many issues of astronomy education, the "journey may be more important than the destination."

Hollow.

Graduate students at University of Western Sydney (Australia) have to take part in outreach programs such as primary school groups/public viewing evenings at our Astronomy Center. Some hesitation but students acknowledge the value of them.

Manning.

Many good ideas have been expressed here, but follow-up will be important. Might it be useful to establish a committee to coordinate, monitor, or encourage the sorts of calls to action expressed here? Perhaps it would be good to have a representative from each of the sponsoring organizations, to maintain communication and interchange.

Pasachoff.

To pick up on Dennis Schatz's earlier statement, let us make a concrete way for graduate students to participate in meetings like this. I propose that for the next ASP meeting on education we invite universities to nominate one graduate student each to have a Fellowship to attend, and to provide that person with free registration and lunch.

Are there any graduate students here? [One hand is raised.]
Alison Procter: I am.
JMP: I am very glad to have you here. In some sense, this meeting is for you. I hope you come to many more.

To make my valedictory statement for this meeting let me associate myself with Darrel Hoff's statement by quoting Robert Louis Stevenson, who said – using the word "hopefully" correctly – "to travel hopefully is better than to arrive."

Richter.

Two general comments:

1) In terms of K-12 education and where astronomy fits in the curriculum: the strength of astronomy is its interdisciplinary nature and integration of science themes, big ideas, and concepts we want kids to learn. Rather than focusing on astronomy as a separate subject or lack of mention in curriculum standards, <u>use the standards to make an argument</u> for teaching astronomy (as integrating subject).

2) Astronomers need to link and learn with and from other sciences active in education – chemists, physicists, biologists, doctors.

Roettger.

There already exists a group whose mission is enabling scientists to become involved in education, do workshops at professional society meetings, collect and disseminate the needed information, etc. In 1994, the National Research Council (NRC) established Regional Initiatives in Science Education (Project RISE) to develop a national cadre of scientists and engineers equipped to participate effectively in K-12 science education reform. (from *The Catalyst*, 1:1, Jan 1995, Ed. Karen Goldberg)

They advertise "Hold a Project RISE workshop at the next annual meeting of your scientific society" to get information on national, regional, and local initiatives, experience activities, learn how to develop partnerships, etc.

Contact information:
Project RISE
National Research Council
2101 Constitution Avenue NW
Room HA 486
Washington DC
USA 20418

Tel.: (202) 334-2110
FAX: (202) 334-3159
E-mail: RISE@NAS.edu

Karen Goldberg, Editor
Jan Tuomi, Director

Rosendhal.

The general issue of training is going to be critical. If a group of astronomers said they wanted to go into brain surgery, the prospective recipients would certainly ask valid questions about credentials. Other professions regard formal training and professional development as a normal part of career progression and that there is a routine need for the development of new skills. Since what we are doing is at least as important as brain surgery, we are going to have to address the issue of training. The astronomical community needs to become more aware of the current trends in education, become aware of what is and isn't effective etc. We must find out how to arrange this if we, as a community, are going to be effective in education and public outreach.

Hemenway.

[response to Jeff Rosendhal] AAS sponsored a one-day workshop at Lawrence Hall of Science following the AAS Berkeley meeting for 40 astronomers. The LHS staff and ASP staff co-presented. Other professional societies offer 1-day short courses connected with meetings to offer training on various aspects; perhaps AAS and/or ASP should consider something similar to educate astronomers about educational techniques.

Mattei.

I participated in the workshop for astronomers at Berkeley that Mary Kay Hemenway mentioned and found it extremely helpful in learning about hands-on-activities, the kinds of things to consider in giving educational workshops and in being informed about resource materials.

Schatz.

Now is the time, with the new AAS Education Coordinator being hired, to write an NSF proposal to the Astronomy Division to develop a staff development workshop for professors and graduate students – possibly focused on TA and professor instructor teams – that occurs in regional sites, where staff from surrounding institutions come to a common site. It would have to be more than one day – three would be nice.

Shawl.

A followup to both Jeff and Mary Kay ... I've been impressed with the number of people who have overcome administrative roadblocks and accomplished great things. Some professional development to teach us how to be more effective would be extremely helpful.

SECTION III. POSTER PAPERS

Participants were encouraged to contribute poster papers, and many did so. Summaries of most of them are included in the following section. (The non-alphabetical position of the papers by Hennig and Hoff is due to an oversight on the part of the editor.) They include many of the most interesting and important astronomy education programs and projects in North America. A more complete list is included in the appendix. The following posters were presented, but not summarized here.

Chaisson, Eric: Programs and Activities at the Wright Center for Innovative Science Education

DeVore, Edna and Stoneburner, C.: SETI Institute Educational Projects

Heinzman, Tony: Remote Access to Astronomical Images

Mechler, Gary: You Teach Astronomy, but Do You Teach Science?

Sadler, Philip et al.: The Micro Observatory Net, and Project STAR: A New Approach to Teaching Astronomy

Because of the time constraints in the symposium, only limited time was available for poster viewing. In future symposia, more time should be allocated for poster viewing, and the posters should be situated in a single, large, comfortable area.

The Angel Ramos Visitor and Educational Facility – Arecibo Observatory

Daniel Altschuler

and

JoAnn Eder
Arecibo Observatory
Cornell University
P.O. Box 995
Arecibo, Puerto Rico
00613

E-mail: daltschuler@naic.edu - jeder@naic.edu

As the world's largest single-dish radio telescope, Arecibo Observatory in Puerto Rico attracts thousands of visitors each year, of all ages and from many countries. Pride in the Observatory has caused local Puerto Rican organizations to contribute the funds necessary for the construction of the new Arecibo Observatory Visitor and Educational Facility (AOVEF).

When the AOVEF is completed, it will be the only facility of its kind serving the general public and the public and private schools of Puerto Rico. At Arecibo, we see an ideal opportunity to positively affect this large, minority population. Currently, without any formal visitor program, approximately 40,000 persons visit the Observatory annually, half of whom are of school age. We anticipate that when the visitor center opens, this attendance will easily surpass 80,000.

The AOVEF consists of approximately 10,000 square feet of building and outdoor program space. It will house about 3500 square feet of exhibits, a 100 person multi-purpose theater, a science merchandise store, and appropriate meeting rooms and work space. In addition, the facility offers a breathtaking view of the primary research instrument of the observatory – the 305 meter diameter radio telescope.

We will establish a bilingual educational program centered on the theme "More Than Meets the Eye", which will reflect the general idea that we can study our world with tools which extend our direct sensory experience. The proposed program will explore the unseen sky and, in particular, the objects that fall under the scrutiny of a radio telescope. Specifically, the program will introduce the visitors to the electromagnetic spectrum as a means of exploration, will offer a framework of basic astronomy and atmospheric science, and will provide understanding of the function and operation of the Arecibo radio telescope. Some of Arecibo's most exciting new discoveries in the fields of radio astronomy, solar system radar astronomy and atmospheric science will be presented.

If we obtain the necessary funds, we plan to establish a "Hyper Exhibit" – a network of workstations which will allow visitors to navigate through several levels of information. A Hypertext platform such as Mosaic will be used. Also, in the theater a 15 minute multimedia program "A Day in the Life of the Arecibo Observatory – Un Dia en la Vida del Observatorio de Arecibo", will endeavor to tell the story of the people who make Arecibo possible, from the guards and kitchen staff to the telescope operators, staff and visiting scientists. Here, the techniques of science will be combined with stories of scientists. The audiovisual piece will move comfortably and creatively back and forth between science, tools and technology and the many people who make science possible at Arecibo.

The general goals of the Educational Facility can be summarized as follows:

- To provide visitors with the necessary background to appreciate the research work in atmospheric science, radio and radar astronomy pursued at the Arecibo Observatory.
- To introduce visitors to the general principles of astronomy at all wavelengths of the electromagnetic spectrum.
- Encourage minority youth to pursue careers in science, mathematics and engineering.
- Help to improve science teaching in the public and private school systems of Puerto Rico.
- Provide local university students with a direct experience at a major research center working as student guides.

There are approximately 650,000 students in Puerto Rico under one Department of Education, the largest school district in the nation. A total of about 37,000 teachers are employed. We plan to use the AOVEF to hold workshops and training courses in the sciences, for educators. We also plan to use the facility for activities such as the organization of scientific workshops.

Funding for the facility has been obtained from private and government sources in Puerto Rico, outside of our normal funding sources. In particular, the *Angel Ramos Foundation* provided a matching grant of one half the total construction cost. The other half was provided by the *Municipality of the City of Arecibo,* the *Puerto Rico Tourism Company,* the *Government of Puerto Rico* and several corporations and individuals. Funds to develop the exhibits were obtained through a grant from the *National Science Foundation's Informal Science Education program.* Total investment will reach $2,500,000.

Construction documents were prepared by the office of Mendez, Brunner Badillo and Associates of San Juan and we expect to start construction in mid 1995. Zalisk Martin Associates of Cambridge MA, and Design Craftsmen of Midland MI have been contracted to design and manufacture the exhibits.

Hands-On Universe: Bringing Astronomical Explorations to the Classroom

Jodi Asbell-Clarke
TERC

Hughes Pack
Northfield Mount Hermon HS, MA

Carl Pennypacker
Lawrence Berkeley Lab

Dane Toler
Robinson HS, VA

For the past three years, with support from the National Science Foundation and the Department of Energy, Hands-On Universe (HOU) has developed and piloted an educational program that enables high school students to request their own observations from professional observatories. HOU students download CCD images to their classroom computers and use HOU's powerful image processing software to visualize and analyze their data. HOU also provides comprehensive curriculum that integrates many of the topics and skills outlined in the national goals for science and math education into open-ended astronomical investigations. Over the next several years HOU will develop activities and tools for middle school students and informal education centers as well as implementing HOU in regional high school networks across the nation.

The HOU Telescopes: Currently HOU students request images from the 30"automated telescope at Leuschner Observatory of the U.C. Berkeley astronomy department. HOU has gathered a group of collaborators with telescopes in Hawaii, Illinois, California, Washington, Sweden, and Australia that are working together to form ATHENA, Automated Telescopes for HOU Educational Network Access. Student requests will be processed by the network to decide which telescope is best suited for the particular request, considering weather, geography, scheduling and equipment. This network will provide fast turn-around for student requests and allow real-time observing in certain cases because of the various time zones of the telescopes.

The HOU Telecommunications: HOU began with teachers and students dialing up the HOU computer over toll-free phone lines and using MS-Kermit to download images and use e-mail accounts. In the past year several of the HOU pilot schools have obtained Internet access either directly or through a local provider. This summer the HOU WWW pages are being released and will be the primary interface between HOU students and the image request and retrieval system. The WWW provides a much cleaner and more efficient mode of requesting and downloading images, as well as providing observatory weather information, a forum for student reports and Q&A sessions. All HOU classes without direct Internet access will connect to the HOU computer using PPP or SLIP. To make this possible, the toll-free phone lines will remain in use for those schools that do not have a local provider available.

The HOU Image Processing Software: The HOU image processing and data analysis software runs under Windows on a 486 (or extended 386) as well as a Macintosh II series or better. Image processing tools include: log scaling, min/max adjustment, resizing, a variety of color palettes, and image manipulation such as rotation, flip, shift, and arithmetic functions (add, subtract, multiply and divide).

For data analysis, counts are displayed for each pixel and slice plots and histograms show brightness distributions. Photometry routines calculate full width half max and sky-subtracted brightness.

The HOU Curriculum: The HOU curriculum integrates mathematics, science and technology in the context of exciting astronomical explorations. HOU addresses many of the goals for mathematics and science education standards by the National Council of Teachers of Mathematics and the National Research Council. Through the investigation of the solar system, galaxies, variable stars, and supernovae students develop problem-solving techniques and critical thinking skills. Along the way students discover the need for algebra, geometry, various data representations and interpretations, as well as physical principles such as force and motion, energy transformations, and properties of light. The classroom computer, along with the image processing software and telecommunications, are regarded as research tools which enable students to pursue their investigations.

The HOU Teacher Training and Support: Over the past three years, HOU has held a one-week workshop each summer in which the staff trained 30 pilot teachers to use HOU and received feedback that was used in the development and refinement of the project. The next phase of HOU expansion will begin this August with two one-week workshops where previous HOU teachers will serve as Teacher Resource Agents (TRAs) to train 20 new teachers per workshop. Over the next several years, we expect HOU to grow exponentially as a few teachers from each workshop will go on to become TRAs after using HOU successfully in the classroom and undergoing further training. Following the workshops, TRA's will give ongoing support during the school year. The HOU WWW pages provide a forum for communication among teachers, students, HOU staff and professional astronomers. Through e-mail, classroom visits, and video-conferencing (where available), professional astronomers mentor teachers and students to help guide their HOU investigations. The TRA method of HOU expansion will encourage regional growth since TRAs will generally train teachers within their own geographical area. HOU has developed a system for local support called Support Teams with Resources for Educational Explorations using Technology (STREET). Each STREET will include members of local universities, museums, amateur astronomy groups, and industry working along with schools to build a strong foundation for projects such as HOU. Contributions from STREETs will include mentorship, equipment, facilities for meetings and training workshops, and technical support. HOU has already obtained commitment from all of these sectors of the community in three potential STREET pilot sites.

HOU's Success: Over the past two years HOU has been studied by an external evaluator to determine the effect on student outcomes in both content learning and attitude towards science as well as the effect of HOU on teachers and the use of technology in the classroom. Results from student surveys show that students are motivated to learn and retain scientific and mathematical concepts because they see the direct application of such material. They enjoy using the computer and see it as much more than a word processor when they participate in HOU. Many students who did not think of themselves as scientifically-inclined, came away from the HOU experience with a much better understanding of what a research scientist does and felt that they could and may pursue science as a career. Startling discoveries and valuable science can occur when high school students are given access to professional telescopes as witnessed by two Oil City High School students in spring 1994. Melody Spence and Heather Tartara requested observations of M51 (The Whirlpool Galaxy) during their investigation of spiral galaxies. A few days later they received a phone call informing them that they had captured the first light of SN1994I, the ninth supernova of 1994. Ms. Spence and Ms. Tartara will appear as co-authors on the SN1994I photometry paper.

Further Development of HOU: HOU has received funds from the National Science Foundation to develop educational materials for use in informal science education (ISE) centers, including museums, science camps and community centers. These materials will focus on themes that incorporate astronomy through various cultures and also serve a large range of academic backgrounds and age groups. The ISE materials will be geared toward ongoing workshops within the centers or drop-in labs. In addition, this summer the Boston Museum of Science will be introducing a museum exhibit they have developed along with HOU. This exhibit will serve as a prototype for exhibits to be developed by other museums and planetaria around the world.

For More Information on HOU: Please e-mail: houstaff@hou.lbl.gov, or write: Hands-On Universe, MS 50-232, Lawrence Berkeley Lab, One Cyclotron Rd., Berkeley, CA USA 94720. the Web site is http://hou.lbl.gov/

Astronomy as a Doorway to Science for Disadvantaged Students

Harry J. Augensen
Widener University
Chester, PA
USA 19013-5792

E-mail: harry.j.augensen@cyber.widener.edu

Abstract: Students with academic deficiencies, especially in the sciences, are becoming increasingly common in college classrooms. Many universities and colleges provide special, often state-funded, services for educationally and economically disadvantaged students, but deficiencies in their pre-college education often hamper higher learning and at the very least limit career options. Most active scientists have one thing in common: their curiosity about nature. Learning about the stars and planets at an early age can help all students, disadvantaged or otherwise, can help develop this natural curiosity. I will describe my experience teaching and advising disadvantaged students and suggest some projects for motivating them to become interested in and perhaps pursue careers in the sciences.

Introduction: It is a disturbing fact that only a relatively small percentage of minority students living in the US pursue college degrees in the sciences. A recent study from the US Department of Education concludes that blacks, hispanics, and native American Indians are underrepresented in science- and math-related fields due to deficiencies in their early education. During their pre-college years, minority students often receive inadequate preparation to take science and math courses, and/or they become disinterested as they fall behind in science and math learning. Only rarely does a minority student receive stimulation early on to pursue a career in science. Yet, the report states, students of all racial and ethnic backgrounds are equally positive about science and mathematics.

Widener University, a private university located in southeastern Pennsylvania, is one of a number of institutions of higher learning which provide special services to assist educationally and economically disadvantaged students under a state-supported program called ACT 101/Project Prepare. Nearly all of the students accepted into the program grew up in light-polluted, urban environments in and around Philadelphia, where the view of the sky is often obstructed by tall city dwellings. Many have never noticed such simple astronomical events as sunrise or sunset, and, because of urban light pollution, are barely able to see the night sky beyond the Moon, a few bright stars, and the planets.

Such deficiences, while debilitating, need not be fatal, and can in fact be partially overcome. In Project Prepare, we have two goals:

1. To provide the necessary remedial training for students with learning deficiencies prior to their matriculation as freshman in science/engineering,
2. To provide direct laboratory experiences to motivate some of these young people to pursue physics and/or astronomy as a career.

This presumes, of course, that the students already have the necessary willingness to succeed, but this is generally true of anyone selected into this program. Details of the Project Prepare program are available from the author at the address listed at the end of this paper.

The Role Of Astronomy And Physics

Most active scientists have one thing in common: their curiosity about nature. Learning about the stars and planets at an early age can help all students, disadvantaged or otherwise, to develop this natural curiosity. It is not even necessary that a student pursue a career in astronomy or one of the other sciences, but simply to develop a foundation of curiosity that will ensure their continued interest in the subject throughout their adult lives.

In a recent AIP survey of new bachelor's degree recipients in physics, one-fourth of the respondents said that it was their curiosity about nature that attracted them to physics. An equal number said that they were inspired by their exposure to the subject in high school. One-tenth said their home environments led to their decision to study physics – often they said their fathers were physicists. The latter two factors are usually not applicable to disadvantaged students, and so it is the first of these, the curiosity about nature, that I try to instill in my students in order to motivate them.

Suggested Projects

The following projects have been used (or are anticipated to be used) in the physics component of the Project Prepare program. The ones presented below are, from my experience in teaching the science-oriented students, the ones which seem to pique the students' curiosity the most in astronomy and physics. These exercises provide useful experience for any students, disadvantaged or otherwise, and may be categorized as primarily either concept-oriented or calculation-orientated. None require the use of computer. It is important to present a variety of experiences, with various levels of difficulty, as many of these students, although disadvantaged, are often quite enthusiastic and capable, and one must ensure that the level of the work is neither consistently too low or too high.

1. Observing the Changing Sky

Objective: To instill in the students the habit of noticing what is happening in the sky, day and night.

Procedure: This project requires nothing more than looking at the sky each day and night. Yet, it is probably the single most important exercise (see Borden, in *Sky & Telescope*, June 1991, p.5). During the day session, we look at the color of the sky, noting that when the sky is a deep blue, the air is relatively dry, with air molecules scattering the blue light very efficiently. When the air is humid, the sky becomes a paler, more whitish blue due to scattering by water molecules. If clouds are present (especially cumulus and cumulonimbus), I try to explain how they form and the different types.

At night, we observe the Moon and its run of phases during the 7-week period. Also, we try to identify some stars and any bright planets. We also note the color of sunset, and the position of the setting sun. I point out any bright planets and stars. I also challenge them to note that different stars (e.g., Vega and Arcturus) have different colors, and that these different colors are due to differences in surface temperature.

Motivating Factor: The students are made to realize that there are many interesting things to observe in the sky, and it costs nothing to see! One need only turn one's gaze upward occasionally to witness an ever-changing pageant.

2. Determining Velocities of Speeding Cars

Objective: To involve the student in planning and executing an experiment to measure the speed of cars on a city street. The concepts of average velocity and uncertainties in measurements are emphasized.

Procedure: Each group of two or three students is given a stopwatch and meter stick and asked to place themselves along a well-traveled street near campus (speed limit 25 mph). They measure out a substantial distance interval Δx and mark the begin and endpoints. The students then, working as a team, record the time Δt that 20 different motor vehicles take to pass through the marked distance interval. Later, they compute the average velocities $\Delta x/\Delta t$ for each vehicle in m/s, convert them to miles/hr, and compare them with the speed limit to determine if each car was speeding or not. We discuss the effect the uncertainty in the timing of the cars will have on the calculated velocity.

Motivating Factor: This lab is an active project, involving coordination and teamwork. The students get the feeling that making measurements can be fun as well as instructive. They also obtain a fairly clear notion of what an average velocity is and how to compute it.

3. Identifying Absorption Lines in the Solar Spectrum

Objective: To show the students that ordinary sunlight consists of a spectrum of colors, and in that spectrum there are dark absorption lines which correspond to known chemical elements. Also, to look at a wide variety of emission-line spectra from gas-discharge tubes of hydrogen, helium, nitrogen, neon, and mercury.

Procedure: Each lab group has a spectroscope with which to directly observe spectra from bright sources. The position angles of the observed lines can be read from a vernier scale near the base of the instrument. The students point the spectroscope at an outside window (not directly at the Sun) and observe the absorption lines in the otherwise continuous spectrum of colors. They then observe the spectra from several discharge tubes and compare them with known lines.

Motivating Factor: The students learn that ordinary sunlight contains an encoded message of the chemical elements present in the Sun's atmosphere.

4. How Lenses and Mirrors Form Images

Objective: As stated in title.

Procedure: The students mount a converging lens, screen, and light source on an optical bench and try to form an image of a distant tree viewed through the window. They repeat this with the much closer artificial light source on the bench. They then repeat this procedure with a converging mirror and an L-shaped screen suitable for forming the reflected images.

Motivating Factor: The students observe first hand how a simple camera or telescope works to form an image. They also learn that imaging can be done with either a lens or a mirror.

5. The Fermi Question

Objective: To instill in the students the notion that even if a precise answer cannot be obtained for a problem, there may be a way to come up with a "ballpark" figure.

Procedure: This is really an outside assignment, based upon physicist Enrico Fermi's way of dealing with finding the number of piano tuners in New York City (see Schwartz, *Prelude to Physics*, John Wiley, 1983). We ultimately apply it in class to the famous Drake equation, which seeks to obtain a crude estimate for the number of planets in the universe with intelligent life.

Motivating Factor: The students learn that often one can (and sometimes must) obtain "ballpark" answers to many different questions.

6. Newton's Law of Cooling

Objective: To show that the temperature of a hot object falls off exponentially as it cools.

Procedure: The students heat water to near boiling in a flask and then pour it into a small copper vessel, insert a thermometer, and cover the top with foil. When the temperature has fallen to 80°C, temperature measurements are taken at two-minute intervals for 30 minutes. The students later plot the temperature versus time to reveal the exponential decay.

Motivating Factor: The students find that objects cool in a definite exponential fashion.

The author welcomes correspondence with other interested colleagues on the teaching of disadvantaged students. Please write to the address given above, or to the e-mail address: harry.j.augensen@cyber.widener.edu

Acknowledgement: The author thanks Dr. Alonzo Cavin, Director of Project Prepare at Widener University, for his helpful suggestions in preparing this presentation.

U N I V E R S O: A Spanish-Language Astronomy Radio Program

Sandi Barnes
The University of Maryland
College Park, Maryland
McDonald Observatory
2609 University, Room 3118
Austin, Texas
USA 78712

Tel.: (512) 475-6765,
FAX: (512) 471-5060
E-mail: sandi@astro.as.utexas.edu.

Production Progress: The University of Texas at Austin McDonald Observatory began forming the Universo program in September 1995. The program is modeled after Star Date, an English-language astronomy radio program.

Initial steps included finalizing a production agreement with KXCR-FM, a radio station in El Paso, where the program is recorded; selecting and auditioning professional voice talents from the El Paso area; forming focus groups to evaluate the program and provide advice in several areas; and selecting a translator and technical editors.

The production team narrowed the list of voice talents to three candidates, and sent demo tapes with all three voices to the focus groups in late 1994. Focus groups also evaluated program content, style, presentation, and other areas. Their feedback was used in the final selection process, in which Teresa Fendi de la Cruz was chosen as Universo announcer. An informal focus group helped the production team select the program name. Copies of focus group results and the forms used to create the focus group results are available.

Rosario Torres was selected to translate the Universo scripts into Spanish. Three scientists were chosen to review the translated scripts for technical accuracy: Dr. Jorge Lopez, associate professor of physics, University of Texas at El Paso; Cecilia Colome, doctoral student in astronomy, University of Texas at Austin; and Carmen Pantoja, assistant professor of astronomy, University of Puerto Rico at Arecibo.

Script translation and production rehearsals began in January 1995, with the first on-air programs recorded in February. These were recorded on cassette tape. Beginning with April 1 programming the master tape was sent to Nimbus Recording for compact-disc production.

While most programs are translated versions of the Star Date episodes, several programs each month are written specifically for Universo. These have covered Mesoamerican astronomy and skylore, the work of Hispanic astronomers, joint research projects between the United States and Mexico, and similar topics. Before the remaining scripts are translated, they are edited for length and to customize the content for the target audience. The number of scripts customized in this way has increased with each month, so that perhaps one-fourth of the July and August programs are significantly different from the Star Date episodes. Yet the overall program retains its emphases on skywatching and the science of astronomy, providing a strong informational and educational focus.

Current priorities for summer 1995 include smoothing out the translation process, refining the announcer's delivery, and improving overall production standards.

Marketing Progress: As of May 22, 1995, Universo was distributed to 25 radio stations in the United States, Puerto Rico, and Mexico. One of the two Mexican stations is in Ciudad Juarez, so it also serves the El Paso, Texas, market.

Initial radio stations contacts were made through attendance at the National Association of Broadcasters conference in October 1994 in Los Angeles, a survey and newsletter article in the Star Date newsletter, and direct mailings and follow-up telephone calls to four groups of Spanish-language stations.

The NAB conference and Star Date notices produced interest among several dozen stations. In addition, KXCR-FM in El Paso, where Universo is recorded, is a member of a network of community bilingual radio stations; contact between KXCR and other network stations produced additional interest.

Mailings were divided into four groups: community bilingual stations, Star Date stations, stations that carried a former Spanish-language astronomy program, and stations in markets that rank 21-40 in Spanish-speaking population. For the first three mailings, station managers and program directors received a demonstration tape with five sample programs and a cover letter; stations in the fourth mailing received a May or June compact disc. The first three mailings and follow-up phone calls have been completed. The fourth mailing is taking place in May, and follow-up telephone calls should be completed by mid-June.

Subsequent mailings will use a four-color marketing brochure, which is in production. Printing and binding will be completed by the first week of June.

Initial long-range planning meetings between members of the Universo production and marketing team and two outside consultants, Estrada Marketing and Hispanic Marketing Group, produced many possible approaches, with suggestions for interim steps to increase program awareness among radio stations, potential underwriters, and the Spanish-speaking population at large.

A proposal for additional underwriting support for years two and three was submitted to Montemayor and Associates in San Antonio, an advertising and marketing agency that handles Spanish-language accounts for several national clients, including Chrysler Corporation.

An expanded marketing plan for years two and three was prepared and is available.

Scaling the Solar System at Lake Afton Public Observatory

John E. Beaver, W. Scott Kardel, Greg Novacek

Lake Afton Public Observatory
1845 Fairmount
Wichita, KS
USA 67260-0032

The solar system is mostly empty space. No picture can really demonstrate this fact. It is difficult enough to draw the orbits of the planets to scale; the orbit of Pluto is more than 100 times the size of the orbit of Mercury. To also put the planets themselves to the same scale as the orbits is an impossible task for any reasonably-sized sheet of paper. Thus, every drawing of the solar system is misleading. Because the sizes of the planets are so tiny compared to the size of their orbits, one cannot really experience both directly at the same time. At Lake Afton Public Observatory in Wichita, Kansas, we have developed a scale model designed to come as close to this direct experience as possible. We do this by having two levels of experience--one for the planets, the other for their orbits. These two experiences are then connected to each other by the physical actions of the student. This is possible because we have chosen the scale to be just big enough to make Pluto, the smallest planet, barely visible to the naked eye. At this small scale the whole model is visible simultaneously and one can easily walk to Pluto. In addition one can take advantage of the fact that angular sizes are preserved in a model that puts the sizes of the planets and their orbits to the same scale; as seen from the scale Earth, the scale Saturn appears the same size as the real Saturn does in the real sky. We feel this is a distinct advantage over large scale models that scatter the planets over an entire city, only one planet being visible at a time.

For our chosen scale (1:45 billion) the planets, placed in random directions from the sun, just fit onto a small field next to the observatory. We have chosen the semi-major axis of Pluto's orbit to be 428 feet. At this scale, the sun is a painted wooden ball of about 31 mm diameter. The Earth is no more than a tiny dot, 1/3 mm across. Mercury has a diameter of 0.11 mm, while Jupiter is 3.1 mm across. An astronomical unit is a little under 11 feet. Jupiter and Saturn are large enough to be represented by small metal ball-bearings; the rest of the planets are represented by appropriately-sized 2-dimensional dots. We have made these tiny dots by first drawing them at five times this scale, then reducing them photographically to the correct size onto transparency film. Also included on the transparencies are circles marking the orbits of the larger satellites, as well as labels and arrows. Since Pluto is at the margin of naked-eye visibility, a clearly-labeled 5-times-enlarged blowup is included on its transparency as well. These transparencies, sandwiched between thin pieces of clear plastic, have been mounted in transparent plastic insect viewing jars. These small plastic containers are transparent on all sides, with a magnifier for a lid. Thus, one can look through the sides of the container and see the transparency at its actual size, but the magnifier allows one to see extra detail. Each transparency has been mounted in its container at the same angle as the planet's rotational inclination. For Jupiter and Saturn a small hole was drilled in the transparency and an appropriately-sized ball bearing placed at the location of the planet. Saturn's rings are part of the transparency, with the ball bearing planet placed at their center.

The insect viewing jars, with the planets inside, have been fixed to red poles roughly 3 1/2 feet in length. These "planet poles" can then be placed in holes already set into the ground at the appropriate locations. The "planet holes" are small pieces of pipe, with a pointed plug on one end and a flange on the other. It is this flange, about 4" in diameter, that is visible at ground level, the rest of the pipe having been driven into the earth. A planet hole, once placed in the activity field, is easily visible from a distance of about 10 feet. They have been placed in random directions from the "sun" so as not to give the impression (as some scale models do) that the planets lie all in a straight line.

If one stands at our scale sun, Pluto is of course too small to see. One can, however, just see the red pole, with the insect viewing jar on top, off in the distance. Having actually walked the distance from the sun to the planet pole, while also having seen firsthand the scale planet inside the insect viewing jar, the student is able to mentally connect the diameter of the planet with its distance from the sun. For any scale greatly different from the one we have chosen, such a direct connection between the size of the planet and the size of its orbit would be difficult to achieve.

The Scale Model Solar System has been used at Lake Afton Public Observatory by groups ranging from 1st grade to college level and with class sizes ranging from 6 to 60. For all of these groups, we generally begin by showing them a typical poster of the solar system. The students are then asked, "What is wrong with this picture?" The poster used is similar to many others in that neither the relative sizes of the planets nor their distances from the sun are shown to scale (let alone to the same scale). Furthermore the poster shows the planets and their orbits with full perspective and 3-dimensional rendering, implying that they would actually look like this from space. We emphasize that this poster is not atypical; an illustration much like it appears, without qualifying comment, in a science textbook used by the Wichita Public Schools.

For a small attentive group a "solar system tour" has been the most fruitful approach. The planets are placed in their appropriate places prior to the arrival of the students, who are then taken from planet to planet on a guided tour. This allows the opportunity to ask many basic questions: How close would you need to be to Saturn to see its rings with the naked eye? How could you see the rings from Earth? (Often we will even set up a small telescope at "Earth" and view "Saturn.") What does the sun look like from Neptune? On this same scale, how far away is the nearest star?

For larger groups it is often impractical to spend as much time on detail, and so the activity simply concentrates on the contrast between the sizes of the planets and their orbits. The activity is begun with only the sun in place. The students must then find the "planet hole" in the activity field for each planet. Once a hole is found, the students must decide for themselves which planet belongs in it; if that planet has already been placed in another hole, they must negotiate with each other. Usually a class will have found most of the planets within 20 minutes or so. At that time they are brought back to the inner solar system to evaluate what they have found. Any remaining planets are found by having the class form an expanding circle, pacing off the proper distance from the sun using the astronomical unit (about 6 steps) as the reference. This "search and discover" approach to the Scale Model Solar System activity, although crude, can be very effective. After running around a field for 20 minutes looking for small holes in the ground, one cannot help but be impressed by how much space there is between the planets.

We have found that a model of the solar system with both the sizes of the planets and their orbits to the same scale can be a valuable teaching aid – especially if the entire model is visible at once and it is constructed in a physical space the students can walk around in.

Tennis Ball Astronomy

Walter J. Bisard
Physics Department
Central Michigan University
Mt. Pleasant, Michigan
USA 48859

This paper is a summary of a major departure from the usual pedagogy combination of lecture and slide presentations utilized in the majority of introductory astronomy classes. What is strongly suggested is an approach built on the conceptual structure of the knowledge of astronomy and the manner in which conceptual learning takes place. This paper describes a successful approach of involving the student in the learning process and also a means of assessing their understanding either before, during, or after instruction at all educational levels.

It is well known that astronomy is filled with potential for the development of incorrect alternative frameworks or "misconceptions" because of the large amount of "connectedness" to other subjects and the abstractness of many of the fundamental concepts. Furthermore, these concepts in astronomy are not independent and learning can be viewed as a progression from incorrect and naive conceptions to more correct conceptions. When one views this progression as an educational process, it has become known as the constructivist model of cognition.

The most successful attempt to improve the conceptual learning of seasons, lunar phases, and planetary motions used cooperative learning and hands-on with minds-on teaching strategies utilizing a set of activities developed and field-tested by the author entitled TENNIS BALL ASTRONOMY. By using a light bulb (Sun) with a tennis ball (Earth) and a ping-pong ball (Earth's moon) in cooperative learning groups, we have found better results in trying to "unteach" the misconceptions associated with planetary motions, seasonal variations, lunar phases, eclipses, and even day and night.

Materials: A small light bulb (25 watts) in a porcelain socket taped in the middle of a large group table represents the sun. The tennis ball and ping pong ball can each be drilled with an appropriate bit to allow a pencil and small wooden dowel to be inserted and represent the spin axis of each. (It is important to realize this process is quite easy - the balls will not explode and is easiest to do with a simple drill press with a quarter or eighth inch bit.) The dowel should be about 20 centimeters long or approximately a new #2 pencil sharpened at one end because this represents the circumference of the 6 centimeter diameter tennis ball and will be useful in scaling the Earth-Moon system.

After the pencil and/or dowel has been inserted into the tennis ball and ping pong ball, the "Earth tennis ball" can be labeled with equator and northern and southern hemisphere (and continents if so desired). Furthermore, the "Earth tennis ball" can have one or more push pins inserted to represent observers in various locations such as the arctic, northern hemisphere, equatorial, southern hemisphere, or eastern/western USA. The "Moon ping pong" ball can be labeled with "Earth-facing" side and "Far-side" (not the "dark side"). The advantage of these materials is they are economically available, accept markings easily, and very easy to use. Furthermore, both the surfaces of these two types of balls show good shadow boundaries for phases.

Assessment of Fundamental Motions and Astronomical Relations: Authentic assessment procedures with these materials are relatively easy and usually very revealing to the instructor. For instance, the above materials can be the equipment used in a formal or informal interview with a student or a group of students. By providing the materials to students and then asking them to simulate and explain to their classmates the concepts of day, night, year, month, axial tilt, seasons, phases, eclipses, and tides is an excellent assessment activity for the teacher. This is very revealing to do this before, during, and after instruction!

1. **Spinning and Orbital Motions of the Earth, Moon and Sun:** This might be called the overhead or bird's eye view of the motions of the Moon around the Earth and the Earth orbiting the Sun. After demonstration by the teacher, have the students simulate it in their groups using related terms such as "orbit", "month", "day", and "year".

2. **Day, Night, and Time of Day:** With just the "Earth tennis ball" and "Sun light bulb", demonstrate day & night and the times of noon, sunset, midnight, and sunrise. Utilize related terms such as east, west, and overhead for the "push-pin observers" on the tennis ball. Again, have the students repeat but insist on other people in the group doing the explaining, spinning, or pointing--i.e. use good cooperative group procedures.

3. **Phases and Eclipses:** Teaching lunar phases inside the classroom or lecture hall with tennis balls, ping pong balls, and a light bulb sun is best utilized in a cooperative group fashion. As they simulate these concepts, note their attention to the previous related concepts of motions and time.

4. **Tides:** The tennis ball earth can be squeezed by the hand to simulate tidal forces of the ping pong moon and light bulb sun. Do this with one light bulb sun for all the groups and expect lots of questions and interactions!

5. **Seasons:** For the student to conceptually understand seasons, the above activities and learning of motions and time must have taken place including the emphasis on the axial tilt toward Polaris during a year. The "push-pin observers" in the northern hemisphere as opposed to the southern hemisphere and Australia offers a large conceptual leap for the students.

Assessment: One can use the same materials or an Earth globe and a bright light to check for understanding of the team or individual members in private or in class sessions. The process of significantly altering the usual introductory astronomy pedagogy of which one part is using cooperative learning groups in large classrooms or "interactions" is the theme of another paper by Zeilik and Bisard summarized in these proceedings.

For a complete copy of the paper presented at the ASP 1995 Summer Symposium, please contact the author at the Physics Department, Central Michigan University, Mt. Pleasant, Michigan 48859.

Learning Theory and Implications

Jeanne E. Bishop
Westlake Schools - Planetarium and High School
24525 Hilliard Road
Westlake, OH
USA 44145

Current cognitive learning theory involves several schools of thought, which agree remarkably with one another and oppose the transmission model of learning still widely in use.

1. Constructivism: Over the last 15 years, this theory has gained major support. It asserts that each person constructs his or her own set of ideas, and the process is lifelong. The status of constructed ideas affects how events are interpreted. The construction takes effort. A person is not a sponge that is ready to accept transmitted knowledge. Misconceptions--totally wrong or fragment-only knowledge--are overcome only with effort. Students taught a mythical or religious reason for objects and happenings in the universe may retain the non-scientific explanation, often to co-exist with the scientific one. (This occurs with college students in Malayasia, reported by Mazlan Othman.) In Philip Sadler's film *A Private Universe*, the students graduating from Harvard who incorrectly explain the cause of seasons with great confidence, are a good example of tenaciously held misconceptions. This year, to learn about seasons, my high school astronomy students had opportunities for discovery with models in small groups, they took data on the earth-based view in the planetarium, they sketched their observations, they wrote essays about their observations, they read text descriptions, and they applied ideas to other planets. I even showed them the Harvard film and we laughed together at the explanations. Although my students almost all had good, correct explanations on the unit exam, only half correctly explained the seasons on the final exam. Misconceptions resurface, and they die hard!

2. Piagetian Theory: For more than 50 years Jean Piaget gathered data on the ways in which children and teenagers learn. His general development theory is put forth in The Growth of Logical Thinking from Childhood to Adolescence (Inhelder and Piaget, 1958). There is a vast literature of mainly supportive studies. These are a monument to the greatness of his theory. Two more of his books discuss his spatial theory, *The Child's Conception of Space* (Piaget and Inhelder, 1956) and *The Child's Conception of Geometry* (Piaget, Inhelder and Szeminska, 1960).

Some basic ideas of Piagetian general theory are: Heirarchization, a fixed order or succession of the different levels which compose a sequence of development. There is no reversal of this, except in cases of brain damage. Structuring, organization of the intellectual behaviors characteristic of a particular level of development. Integration, organization of the structures representing a level of thinking. Equilibration, re-named self-regulation, the process by which individual's levels of thinking evolve. This older idea of Piaget's is the basis of the idea of the constructivist movement.

Many believe that Piaget general stages are helpful in classifying reasoning patterns, but that individuals vary at the rate at which they pass through the levels. In elementary school, most students are at a concrete operational level. This means that a child can serialize, combine, differentiate, subdivide, and conserve. The child requires reference to familiar objects and actions, needs step-by-step instructions in a lengthy procedure, and makes contradictions which reveal that she is unaware of her own reasoning.

The formal operational level which succeeds the concrete level are a set of interdependent reversible mental operations, which Piaget called "transformations." An adolescent or adult with formal structures can think about thinking. The formal thinker can deduce the general laws behind a set of instances. This student can consider hypotheses and figure out what should follow if they are true. He/she can plan a lengthy procedure when provided with overall goals. He/she can use symbols to express ideas. He/she is aware and critical of his own thinking. This person has good problem-solving ability with an internally well-organized set of ideas about a topic.

Although understanding many astronomical concepts require formal operations – such as the H-R diagram – a large number of basic astronomy ideas, such as seasons, also have a strong spatial component. Perhaps spatial ability accounts for one-half of astronomy aptitude. Piaget's spatial levels are also important for knowing when certain astronomical concepts can be understood by most.

Piaget notes that most individuals possess an understanding of relationships such as an object's closeness to other things, its order in a group, and its isolation or enclosure (the stage of topological relations). The next stage of spatial ability is usually reached in adolescence, projective representation. Projective representation means an ability to visualize what an object will look like from a different point of view. Of course, projective ability is necessary to completely understand topics such as seasons, lunar phases, and planet positions and motions. One has to be able to switch back and forth between an earth-based view of the sky and the bodies in space. The advanced spatial stage declared by Piaget is Euclidean abilities. One who has reached this level can conserve and measure distance, length, area, volume, and angle. Piaget found that such abilities become integrated with projective representation in the teen years. The concepts of magnitude, H-R diagram and mass-luminosity relationship require this spatial level, as well as formal operations.

There is a documented general difference in spatial ability between males and females. And it seems to be most different in adolescence. My own research in astronomy learning and research of others has shown that students show no spatial difference in upper elementary school. But boys outdistance girls in eighth grade. Further, boys in eighth grade significantly improve in spatial ability with astronomy activities using discovery and modeling, while girls in eighth grade do not.

Those who have repeated Piaget's experiments in the United States have found both general and spatial levels reached at later ages than Piaget found. Also, levels are not necessarily the same for each subject or topic. Primarily, this is due to experience, or lack of it, with the particular ideas. For example, a student might think in abstract terms about baseball or have projective representation abilities in building model airplanes, but not be able to apply the same type of critical thinking to astronomical positions and motions.

The Karplus learning cycle, currently known as the learning cycle, is based on Piagetian Theory and is in complete agreement with constructivism. Robert Karplus, who left a very successful career as a theoretical physicist, is its author. He implemented it in the elementary Science Curriculum Improvement Study (SCIS) in the 1960's. Experience with the physical environment, social transmission of ideas, and opportunities for self-regulation or constructing ideas led to his three step cycle. 1) Exploration, in which students in groups explore cause-and-effect relationships with materials in a situation where minimum guidance is offered and they are likely to discover relationships. 2) Concept introduction, a formal discussion of the principle from a textbook, teacher discussion, film, or other source. The student must see the relationship between the exploration and this step. 3) Concept application, in which students apply the principle to similar but new situations. Concept application is considered particularly helpful to students whose conceptual reorganization is slower than average or who did not relate the exploration experience to the introduction. Self-regulation can occur throughout the steps, different amounts of the total for different students in each step.

3. Cognitive Style Mapping: There has long been an awareness that different people learn in different ways. The experiences which lead to development of different levels of thinking can be different, based on genetic or cultural predisposition. Mapping is the process of finding an individual's style. There are different style models. Most include a prescription of varied learning experiences for most learning styles. Cognitive style mapping is closely related to Aptitude Treatment Interaction (ATI).

4. Brain Theory (based on physiological research): Physiological research has shown that there are brain growth spurts which coincide with the times found by Piaget in his development of thinking ability: 2 to 4, 6 to 8, 10 to 12, and 14 to 16.

The brain processes language primarily on the left side through three sets of interacting structures. The left side processes information in sequence--it is the side responsible for logical thinking. The right hemisphere processes spatial information, and it does so simultaneously or holistically. Most people have a tendency to use one hemisphere more. The majority prefer the left, and are called "left-brain dominant" or "left-brained." Some prefer the right side, and are said to be "right-brained." A few use both brain hemispheres about equally and are known as "mixed dominant."

There are gender differences in brain use for different tasks. Apparently sizes of areas or connections to or within these areas vary between men and women, as women and men generally have different "best" skills. Problem-solving tasks favoring women are: tests of perceptual speed, memory of displacement of objects, identification in word lists, fine-motor coordination tasks, and mathematical calculation. Problem-solving tasks favoring men are: mentally rotating or manipulating an object shown in a drawing, target-directed motor skills, finding shapes hidden in figures, and tests of mathematical reasoning. Of course, some individuals are exceptions; some women beat almost all men in the tasks that usually favor men.

Some implications for teaching/learning specific astronomy concepts (average-ability students):

1. Up-down on earth and other space bodies will not be fully comprehensible until about fourth grade.

2. Orientation of a star or object compared to another (such as the Pleiades relative to Orion) will not be correctly predicted for a different sky view until about seventh grade.

3. Correct succession of lunar phases will not be identified until about late first grade. This means just the pattern, without reference to cause.

4. Understanding of the Andromeda Galaxy's elliptical appearance will not be related to the inclination of a round disk and the understanding of eclipse relationships will not be achieved until the late elementary grades (developing

third-sixth grades).

5. Comprehension of the projective concepts (combining earth and space views) of earth rotation and the celestial sphere, seasons, lunar phases, and planet positions and motions will not be mastered until sometime after eighth grade.

Some implications and recommendations for teachers, textbook writers, and those who design curricula:

1. Be aware of cognitive theory. Topics such as seasons cannot be understood as projective concepts by most elementary students. This is not to say that we should not focus on their learning the two views-from earth and from space--for the equinoxes and solstices. Later the two can be put together and probably sooner and better due to prior learning.
2. Know the astronomy content well. Do a task analysis for a topic before deciding on the appropriateness of the lesson at the level and the skills that are needed for full comprehension. A task analysis is the determination of specific perceptions, vocabulary needed to internalize perceptions, an analysis of earth view ideas, and an analysis of space view ideas required for a larger concept.
3. Utilize a variety of methods in teaching. Know many strategies. Strongly consider the learning cycle.
4. The learning cycle, with opportunities for real discovery (not just rote experiments), followed by concept introduction with important terminology, should be employed by teachers and textbook writers alike.
5. Teachers should look for and be alert to student misconceptions. One set of recommendations for this given by Jose Mestre (*Physics Today*, September, 1991):
    - Probe for misconceptions.
    - Ask questions to clarify student beliefs.
    - Suggest events that contradict student beliefs.
    - Encourage debate and discussion.
    - Guide students toward constructing scientific concepts: synthesis of responses, discussion of thought experiments, design experiments to test hypotheses, reevaluate student understanding – he advocates a Socratic Method in discussion.
6. Exercise care in testing procedures. In the United States there is a growing emphasis on "authentic assessment" – the recognition that some tests reveal memorization only.

Problems for following these recommendations:

1. Often teachers lack understanding of astronomy concepts. They cannot determine student misconceptions, determine what concepts are appropriate or inappropriate for their students, or design lessons for discovering the concepts. (Also, elementary and middle school teachers are often overworked and do not take time in a very busy schedule to learn ideas and work out best discovery lessons for a subject with which they feel uncomfortable.) Also, not many teachers are very knowledgeable about learning theory.
2. A number of current textbooks are not very helpful in promoting inquiry science (including astronomy) learning. For elementary and middle school, there is an emphasis on terms, and the transmission model of learning is followed. The

special programs for many grades, written by scientists and teachers in the 1960's and 1970's and which have strong critical-thinking components, currently are not used much. Fortunately, there is an inquiry-based elementary astronomy program, SPICA, and a high school astronomy program, Project STAR, both of which originated at Harvard.

3. Astronomers who might like to help, and know the astronomy content well, may lack understanding of the pressures on teachers and problems with school schedules, knowledge of different instructional strategies, and knowledge of learning theory.

For Further Reading

Bishop, J. E. "Planetarium Methods Based on the Research of Jean Piaget." Most recently reprinted in *Planetarium, A Challenge for Educators*: A Guidebook Published by the United Nations for the international Space Year. Edited by Hans Haubold and Dale Smith. UN Press, 1992.

Bishop, J. E. "The Development and Testing of a Participatory Planetarium Unit Emphasizing Projective Astronomy Concepts, Utilizing the Karplus Learning Cycle, Student Model Manipulation, and Student Drawing with Eighth Grade Students." Unpublished Ph.D. dissertation, University of Akron, Ohio, 1980.

Inhelder, B., and Piaget, J. *The Growth of Logical Thinking from Childhood to Adolescence*. New York: Basic Books, 1958.

Mestre, J. P. "Learning and Instruction in Pre-College Physical Science." *Physics Today*, September, 1991, pp 56-62.

Othman, M. "Influence of Culture on Understanding Astronomical Concepts." *The Teaching of Astronomy*: Proceedings of IAU Colloquium 105 held July 26-30, 1988. Edited by J.M. Pasachoff and J.R. Percy. Cambridge: Cambridge University Press, pp. 239-242.

Piaget, J., Inhelder, B., and Szeminska, A. *The Child's Conception of Geometry*. New York: Basic Books, 1960.

Piaget, J., and Inhelder, B. *The Child's Conception of Space*. New York: Humanities Press, 1956.

Scientific American Special Issue: "Mind and Brain." September, 1992.

Swartz, Clifford. "The Physicists Intervene." *Physics Today*, September, 1991, pp 22-28.

Integrating Astronomy With Elementary Non-Science Curricula: A Workshop Presented on May 24-25, 1995

Matt Bobrowsky
CTA Incorporated
6116 Executive Blvd. #800
Rockville, MD
USA 20852

Tel.: (301) 816-1281
FAX: (301) 816-1429
E-mail: mattb@cta.com

Introduction: Elementary school is a critical time in a person's education when lasting impressions get formed, attitudes about school are developed, and preferences for certain subject areas become established. Therefore, it is the best time for using astronomy to increase the interest of students in a variety of subjects while enhancing their knowledge of science as well. Students who thought they were interested only in science will find out that the non-scientific subjects are also of interest. Students who thought that they had no interest in science will discover that science actually relates to the non-scientific subject in which they are most interested. If more students are exposed to astronomy at an early age, there will be a greater interest in later years leading, perhaps, to scientific careers or simply to greater scientific literacy as adults. It is also the hope of this author that, as the students who benefit from this initiative reach adulthood, more women and under-represented groups will pursue a scientific career at least partly due to knowledge and confidence gained from this project.

Social Studies and History: Social studies and history are usually taught with the emphasis on the *political* aspects of these subjects. These topics could have a wider appeal if scientific discoveries were covered as well, including their relation to the more political issues under discussion. A more specific problem is that the globe of the earth in most classrooms does not show the earth as it appears from space. Workshop participants got to see what the earth really looks like. Without the distraction of labels and boundaries that do not physically exist on the earth, teachers can lead a fascinating discussion about the voyages of early explorers and the sociological reasons for populations being clustered in the areas where they are presently located. In elementary school, there are many discussions about other cultures. Referring to the realistic earth globe, teachers can explain how the location of people on the earth (e.g., through climate considerations, proximity to water and other natural resources) is often a major factor in how they dress, what products they import and export, and even in their physical appearance.

Music and Art: Both music and art can be related to astronomy through waves. One way that astronomers study celestial objects is by analyzing the waves of light that come from those objects. The different wavelenghts appear as the different colors that are used by artists. In fact, there is a considerable amount of art (especially in museums, planetaria, etc.) having astronomical subjects. In recent presentations to a kindergarten class, waves were related to light and sound, as well as other types of waves such as water waves. The students were allowed to get their hands on a variety of wave-producing gadgets and make their own waves. They could, for instance, see (and feel) what was required to make a short wavelength as opposed to a longer wavelength.

Mathematics: The applications of astronomy to mathematics are innumerable. Different activities in the workshop were appropriate to different grade levels. In the higher grades, the students can, by use of ratios, calculate the correct sizes of the planets given the size for a model earth. The students can actually construct model planets. (Correctly painting the surface of each model planet provides another connection of art to astronomy.) Even in the lower grades, many of the usual math activities can also be presented with astronomical themes.

Language Arts: At all levels, the reading material can have astronomical content. Kids love to look at astronomical pictures in books and they love to hear or read about astronomy. The inclusion of astronomy is an excellent way to get students to read more.

Astronomy can be incorporated in activities as simple as penmanship exercises. For the lower grades, each teacher received a penmanship activity consisting of a different letter of the alphabet on each page, along with a space-related word beginning with that letter, and a picture to go with the word. For example, the kindergarten or first-grade students may be on the page for "H". They will practice making the letter "H", then they can trace over the letters in "Hubble Space Telescope." Finally, they get to colour in the picture of HST. (More art!).

Health and Safety: The students who have heard that different waves carry different amounts of energy can understand that the large amount of energy in very short-wavelength waves like UV and X-rays makes these waves harmful. These older students can also learn that warm objects (like planets) emit infrared waves. Many children visiting their physicians now get their temperature taken with a thermometer that is inserted in the child's ear. This aural thermometer works by detecting the infrared radiation emitted by the eardrum.

A Role for Planetariums in Science Education Reform

Stu Chapman
Harford County Public Schools
Southampton Middle Moores Mill Rd.
Bel Air, MD
USA 21014

E-mail: schapman@umd5.umd.edu

In the 1960's and early 1970's the National Defense Education Act (NDEA) provided federal funds to local governments for science education purposes. Largely, due to the aid provided by these funds, planetarium instruments were built by many local governments and school systems. Unfortunately, due to the lack of trained personnel to operate them and the lack of local funding for their maintenance, many of these instruments were (and are today) not used to their fullest potential. Throughout the years of 1970-1995, however, several excellent school system based planetarium programs flourished in various parts of the country and were devoted to science education as it related to the field of astronomy. Most featured the introduction of astronomical content in a chronological approach on a yearly basis which met the needs of the learner best. Many, though not all of these, involved instruction in the smaller, inflatable, and portable planetaria. A factor that each of these school based programs had in common was that they promoted a high degree of student and attending teacher participation. However, in the 1990's a new aspect of science educational reform began to emerge which necessitated the restructuring of much of science curriculum in the K-12 level, including that related to astronomy. A major document to emerge in recent years with some excellent suggestions reforming the science curriculum is Project 2061: Science for All Americans which was published in 1989 by The American Association for the Advancement of Science (AAAS). Project 2061 was designed to occur in three phases. The first phase of Project 2061 culminated with the publication of the Science For All Americans manual, where the conceptual base for reform was established by recommending the knowledge, skills, and attitudes all students should acquire as a consequence of their total school experience. The second phase, still occurring in many local school systems, involves the cooperative effort of scientists and educators to transform the recommendations of the first phase into new and alternative curriculum models. This paper describes the implementation of one of those models in the form of a school based K-12 planetarium program which is interdisciplinary in nature and which supports the total school science curriculum utilizing these recommendations. Four planetarium lessons from the program are highlighted which actively involve the attending teacher in the lesson delivery through well planned pre and post planetarium lessons. Each lesson also promotes active student participation and stresses the mastery of specific scientific habits of mind as a primary focus instead of merely the delivery of more astronomical content. One lesson each has been selected from the elementary (K-5), middle (6-8), high school (9-12), and university (13-14) level. The focus is on utilizing the planetarium as a teaching laboratory rather than a passive recreational device. Suggestions for improving and refining the delivery of science content have been adapted from the American Association for the Advancement of Science's Project 2061 and its related manual, Benchmarks for Science Literacy. These planetarium lessons are available in DOS text and WordPerfect 5 format via anonymous ftp from: ftp://magnus1.com/planetarium.

Worksheets and student activity sheets are available from the planetarium at the address noted above. Lesson descriptions follow:

ELEM (G 2-5) A First Data File of the Solar System: Students pretend to "discover" the nine planets as they enter the Solar System from the outside. As they visit each of the planets they classify each on the basis of size (relative to Earth), solid surface (yes or no), rings (yes or no), moons (yes or no), and color. At the end of the trip each student has his/her own planetary database! Skills focus on collection of data and data classification.

MIDDLE (6-8) A First Analysis of Stellar Data: Students collect data and classify data on selected stars of the winter sky according to several traits including surface temperature, size, and luminosity. Then they solve comparison problems involving the brightness, distances, and surface temperatures of stars.

HIGH SCHOOL (9-12) How Can The Distances of the Stars be Inferred?: Given the starting information, students construct a H-R Diagram prior to the planetarium lesson. In the planetarium, they are presented with a list of 15 stars along with their spectral types and luminosity classifications. Students must work in cooperative groups to 1) find each star on the dome, 2) estimate its apparent magnitude to the nearest 0.5 magnitude unit compared to some known standards, and 3) estimate its distance by using their H-R Diagram to compare the star's apparent and absolute magnitudes.

COLLEGE (12-14) The Mass of a Binary Star System: Students observe a "variable star" and graph its apparent magnitude compared to some known standards over about a forty day simulation. Then, after plotting the star's light curve, which turns out to be a classic eclipsing binary of the Algol type, they are presented with simulated spectra which display doppler shifts toward and away from Earth at different phases of the orbits. Students then calculate the star's orbital velocity from the doppler shifts. Using the time for one complete revolution obtained from their light curve, they calculate the orbit's circumference and radius and use Kepler's third law to calculate the mass of the binary system.

The third and final phase of the recommendations of Project 2061 is the current collaborative effort in which many groups active in science education reform such as the Astronomical Society of the Pacific use the resources of the first two phases to move the nation toward scientific literacy. Today's students are preparing to function in the complex, technological society of the 21st Century. It is essential that all teachers assist them in acquiring the skills and knowledge necessary to improve the quality of life by understanding the interrelationships between the Earth and its surroundings in space. It is equally important that students develop attitudes of open-mindedness as well as skills of careful investigation. A goal of this planetarium program is to support science education reform through promoting scientific literacy in its citizenry. A scientifically literate person is aware that science, mathematics, and technology are interdependent human enterprises with both strengths and limitations; understands key concepts and principles of science; is familiar with the natural world and recognizes both its diversity and unity; and uses scientific knowledge and scientific ways of thinking for individual and social purposes. We believe that through the implementation of this planetarium program, we can demonstrate our commitment to achieving this goal.

The Mystery of The Nazca Lines

Kathleen Cochrane
Our Lady of Ransom School
Niles, IL
USA 60714

The poster display is based on student activities from one of THE NEW EXPLORER episodes about the Nazca Lines. The videotape, called The Mystery of the Lines, sets the stage for a student investigation into the purpose of the Nazca Lines. Often called the strangest messages ever left by man, the immense figures are created in pebbles using the Nazca desert as a background. Enormous figures of birds, animals, insects, flowers and geometric shapes fill the desert floor. These figures which are visible only from the air would be an enigma unto themselves. How did they get there? What is their purpose? Who was their intended viewer? The mystery is compounded, however, by a series of lines constructed to crisscross them in the desert.

After viewing the videotape, the student teams are challenged to see if they can devise a method to reveal the purpose of the figures and lines to solve this 2000 year old mystery. They may choose to recreate the Nazca desert and see if there are any clues in the figures themselves. This activity is given in the teachers guide that accompanies the video. The student investigators will use small stones or gravel to etch the figures on a paper desert floor. Above the desert scene, the students recreate the constellations that would have been visible to the ancient inhabitants of this Peruvian desert. When the lines are added, students realize that some lines point directly to constellations. What they have learned to recognize as Orion the hunter is directly over the figure of a spider. The Pleiades rise over a drawing of a flower. They confer to decide if there is a connection. Another clue is found carved into a mountain. A figure of a person stands against the mountain pointing with one hand to the desert floor, and with the other he points to the sky. The answer to the mystery before then may be found in the sky above.

Student investigators soon realize that although we recognize these constellations as representing one thing, other cultures may have had different ideas, different myths. After conferring with their team, or with other teams, students also realize that we do not see constellations in the same place every night of the year. Through questions, they are led to conclude that the lines only point to those constellations at certain times of the year. In researching the activities (and aided by the clues given in the videotape) students conclude that the lines and figures were actually a gigantic astronomical calendar whose purpose was to tell the Nazcan people when to plant and when to harvest. The lines can be read like the hands of a clock, pointing to rising and setting constellations that correlate to a particular time of year. Some students have used the computer program VOYAGER II to check to see at what time of the year each constellation rises and sets.

Some student teams chose to do the activities by recreating the pictures in the parking lot of the school and using a flashlight to follow the lines up toward the horizon. Others may create the lines and figures on black cloth or plastic in fluorescent paint and view them in a Starlab. Using a laser pointer, they can follow the lines straight up to the corresponding constellation. Students may even devise a completely different method of solving the mystery.

Teams present their findings to their peers at an Archaeoastronomy Convention. They may use pictures, transparencies or models to substantiate their findings. Samples of this activity and order forms for the video series The New Explorers can be requested from Kathleen Cochrane, Our Lady of Ransom School, Niles, IL 60714 USA

Addressing Common Astronomy Misconceptions In The Classroom

Neil F. Comins
University of ME
Bennett Hall
Orono, ME
USA 04469

E-mail: galaxy@maine.maine.edu

Abstract: I discuss elements of teaching that I have found useful in helping students realize that they have misconceptions and in dealing with the misconceptions. In the poster paper I also present the background for this research, including the concepts of a conceptual framework and metacognition and an extensive bibliography. I then present 33 of the most common misconceptions about astronomy that I have identified and their causes. While there is no quick fix to helping students overcome their misconceptions, there are practices that I have found very successful.

Classroom Techniques

Getting student attention Students are much more successful in changing their beliefs if they are aware that they harbor misconceptions. This **metacognition** on the students' part is at the heart of my classroom activities in helping them deal with misconceptions. My first task each semester before a diverse class of 260 students is to convince them that they have misconceptions. Just telling them so is ineffective - it creates irritation without illumination. Therefore, before saying anything on the subject, I do a demonstration of levitating a beach ball using a high speed air gun like those used to clear driveways. After asking what they expect will happen (The ball will fly away), I throw the ball straight up, turn the air gun straight up, and wait for the ball to hover several feet over the air gun. I then ask, What will happen if I tip the air stream over? (Now it will definitely fly away.) I tip the air gun over and the ball hovers over empty space, several meters to the side of the air gun. Now I have their attention.

Turning on student metacognitive radar After the demo, I briefly explain that everyone has misconceptions about astronomy and that they shouldn't feel inferior or inadequate as a result. However, I assert that they will have great difficulty fully understanding and remembering much of what I teach if they don't address their incorrect prior beliefs. I note that I will be helping them identify and deal with the misconceptions throughout the course. I then discuss my version of conceptual frameworks and how these structures make it hard to unlearn incorrect ideas.

Finally, in this initial stage, I hand out a course pretest (available upon request) to be returned at the next class. Students are instructed that the test will not affect their grade and that they must do it entirely by themselves. I explain that the pretest is designed to help them "see" what they think about a variety of astronomical ideas. I keep student attention on the pretest questions throughout the course by highlighting the correct science relating to most of them during the appropriate lectures.

I now address misconceptions in my introductory college astronomy text, **Discovering the Universe**, 4th ed, by William J. Kaufmann and Neil F. Comins. At the beginning of each chapter, students are asked questions based on common misconceptions. Correct information is identified by a tab at the appropriate paragraph.

Keeping the search for misconceptions alive It is part of human behavior to try to make new information conform with our present beliefs. Therefore, in class, I have found it necessary to keep student awareness focussed on the prevalence of misconceptions and the need to address them. To this end, I offer extra credit (an increase of grade by one segment, e.g., B to B+) to any student who turns in a list of 42 misconceptions I have corrected for them at the end of the semester. The number 42 corresponds to the number of lectures. I instruct students to write two sentences for each misconception: State the misconception, and tell me where you first remember learning it.

Disassembling conceptual frameworks By far the hardest, yet most important, facet of addressing misconceptions is helping students to disassemble incorrect conceptual frameworks and then to build correct ones. Once they are aware that a fact is in conflict with a prior belief, it is necessary to engage students in discovering how their prior belief has affected their conceptual framework(s). Otherwise, they will just distort the correct information to conform with the incorrect framework or they will never get the new information into long-term memory at all. That is, they will memorize it long enough to use it on exams and then forget it.

I strongly encourage questions, which I often use to draw out students about common misconceptions. The important point here is for teachers to have some understanding of the relevant conceptual frameworks relating to common misconceptions.

Exploring misconceptions by assuming they are correct This approach enables students to express the background for some of their incorrect ideas. For instance, I have asked, "suppose the asteroid belt is as littered with asteroids as we see in the movies." This is a virtually universal belief. "What would happen to the asteroids under those conditions?" (Gravity would quickly pull them together.)

Acknowledging student discomfort Addressing their own misconceptions is discomforting to students in at least two ways. First is the discomfort associated with accepting that long-held ideas are wrong and the struggle required to change them. Indeed, it is normal to change the new information to fit old, incorrect conceptual frameworks. Second is the ubiquitous need to find someone to blame for the old, incorrect information. Several times during the semester I urge students not to let these discomforts prevent them from changing.

Giving students closure During the penultimate class of the semester I hand out a posttest. Besides the student information questions I asked on the pretest, I also ask what grade students expect to earn in the course and how frequently they come to class. The astronomy questions on the posttest are identical to those on the pretest. This test gives students a chance to re-evaluate their ideas.

Conclusions

1. Our sources of information and our abilities to process this information are so numerous and varied that it is impossible to prevent people from developing misconceptions. (Also, it is impossible to correct all misconceptions with one technique.)
2. Simply telling students correct science, without addressing their prior beliefs, rarely leads to long-term acceptance of the correct information.
3. Once a correct concept is incorporated into a conceptual framework, it is very hard to remember how we could have thought otherwise. Therefore, to help students overcome misconceptions it is important for teachers to understand (perhaps to relearn) the reasons for common incorrect beliefs.
4. Metacognition appears to help students deal with their misconceptions. When they are actively engaged in comparing the science they are learning with their prior beliefs, they are better able to disassemble incorrect conceptual frameworks.

GNAT: A Global Network of Small Astronomical Telescopes

David L. Crawford
NOAO
Box 26732
Tucson, AZ
USA 85716

E-mail: crawford@gnat.org

and

Eric Craine
Western Research
Tucson AZ
USA 85719

E-mail: craine@gnat.org

Abstract: There is a pressing need for quality small telescopes for astronomical research and for education. Such telescopes, with good instrumentation and software, can do frontier research in many areas, and they can and do make a major positive impact on science education. This need is global: the interest in astronomy is universal, both in students and in amateur and professional astronomers. Technology is such today that quality small telescopes can be produced for a reasonable cost. We believe that such new generation small telescopes are an essential part of a balanced and rational approach for astronomical facilities. We have proposed, and are beginning to develop, a global network of such telescopes. A number of universities and individuals are already formal members. We invite you to participate.

I. Introduction.

There are many areas of astronomy where relatively small telescopes can make a major impact on research. For a number of these, they are in fact better than trying to use larger telescopes. A few examples only: monitoring variable objects of all sorts, from lensing galaxies, to variable quasars, to all types of variable stars; surveys of all sorts; systematic imaging photometry on open and globular clusters; standard star networks; and the search for extrasolar planetary systems. Certainly all the traditional areas where small telescopes have been used are still valid, and much needed. Many, most, areas of astronomical research are data poor, and more telescopes are needed to effectively attack the research problems. Not only are there many more stars and galaxies than there are telescopes, but there are many more astronomers than there are telescopes. Few if any of us have the amount of telescope time we need for our research.

Since the cost of telescopes goes with a relatively high power of the aperture, the only way to supply enough telescope time for astronomers is with relatively small telescopes. Fortunately, as noted above, such telescopes are still frontier instruments for research when equipped with modern detectors and software analysis tools.

Likewise, such telescopes can be powerful tools for helping with science education. There is no substitute for real data, especially when it deals with exciting topics. Astronomy is interesting to students and to the public, not just to professionals. It can be the hook to attract interest in science and in the study of the universe. As with research, such interest and needs are global, not just in one country. Creative, fertile minds exist everywhere, and need nurturing. A viable network of small telescopes, accessible by many, could well be making a positive impact on science education for many more than 100 million students worldwide within a decade.

II. The Need

It seems today that almost all of us are "have-nots" relative to adequate telescope time for our research needs. While a few, mostly very large, new telescopes are being built, it is also a fact that a number are closing down, mostly small ones. It is also a fact that many of our observing sites are becoming seriously compromised by light pollution.

We like to think of this dilemma as "Second Class, First Class Astronomy." By that we mean that one has a very good research idea, so good that it actually gets telescope time assigned in spite of the severe pressures for such time. However, one needs five nights to do a good job on the problem, but one only gets three nights, due to the extreme pressure. One of these turns out to be cloudy, on another the equipment is not working well, and on the third one suffers from the learning curve, since things are different (detector, software, and such) from the previous run (if any). So one actually gets only a little good data, and naturally such things as dark frames, standard frames, and such necessary items get compromised too. So results are meager, maybe even unusable for the task, but one publishes anyway or one does not get any more time in the future. Such is life. And it is not getting any better, but worse. The pressure for telescope time is up everywhere, as there are more astronomers, very good ones, competing for the limited hours. Furthermore, the complexity of the astronomical instrumentation is increasing (not to speak of the cost). We are getting behind. Only by increasing the number of first class, relatively small and hence relatively inexpensive small telescopes, accessible to many, can we hope to recover. And these may well be the only ones where we can spend the time to experiment, to allow serendipity to give us new breakthroughs.

The same holds for using telescopes as powerful tools for education. Real data, on real objects, in nearly real time can hold a great fascination for students (not just for the professionals). Astronomy is perhaps one of the very few scientific fields where the number of amateurs greatly exceeds the number of professionals. Consider the amount of media coverage astronomy gets for being such a relatively small field (number of professionals doing research). After all, we cover the universe, and we are the Science of the Extreme. Mind boggling is a definition for "astronomical." As such, astronomy, and the use of small telescopes and the data produced by them, can be a fascinating hook for grabbing and holding the attention of students to science. Not just those who will go into science, but all students, including those who become librarians, lawyers, doctors, politicians, and homemakers. They all need to know what science is about and why it is useful and important!

III. The Technology

Many changes in technology have been occurring, and many of these have had a great impact on astronomy, allowing the new generation of large telescopes to be designed and built offering very high quality and potential at a rational cost. But such advances also allow a new generation of small telescopes too, of very high potential and at relatively low cost. Since the cost of a telescope goes with a rather high power of the aperture (perhaps 2.6 or so), small telescopes are relatively low cost, offering a very high Value Per Cost ratio. It is rather clear that the only way we can supply a significant increase in the number of available telescope hours is to build a significant number of these new generation small telescopes, and equip them with modern detectors and instrumentation.

By using computers at all phases of the design and operation of such telescopes, we can insure that they operate efficiently, even automatically and remotely. We can then conceive of a global network of such telescopes, many of them located at excellent observing sites, thus giving us full coverage of any objects (not limited by longitude or latitude issues) and many hours of first class clear hours. With the increase in our communication abilities (e-mail, Internet, the Web, for example), we can "easily" operate such a system efficiently and remotely. Users can be anywhere, so can the telescopes. We thus have a multi-user, multi-telescope observatory. Such an observatory can and would have very many part-time, unpaid staff members, essentially anyone, anywhere who is interested in the issues and wants to contribute. A small core of paid staff would be the "oil-can" keeping the hardware, software, and communication systems working well.

IV. GNAT

Such a system seems to make a great deal of sense, if it can be developed and operated efficiently. There are now a number of small telescopes operating automatically or remotely, some others coordinated globally for special programs. But we think that the potentials cry out for a broader scope and considerably more coordinated operation. We have therefore incorporated a new non-profit organization: GNAT, Inc. The Global Network of Astronomical Telescopes.

GNAT's goal is to be a catalyst for those interested in small telescopes in astronomical research and education. Specifically, GNAT's long term goal is to establish a global network of new generation astronomical telescopes, accessible to most anyone, anywhere. It will be a membership based organization, hopefully with most institutions and most individuals who are interested in such issues as members. Several such have already joined.

We have been in contact with many commercial companies who are active in producing cost effective small telescopes and/or associated instrumentation. We have hosted a meeting to discuss the viability of a "standardized" design for new generation small telescopes of several sizes. It appears that one or more such standard designs is a viable potential. We will be hosting several more such meetings, in various locations, over the next year, to continue the discussions. Our Web page contains details of the past and of the upcoming meetings (and other items of interest, such as references and links to other entities involved in the issues). While we are trying to communicate regularly with such companies, GNAT does not have any formal connection with any such company, nor do the individuals currently active in GNAT affairs. GNAT must remain independent of any such connection with any one company if it is to meet its goal of being a catalyst for all. In addition, GNAT has no connection in any way with KPNO or NOAO, except that a few of the individuals involved with GNAT as volunteers are staff members of NOAO.

We expect to have one or more GNAT telescopes operating by the end of 1995, and to have established links with several non-GNAT telescopes. The goal is to have a proposal ready by the end of 1996 so as to solicit funding for the full GNAT network. Advice and feedback by institutions and individuals is essential if GNAT is to become a viable entity.

We welcome anyone who is interested to contact us and get involved. Our Web page address is currently: http://www.csn.net/~jls2/gnat/ but it will be changing soon to: http://www.gnat.org/

Astronomy And Teacher Training At The University Of Waterloo The Waterloo – Queen's Science Teaching Program

Gisèle Dagenais:
94 Marmora Street
Trenton, Ontario
Canada K8V 2J3

E-mail: gisele@astro.uwaterloo.ca
home page: http:\\astro.uwaterloo.ca\~gisele

The Waterloo - Queen's Science Teaching Program (contact: Hugh Morrison, University of Waterloo, Waterloo, ON N2L 3G1, (519) 885-1211) combines a Co-op Honours B.Sc. from the University of Waterloo with a B.Ed. from Queen's University. It is a concurrent education program. Through the Co-op program at the University of Waterloo, this program provides its graduates with opportunities to gain teaching experience in both classroom and less traditional settings. This provides many opportunities to test out a variety of astronomy activities with many different age groups. The following are just a few successful experiments in astronomy that were popular with students in varied settings.

Outline of the Waterloo-Queen's Science Teaching Program

Term	Activity
Fall - 1A	First half of first year.
Winter - 1B	Second half of first year.
Spring - W1	First work term - in industry.
Fall - 2A	First half of second year.
Winter - W2	Second work term - in industry.
Spring - 2B	Second half of second year.
Fall - T1	First teaching placement.
Winter - Queen's	B.Ed. program at Queen's University.
Spring - 3A	First half of third year.
Fall - T2	Second teaching placement.
Winter - 3B	Second half of third year.
Spring - W3 or T3	Work term, teaching placement or holiday.
Fall - 4A	First half of fourth year.
Winter - 4B	Second half of fourth year.

A Micrometeorite Hunt at day camp: Engineering Science Quest is a summer day camp program at the University of Waterloo, for students going into grades 5 to 8 (9 to 13 years old). The following activity was run following the Perseid meteor shower and was very popular with the participants. Adapted from "À la cueillette des météorites" in *Encore des Expériences*, by Professeur Scientifix and Bernard Larocque, Québec Science Éditeur: Québec, 1985. Set containers outside before the first rainfall following the meteor shower. After the first rainfall, bring in the containers with the rainwater. Pour the water into beakers and place the beakers on hot plates to evaporate the water. This must be carefully monitored, since the hot plates will have to be on high, to boil off the water. The containers can be left on a warm heater, for the water to evaporate more slowly, but this can take a couple weeks. When the water has evaporated, use a paintbrush to brush loose all the sediment on the bottom of the beakers. The same should be done to the containers once they have dried out. Brush all of the sediment onto a white sheet of paper. Pass a magnet very closely over, or through, the sediment. Look closely for any small pieces of sediment that stick to the magnet. Use a card or a small piece of cardboard to brush them off onto a fresh white sheet of paper. Double check that they are indeed magnetic. This indicates that they likely contain iron and are thus very small meteorites, not just pollution. The students then taped the micrometeorites to the paper and took as many as 16 meteorites home!

Design An Alien In Grade 10 Science: The curriculum for grade 10 Advanced Science (this is a streamed level) includes an optional unit in astronomy. The focus of the unit is on the solar system and an introduction to some observational astronomy. The rest of the course has a large biology and ecology component, covering topics such as food chains, sensory systems, internal transport systems, competition and adaptation. The focus is on what animals need to survive and how they adapt for survival. The following project was assigned to the students to complete in partners. The quality of some of the submitted work was astounding, even from students who were not usually very motivated by science. Choose a planet, moon or asteroid in the solar system and invent an alien that could survive on that planet. Take into acount the properties of the planet (composition, temperature, atmosphere, etc.) and describe the physical and behavioural adaptations that this alien has in order to survive in its environment. Include a sketch or picture of your alien.

Find Constellations Without Star Charts At An Outdoor Education Centre: Noisy River Outdoor Education Centre is run by the Etobicoke Board of Education, just outside of Toronto. The centre is in a very isolated area, about one hour north of Etobicoke. Students in grades 6 to 8 come up to the centre as a class, with their teacher and stay at the centre for a full week. The advantage of this program to teaching astronomy is that a night sky is actually available for stargazing. However, many students have never done any stargazing before, most have never seen the night sky outside of Toronto and star charts are rather unwieldly tools for the beginner. Star-canisters can be made with a hammer, some different size nails and empty drinking crystal tins. The hammer and nails are used to make holes in the aluminum bottom of the tins, following the patterns of constellations. Different sizes of holes correspond to stars of different magnitudes. A flashlight is then placed into the top of the tin, shining out the bottom, through the holes.

These star-canisters were used to show students an accurate image of the constellations they were trying to find, including their orientation. In this way, students do not get frustrated trying to read a star chart on their first star gazing escapade. This technique was tried on a group of 12 students, with canisters for 10 prominent constellations. Only one student was unable to find all 10 constellations. During the stargazing session, the mythology of the constellations was also discussed, which kept the students very interested.

Resources:

365 Starry Nights, by Chet Raymo

The Beginner's Observing Guide, by Leo Enright, The Royal Astronomical Society of Canada, 1992.

Space Telescope Elementary School Outreach Program (STESOP): Teaming Up STScI staff with Future Educators

Laura Danly
Space Telescope Science Institute
3700 San Martin Drive
Baltimore, MD
USA 21218

Tel.: (410) 338-4422
E-mail: danly@stsci.edu

and

Flavio J. Mendez
Space Telescope Science Institute

E-mail: mendez@stsci.edu

The objectives of the Space Telescope Elementary School Outreach Program (STESOP) are two-fold: (1) to assist the Education schools of local universities in the training of science for elementary school level pre-service teachers, and (2) to motivate and educate elementary school children in the Baltimore City and neighboring counties areas about science and math through results from the Hubble Space Telescope.

The STESOP staff is composed of STScI employees and regional education school students. The synergistic partnership permits the education grad-students to learn more astronomy content while the STScI staff learn more about effective classroom presentation. The presenters are trained in a workshop on effective classroom presentation prior to going into the field.

The STESOP presenters visit schools in pairs, one woman and one man, for role modeling purposes. Each presentation lasts about 60 minutes, and each addresses no more than 40 students at a time. The classroom visits consist of talks, demonstrations, hands-on activities and dialogues using materials such as video-tapes, desktop models, and slides. The teachers participate as evaluators of both content and presentation. The school is left with educational materials including posters, HST-related literature, lithos, coloring sheets and information about the Teacher's Resource Laboratory located at Goddard Space Flight Center.

The planning and scheduling of the STESOP visits is done in close collaboration with the Baltimore City and county Boards of Education. The science supervisors for each district help identify schools and teachers to contact. The aim is to coordinate the STESOP visits with the timing of the astronomy/space unit being taught in the schools.

In the 1994-1995 school year, STScI staff paired with students from the Johns Hopkins University graduate Education program. Roughly 5000 students were reached in 50 schools in 7 different counties in Maryland. STESOP has also visited a few classrooms out of state, including presentations in Pennsylvania, Puerto Rico, and even Trinidad. STESOP has also supported the educational activities of the 4-H program in Maryland. Of the presentations directly in schools, 41 were in public schools and 8 were in private schools.

Evaluations from the 1994-1995 school year were excellent. The project scored between 89% and 95% on questions such as

- Was the presentation at the students' level?
- Was the topic covered clearly and completely?
- Were the materials helpful?
- Did the presenters adequately answer the students' questions?

In the 1995-1996 school year, the STESOP staff will be augmented by students in Education from Morgan State University, an historically black college in Baltimore. Roughly 60 presentations are scheduled.

An expanded evaluation sheet has been developed for this year. Some of the additional questions include:

- Did you prepare you class prior to the presenters' visit?
- Was the length of the presentation appropriate?
- Did the presentation match your curriculum?
- Would you recommend STESOP to other teachers?
- How can we improve our presentation?

In addition, we are compiling responses to the question:

- What is the one question your students ask the most about astronomy that you would like more information on?

Our aim is to collect a set of the most frequently asked questions by Baltimore City and County 4th graders which, together with the answers, will be provided in a publication to local school teachers to assist them in answering their students' questions. We also plan to publish the result on our World Wide Web site (www.stsci.edu).

Finally, we have developed a closer relationship with one elementary school in particular to develop a more expanded program which involves the whole school. The expanded program includes a pre-visit meeting with teachers from all grade levels, an HST-related display on-loan to their school library, and additional classroom meetings plus school assembly. In this way, we hope to reach a wider audience of students, while permitting teachers from each grade to develop their own ways of incorporating lessons from HST into their science curricula.

STESOP is funded jointly by the Maryland Space Grant Consortium and STScI.

Hands-on Learning at Northern Arizona University and Lowell Observatory: The National Undergraduate Research Observatory, and Research Experiences for Undergraduates

Kathy DeGioia Eastwood
Northern Arizona University
Department of Physics and Astronomy
PO Box 6010
Flagstaff, AZ 86011-6010

E-mail: Kathy.Eastwood@nau.edu

The National Undergraduate Research Observatory is a consortium of primarily undergraduate institutions which have joined together to provide hands-on training and research experiences for undergraduate students. NURO students learn about science by participating in the process rather than by hearing about it.

All of the state-of-the-art facilities at NURO are available for undergraduate training and research. Students help plan the observations with their professors, take the data, reduce it, and help with the writing of publications. NURO also provides students with a chance to meet students and astronomers from other institutions, which contributes to their sense of belongingin the astronomical community. In addition to the NURO facilities and astronomers, students have opportunities to interact with scientists from the Lowell Observatory, Northern Arizona University, and the US Naval Observatory in Flagstaff.

The NURO Consortium includes both public and private universities from across the country. Together they share in 60% of the observing time on Lowell Observatory's 31-inch telescope. Astronomers and their undergraduate students at the member schools collaborate on key research projects, as well as conduct their own private research. The key projects are chosen to take advantage of the instrumentation and the large blocks of observing time available. Current key projects include photometric monitoring of chromospherically active binary stars, and light curves for extragalactic supernovae.

Benefits of membership in the NURO Consortium include guaranteed observing time on the 31-inch telescope, technical assistance while observing, a voting seat on the Steering Committee, and participation in NURO meetings. Responsibilities of membership in the NURO Consortium include an annual membership fee of $2,300, supplying travel expenses and housing in Flagstaff (some travel funds are available through the Consortium), occasional collaboration on key observing projects, contributions of expertise, sending a representative to the annual Consortium meeting, and the writing of funding proposals as needed for further support.

The current NURO Consortium members are Alma College, Ball State University, Benedictine College, Central Michigan University, College of Charleston, Denison University, Dickinson College, Franklin and Marshall College, Gettysburg College, Maria Mitchell Observatory, University of Nevada at Las Vegas, Northern Arizona University, University of Oklahoma, Western Connecticut State University, and Widener University.

The facilities include the use of Lowell Observatory's 31-inch telescope, which was completely refurbished in 1990 and is fully computer controlled with a "warm room." The site of telescope is Anderson Mesa, a dark site about 13 miles from Flagstaff at an altitude of 7200 feet. The atmosphere is very transparent, and about one third of the nights are photometric with another third spectroscopic.

The primary instrumentation consists of an imaging CCD camera made by Photometrics. The 512 X 512 Tektronix chip is back-illuminated, with a quantum efficiency of over 70% at V and over 40% at U. The camera is cooled with liquid nitrogen, and networked to an on-site SUN workstation. Also available are a computer-controlled photoelectric photometer and additional SUN workstations at NAU. A Boller and Chivens spectrograph with a CCD detector is under development.

In addition to NURO, the NSF-sponsored Research Experiences for Undergraduates (REU) program is in its fifth year at Northern Arizona University. The program consists of eight nationally recruited undergraduate students per year doing one-one-one research with a mentor from Northern Arizona University, the Lowell Observatory, or a visitor from a NURO institution. Student activities include a lecture series given by local and visiting astronomers for which the students receive university credit. The students are housed together in NAU dormitories and participate in extra-curricular activities together. Each student receives a stipend, housing, and travel support. To be eligible, students must be enrolled in an undergraduate degree program and be citizens or permanent residents of the United States or its possessions.

For further information on either program, contact the Director of NURO, Dr. Kathy DeGioia Eastwood, at the address shown above, or see our World Wide Web page at http://nuro.phy.nau.edu.

Developing Education and Outreach Programs at Research Institutions: Lessons Learned at the Harvard-Smithsonian Center for Astrophysics High Energy Astrophysics Division

K. Dow, E. Mandel, C. Jones, S. Murray and T. Ruiz
Smithsonian Astrophysical Observatory
Harvard-Smithsonian Center for Astrophysics
60 Garden Street, MS-83
Cambridge, MA
USA 02138

Tel.: (617) 496-7586
FAX: (617) 496-7577
E-mail: kdow@cfa.harvard.edu

Like many research institutions, the Harvard-Smithsonian Center for Astrophysics(1) (CfA), has been actively engaged in education and public outreach activities for many years. The Harvard University Department of Astronomy, the formal higher education arm of the CfA, offers an undergraduate concentration and a doctoral program. In the CfA Science Education Department, educational researchers manage ten programs that address the needs of teachers and students (K-12 and college) , through advanced technology, teacher enhancement, and the development of curriculum materials. In this environment of successful programs, we decided to initiate a series of education and outreach projects and partnerships in the High Energy Astrophysics (HEA) division.

HEA is one of seven research divisions at the CfA, and concentrates on X-ray astronomy and the development of advanced X-ray instrumentation. For several years, our researchers have been working individually on education/outreach activities such as presenting public talks, school visits, the development of slides sets and the publication of popular articles. With management support, this long-lived tradition of informal public outreach recently has coalesced into several formal education and outreach programs targeting three specific groups: college students, elementary students and teachers, and the general public. This paper describes three of these projects, namely the Smithsonian Astrophysical Observatory Summer Intern Program, Everyday Classroom Tools, and Astronomy in Motion. All of our formal programs complement the ongoing and informal activities of individual researchers.

The SAO Summer Intern Program is one of 15 National Science Foundation Research Experiences for Undergraduates (REU) program sites (in astronomy). Now beginning its third year, our program offers a dozen undergraduates interested in pursuing a career in the physical sciences the opportunity to spend 10 weeks at the CfA working on individual research projects with staff scientists. We have actively encouraged participation by members of under-represented groups through direct mailings to placement offices and physics professors at women's colleges, and colleges with high African-American, Hispanic, and Native American student populations. Fifty percent of our participants in the first two summers were women and 12 percent were minorities.

In addition to conducting their research project, our REU interns visit other students at the Maria Mitchell Observatory in Nantucket, MA, and at Haystack Observatory in Westford, MA. Students also attend weekly colloquia, "ethics evenings," and weekly intern lunches. Although each student is working on an individual project, we have found that it is important to encourage a collaborative learning community. This type of set-up fosters peer reinforcement, group problem solving skills, and builds self confidence.

We feel strongly that the interns spend time talking about their research with other interns as well as staff researchers. Speaking and writing about their projects is challenging for the interns, but their clarity increases when they have to present their ideas to others. Opportunities to explain their research invariably leads to new questions and further knowledge. The process of preparing their talks and papers teaches them self-reliance, allows them to learn from one another, and provides them with an environment where their peers, scientific advisor, and program staff take their ideas seriously. To further these goals, our program includes frequent informal group and individual meetings, writing of a journal-style paper, and an "Intern Symposium" at the end of the summer. During the Symposium, the students present a 15 minute talk about their project, followed by questions from Observatory scientists and support staff.

At the conclusion of their internship, students have the opportunity to present a poster paper at an American Astronomical Society meeting. Additionally, interns often use their CfA research project as the basis of an undergraduate thesis, and therefore continue to work with their research advisor during the academic year. In the first year, three of these collaborations produced Astrophysical Journal papers. As of this writing, nearly every past participant who has graduated from college is enrolled in a graduate program (in Physics, Astronomy, and Engineering).

Everyday Classroom Tools is one of 18 projects in 19 states funded through NASA's Cooperative Agreement Program: Public Use of Earth and Space Science Data Over the Internet. Working with a dozen teachers in three Massachusetts elementary schools, we are developing a "hands-on" astronomy curricular theme that integrates science and Internet/computer activities into the daily life of the classroom. An important part of this project is the adaption of research quality software, such as the SAOimage display program, for use by elementary students.

One of the key aspects of this project lies in communicating the process and excitement of scientific inquiry to teachers so that they can translate their understanding into effective classroom tools. We are concentrating on applying the scientific method to a small number of situations directly related to the world around us (Sun, moon, shadows, etc.). By learning and applying observation, data acquisition, modeling, and interpretation skills to these everyday occurrences, we aim to go beyond the development of a collection of activities to attain a way of looking at the world. Our expectation is that these skills will find their way into many other areas of the elementary school curriculum, and in particular, will allow schools to become more comfortable and capable in the use of computer technology.

Our project presents a unique set of challenges both for teachers and science researchers. On the one hand, we as researchers need to find a way to distill our experience into a set of activities and concepts that can provide the forum through which the spirit of scientific inquiry can make itself known. On the other hand, teachers need to find a way to refine, expand, and integrate these activities in the "real world" of the classroom. Together, we are exploring a new type of scholar-teacher partnership that is long-term and that goes beyond interfacing through a static collection of science activities.

In addition, there are many opportunities and challenges inherent in forming a partnership between a research institution and an elementary school. For example, we are exploring cost-effective ways to share our institutional computer expertise with our partner schools. By setting up Internet connectivity in those schools based on the already-existing network at the CfA, we are charting new ways in which a research institution can serve the local community.

Astronomy in Motion is one of three HEA division projects funded by the NASA IDEA program. We have produced two posters on astronomical themes ("The Sun" and "Jupiter") for display on Massachusetts Bay Transit Authority's subway lines. The intent of these posters is to heighten public interest and appreciation of astronomy among general audiences. Toward this end, we use full-color NASA photos and high-quality graphics to engage the viewer's attention. Each poster poses a basic question about astronomy, the answer to which is available through a recorded phone message or e-mail.

During the autumn of 1995, 150 copies of each of our posters (both in English and in Spanish) were featured on subway trains and simultaneously were distributed to Boston Public Schools. The schools also received supplementary information such as suggested field trips, local astronomy-related events, and age-appropriate astronomy literature. The success of this pilot project has prompted much interest by other research institutions and mass-transit authorities who would like to expand this program to their cities.

The programs described above are but a part of an overall effort in the HEA division of the CfA to make our expertise, our facilities, and our excitement about science more accessible to teachers, students, and the general public. More information about these and other programs is available from the authors or from the HEA division Education and Outreach homepage: http://hea-www.harvard.edu/scied/.

The Smithsonian Astrophysical Observatory (SAO) is a research bureau of the Smithsonian Institution. Research at the Harvard University Department of Astronomy is carried out at the Harvard College Observatory (HCO). Together, the two observatories constitute the Harvard-Smithsonian Center for Astrophysics.

Undergraduate Research for Majors and Non-Majors

Robert J. Dukes, William R. Kubinec, Harold L. Nations

Department of Physics and Astronomy
The College of Charleston
Charleston, SC
USA 29424

E-mail: dukesr@cofc.edu

The Importance of Undergraduate Research

It is becoming well recognized that undergraduate participation in the research process is an important part of the educational experience of students in the sciences including astronomy. In the past, some colleges ran outstanding programs of involving their undergraduates in astronomical research by using the facilities of the national observatories. Unfortunately these facilities are no longer available. Institutions desiring to involve undergraduates in astronomical research have adopted several means of replacing these facilities including refurbishing telescopes at Kitt Peak and other observatories for use by consortia of smaller institutions. Another approach which we have adopted is to augment campus facilities with both an automatic telescope at a prime site and membership in the National Undergraduate Research Observatory.

The Triad of Facilities

Undergraduate students should be exposed to all components of research. In observational astronomy these include hands-on experience at the telescope, experience in reducing and analyzing data, and (ideally) experience in observing at remote facilities. While this might be impossible for every student we still have assembled the components required to afford our students the opportunity for each. These include an on-campus 0.4-m DFM Cassegrain Telescope with a CCD camera and spectrograph, 30 0.2-m Celestron and Meade reflectors, and six ST6-CCD cameras with mobile computer workstations. We are a member of the Four College Consortium which operates an 0.8-m Automatic Photometric Telescope at the Fairborn Observatory Station located at the Whipple Observatory on Mt. Hopkins in southern Arizona. This telescope is an 0.8-m computer operated telescope which can operate without on-site attention for several weeks at a time. It is capable of uvby, UBVRI, H-alpha, and H-beta photometry and observes using a preloaded list of stars and internal rules governing target selection. Finally, we are a member of the National Undergraduate Research Observatory which is a consortium of 14 colleges and universities and which operates an 0.8-m reflecting telescope at Anderson Mesa near Flagstaff, AZ. This telescope, equipped with a Tektronics 512 x 512 liquid nitrogen cooled CCD, is maintained by Lowell Observatory.

Undergraduate Research for Science Majors

There are a number of problems which have to be considered in designing a viable research program for undergraduates. For example, undergraduates have much less time available for research, and far more varied demands on this time than do graduate students. Observational astronomical research is a difficult field for undergraduates to participate in, since a period of poor weather which might be an inconvenient delay to a graduate student's research, might destroy an under-graduate's project which often must be competed in a semester. Also, undergraduates have greater difficulty adjusting to night work since they usually have a much fuller day schedule. Many undergraduates also lack some of the maturity of graduate students. Supervising a senior undergraduate research project is usually more time consuming than supervising a graduate student. This is especially true since we consider it imperative that the students have the same kind of "ownership" in their research that a graduate student does. One way we approach this problem is to involve the students in on-going research in the area of variable and binary stars. Here the student collects photometric data through one or more components of our triad. They may use the on-campus facilities to do CCD photometry. They may use APT data or data collected on an observing run at NURO to supplement their own data or, depending on the student and the project, they may rely solely on APT data. In any case, the student must become familiar with the basic astrophysical properties of the object or objects they are studying, reduce data, analyze the data set, and present their results at one or more scientific meetings. Students can participate in an ongoing project which has been worked on previously by other students. For variable star research this means that each new student will have an additional seasons worth of data to reduce, combine with the previous data, and perform a new analysis. This is wonderful for the students and is not a great disadvantage in some of our projects which require relatively long-term monitoring. Other students may choose to work on a more instrumentally based project using on-campus facilities. One of us is in the process of adapting an inexpensive, commercial spectrograph to be used with a CCD. The spectrograph and CCD will be housed in a stationary climate controlled chamber and coupled to our 0.4-m telescope by means of an optical fiber bundle. This spectrograph will be used to obtain radial velocities of brighter stars in a search for undetected binaries. Students will gather and analyze the data as well as perform follow-up photometry either on-campus, at NURO, or with the APT.

Undergraduate Research and Science Literacy

It is becoming well recognized that many students graduating from college today are not science literate. Science education in the Liberal Arts Curriculum is one of the prime means of developing such science literacy. However, many students complete a college sequence in science with little or no appreciation for the methodology of science. Instead they view science as a body of facts to be memorized and given back on examinations. Typical laboratory science courses in many cases fail at developing this appreciation and consequent understanding. It has been suggested, that an approach to improving this situation is to more directly teach science to non-science majors as it is practiced. In our program we have tried to do this in two ways:

1. For the larger number of students we have used computer simulations of astronomical research as components of both our lecture and laboratory program.
2. For the most talented and most interested non-science major we have attempted to create an undergraduate research component designed for non-science majors.

This begins by recruitment during the first semester of a one-year sequence. Students who express an interest and demonstrate superior ability are invited to participate in a special lab section during the second semester. These students conduct research projects under the supervision of both faculty and undergraduate teaching assistants. Generally, these students complete a short research project during the course of the Spring semester and present their results at a School wide poster session during the last week of classes.

Acknowledgments We would like to thank the many students who have worked with us over the years. Particular thanks goes to Georgia Richardson, Rose Forsythe, Allan Espano, Lars Omberg, Ron White, and Brittany Camper. This work has been supported in part by NSF Grants AST86-16362, AST91-15114, and USE-9156184 to the College of Charleston. Additional funding has been provided by the College of Charleston Research and Development Committee and South Carolina Space Grant.

Bridging the Cultural Gap with Science

Pam Eastlick
Planetarium Coordinator
University of Guam
UOG Station
Mangilao, Guam
96923

E-mail: pameastl@uog9.uog.edu

The loss of traditional culture and values is of increasing concern to much of the world's population. The small island states of the Western Pacific are no exception to this disturbing trend. This paper presents several ideas and teaching methods to integrate the teaching of 'Western' science (specifically Astronomy) with more traditional concepts and materials.

The island nations of the central and western Pacific Ocean are in a state of cultural flux. Traditional values are giving way under the influence of 27 cable TV channels and other trappings of 'Western civilization'. The suicide rate in Micronesia among young men is one of the highest in the world, a fact that is usually attributed to an inability to reconcile the new ways with the old. Although some of the islands still cling to traditional ways (notably Yap State) most of the island states of the Federated States of Micronesia are a clashing mix of throw-net fishing and outboards, traditional music and boomboxes.

The Federated States are nations of the young. Three-quarters of the people in Pohnpei State are under the age of twenty-five and half the island's inhabitants are under the age of fifteen. Teaching these children and equipping them for their rapidly changing world is one of the major challenges facing the governments of the Federated States. There are few school supplies and most of Pohnpei's teachers have a high school education or above. Many have two-year degrees. The Pohnpeian government holds many workshops to help upgrade teachers' skills and I presented ***Stars Over Pohnpei: A Heavenly Workshop*** to island teachers, using the skills and hands-on activities I acquired in the AASTRA (American Astronomical Society Teacher Resource Agent) program of the American Astronomical Society (AAS).

Stars Over Pohnpei was presented over a five-day period for three hours each morning with optional sessions of observing and stargazing at night. A total of fifteen activities were presented. The activities were chosen with particular care to be relevant in an island context. All activities were based on concrete concepts and easily observable phenomena. No black holes, Big Bang theories or objects visible only with a telescope, were discussed. The activities were also chosen with an eye toward the integration of 'Western Science' with more traditional materials and concepts. The workshop was planned and the activities chosen with the help and advice of Mr. Mariano Mesngon, a Chamorro (native Guamanian) employee of the University of Guam. He advised choosing activities that featured materials that were easily obtainable on the islands or had easily obtainable substitutes.

For instance, the materials required for the activity *Make Your Own Quadrant* include a drinking straw, string, a hexagonal nut or other small weight and the quadrant template. Mr. Mesngon pointed out that bamboo or a papaya stem could replace the straw, a shell or small rock could replace the nut, and traditionally-made string from coconut fibers or the bark of the sea hibiscus (*Hibiscus tiliaceus*) could replace the purchased variety. Although the 'Western materials' were provided to the workshop participants (courtesy of the Pohnpeian government who paid for the workshop supplies) we also made a quadrant using 'traditional materials'. It worked just as well as the 'Western' version. The moon-phase activity requires a small ball. The activity instructions call for a Styrofoam ball or a Ping-Pong ball, neither of which are readily available on Pohnpei. However, several trees grow there that have large round seeds that work just as well. Alhough I took Styrofoam balls for this activity, the teachers told me they wished that I had brought ping pong balls since they are about $3.00 each on Pohnpei!

The workshop participants really enjoyed the activity *Create Your Own Constellation*. It involves dropping seven or eight 'stick-on' stars onto a sheet of construction paper; sticking them down where they land, and then drawing lines between them to make a constellation-like picture. The teachers were told that if 'stick-on' stars were not available, the children could cut out their own stars and glue them to the paper, using traditional paste made from taro starch. After the constellation picture is drawn, the participant then makes up a 'constellation story' about the picture they have just made. The Pohnpeian people still have rich oral traditions and storytelling is a favorite pastime. The participants made up funny stories, tragic stories and epic sagas. One teacher even used traditional Pohnpeian story characters in his 'constellation tale'.

Much of the traditional star lore of the island states has never been quantified. Although there has been much scholarly and popular interest in the navigation stars and sailing techniques of the island's inhabitants, many of the island cultures have an extensive body of star lore associated with farming and fishing that has never been written down. Traditional stories are also told about stars and star patterns.

In an attempt to begin the process of quantification of some of this untapped knowledge, I wrote an activity called *Traditional Skies*. *Traditional Skies* features a blend of traditional and modern ideas and techniques. The activity is based on two sets of seasonal whole-sky star maps. These sky charts are computer-generated for the latitude of the workshop location. The first set of four (spring, summer autumn and winter) has some prominent 'Western' constellations drawn in, such as the Big Dipper and Orion the Hunter. The brighter stars are labeled with their official names. The second set of sky charts is identical to the first, but unlabeled so that only the stars are visible. The first set with its 'Western' constellations and names is used to help the participants orient this unlabeled 'spatter of dots' to the stars and star patterns in the real night sky. The unlabeled set of sky charts can be used in two ways. For the lower grades, the instructor can label the maps with traditional star names and draw in traditional constellations, if this knowledge still survives. In the upper grades, these maps can be sent home with the students so that their elders can teach them the traditional star lore. This activity was extremely well received by the teachers in the workshop and most of them listed it in the workshop evaluation as **the** activity they would definitely use. Although Pohnpeian star lore has traditionally resided with a limited number of community elders, there is a growing awareness that such knowledge should be more widely disseminated to prevent its loss.

Stars over Pohnpei: A Heavenly Workshop was deemed a great success by all who participated in it. The activity *Traditional Skies* has been requested by several student teachers from other islands in the Federated States as an aid in quantifying their island star lore. Preserving that traditional star lore and presenting scientific concepts in a culturally acceptable manner are two of my primary objectives and I hope that ***Stars over Pohnpei: A Heavenly Workshop*** is only the first of many astronomy workshops that will help me achieve these goals.

Periodical Publications on Astronomy for Children

Julieta Fierro
Instituto de Astronomia, UNAM
Apartado Postal 70-264
C.P. 04510
D.F. Mexico

E-mail julieta@astroscu.unam.mx

A way to convey the latest astronomical discoveries and their basic facts, to a large population, is with periodical contributions for children in magazines and newspapers. One of the advantages of simple publications is that they are easier for an adult to understand, especially if he is from a developing country.

We have been writing small articles for children in one of the major newspapers in Mexico, Excelsior. The articles feature general astronomy topics, news, special events, history and Mexican science. Once in a while experiments are suggested, like building a volcano with lava made with dyed vinegar and Alka-Seltzer and bread crumb ashes, or a planetarium out of a shoe box. The articles are usually half a page long so children feel encouraged to follow them. If one gets used to reading a certain newpaper or magazine at an early age it can become a lifelong habit.

As mentioned previously, sometimes these articles are read by parents and instructors; the latter convey the information to children (we have this information from people that write to the newspaper).

These newspaper astronomy articles are part of a page for children, with different topics including human rights and stories. For the second anniversary a public gathering was organized, where children could meet the authors; it included a science fair, workshops and a musical play; it was very successful.

Another way we have published astronomy articles for children is in specialized magazines on science. In Mexico "Chispa" is the leading one; it carries monthly articles on astronomy. This magazine covers one theme per number, so when it is, let's say, about water, language, or children's rights, topics on "Europa and Uranos", "Extraterrestrial Communication" and "Who the Moon belongs to" are addressed.

One of the problems of science popularization for children in Mexico is that projects for children are easily eliminated. Only three newspapers in the country carry special educational pages for children. XERIN, the only radio station for children, was closed two years ago. "Chispa" has to be government sponsored in order to survive. It has serious distribution problems and has very few subscriptions, mainly to public libraries (300 for the whole country).

We feel it is extremely important for developing countries to encourage children to read at an early age. Simple, interesting and challenging articles can be a way to capture their mind. So the more high quality written material that is available for them, the easier it will be to have them learn how to enjoy and use knowledge and how to self educate themselves for the rest of their lives.

Project ASTRO: Partnerships Between Astronomers and Teachers

Andrew Fraknoi, Jessica Richter, Scott Hildreth
Astronomical Society of the Pacific
390 Ashton Ave.
San Francisco, CA
USA 94112

E-mail: fraknoi@admin.fhda.edu

We want to report on an on-going two-year pilot program (supported by the Informal Science Education Division of the National Science Foundation) that brings professional and amateur astronomers together as partners with 4th through 9th grade teachers to improve the teaching of astronomy in our schools. Over 120 partners are participating in this experiment at 45 sites in the state of California.

The Project

While many professional and amateur astronomers make one-time visits to a class, this project is designed to set up a *continuing relationship* between the astronomer and the teacher. We trained astronomers and teachers together at special workshops: our evaluation showed that proper preparation of both partners was the key to a successful experience. We focused on basic, age-appropriate astronomy activities on such topics as the phases of the Moon, similarities and differences between planetary features, and making a constellation finder (not necessarily what each astronomer or teacher initially wanted to talk about, but those topics and approaches that the students of this age are ready to hear.)

We provided the partners with a prototype loose-leaf notebook of detailed activities, resource lists, and teaching suggestions, as well as a range of audio-visual materials. For some of the suggested activities, we also provided hands-on materials, such as a kit for making the star finder or styrofoam balls for demonstrating moon phases.

We focused on grades 4 through 9, because it is in these grades that children are old enough to understand astronomical concepts, but still not so old that their negative attitudes toward science have hardened. In setting up the project, we did not re-invent the wheel, but instead: 1) reviewed research on the teaching and learning of astronomy; 2) evaluated (using our staff and a panel of mentor teachers) hundreds of hands-on activities created at the A.S.P., the Lawrence Hall of Science, the Pacific Science Center, and other centers for astronomy education around the country; and 3) reviewed hundreds of children's books, videos, and slide packages for appropriateness for these grade levels.

The Notebook

A key product from Project ASTRO is the revised, updated, and expanded resource and activity notebook, entitled *The Universe at Your Fingertips*. In its final form, the 813-page notebook includes over 90 hands-on activities (including a number that are interdisciplinary and involve looking at social issues, like the role of women in astronomy). There are also dozens of resource guides, including introductions to astronomical software, observing guides, organizations, reading materials, and vocabulary. The notebook was extensively reviewed by both teachers and astronomers and their suggestions are incorporated in the final version, which is being distributed through the mail-order catalog of the Astronomical Society of the Pacific (address above.) The Society expects to keep the notebook in print and to issue periodic updates and additions.

Project Results

Formative and summative evaluation for the project was provided by an outside evaluator, Science Learning Inc. Their findings, which provide an independent assessment of the project. include the following:

1. The participants committed to at least 4 visits to the classroom, but about 60% made more than this number. Some of the astronomers made more than 10 visits to their school and got involved in a number of projects in and out of the classroom.
2. 91% of the teachers were teaching more astronomy and 48% more science overall as a result of the project.
3. Amateurs were as successful, and in some cases more successful than professionals at this level. Amateurs were often able to organize evening star parties for the class or the school which were very effective in drawing families and school administrators into the project.
4. 76% of the astronomer partners got other astronomers involved, usually local amateurs for a star party.
5. 64% of the partnerships included families of students in the activities of the project. This was one of our major aims, since we believe school learning is most effective when supported by family involvement.

As an example of the sorts of activities the most ambitious partnerships undertook, in Castro Valley, fifth grade students made a town sized scale model of the solar system, with different planetary globes and explanations located in a bank, shops, and the town library. Community members then helped publish a tour guide to the Castro Valley Solar System.

Two of the partnerships (in Santa Barbara and near Stockton) have "spun off" their own mini-Project ASTRO – where the existing partners are training new teachers and astronomers on setting up their own partnerships. We have had inquiries about the project from around the country and are keeping a database of interested scientists and educators.

Future Plans

NASA has given the A.S.P. a transition grant to help disseminate the results of the pilot project and to allow *The Universe at Your Fingertips* notebook to be distributed to all the NASA teacher resource centers. We are initiating a second round of partnerships in the San Francisco Bay Area, with the help of a coalition of astronomy and science education institutions. NSF has funded the expansion of the program to five other areas of the country, forming coalitions of astronomical and educational institutions to run each local Project – with assistance and support from the national Project ASTRO office. A second strand of the expanded project will offer training at A.A.S. and A.S.P. meetings for astronomers interested in doing such partnerships in other areas on their own.

Interdisciplinary Approaches to Astronomy Education

Andrew Fraknoi
Foothill College
12345 El Monte Rd.
Los Altos Hills, CA
USA 94022

E-mail: fraknoi@admin.fhda.edu

These days, when both our high school and college science classes are filled with many students whose interests lie in other fields, making occasional connections between astronomy and these other areas can be an effective way of drawing them into the course. Luckily, there is now a rich literature connecting astronomy with such fields as literature, music, art, psychology, politics, the environment, folklore, archaeology, law, history, and even stamp collecting.

In my college-level introductory astronomy class, I assign a brief paper in which each student is asked to report on some connection between astronomy and another subject, preferably one that is of interest to the student. Among the topics to which students have managed to connect astronomy are surfing (the tides), religion (Galileo and the current Pope), and games (pinball machines that use astronomical images). My new book *Cosmos in the Classroom* (1995, Saunders College Publishing) lists detailed references to some of these and many other interdisciplinary topics.

One of the most accessible areas for students (or instructors) to investigate is science fiction. If your last exposure to science fiction has been some fantasy film like *E.T.* or *Star Wars,* or some of the *Flash Gordon*-type "space operas", you may have a pleasant surprise in store. Not only do a number of science fiction authors now have degrees in physical science, but a number of astronomers and physicists have started writing science fiction as well. The most famous novel in this genre is Carl Sagan's *Contact,* but some other astronomer-physicist authors include: Fred Hoyle, Donald Clayton, John Gribbin, Robert Forward, Paul Davies, Charles Sheffield, Gregory Benford, John C. Wheeler, and David Brin.

Many so-called "hard science fiction" stories today use ideas from modern science and apply or extend them in very clever ways. Some introductory astronomy students are already avid science fiction fans, and enjoy analyzing a science fiction story from a scientific perspective, while others may get interested in science fiction through your course. You can discuss black hole event horizons in the abstract, but a novel like Fred Pohl's *Gateway* (1977, Ballantine), in which the protagonist has to make a split-second decision which sends the great love of his life into an event horizon, can make the concept come alive (pardon the expression) for the reader.

A more complete interdisciplinary resource list was distributed at the conference, and can be obtained by writing to the author at: A.S.P., 390 Ashton Ave., San Francisco, CA, USA 94112. I would be grateful if the request could include a legal-size self-addressed envelope with two first class stamps.

Some Astronomers & Physicists Who Have Written or Write Science Fiction

1. Scientists of the Past

- Johannes Kepler wrote what many consider the first science fiction story, *Somnium.*
- Simon Newcomb (U.S. Naval Obs.): wrote a novel, *His Wisdom the Defender* (1900, Harper)
- Robert Richardson (Mt. Wilson Obs.): wrote short stories under the pseudonym of Philip Latham.
- Leo Szilard (U. of Chicago): wrote short stories

2. Living Scientists

- Doug Beason (Air Force Plasma Phys. Lab): novels such as *Lifeline*
- Gregory Benford (U. of California, Irvine): stories & novels such as *Timescape* and *Furious Gulf*
- David Brin (has a PhD in astrophys.): stories & novels such as *The Uplift War* or *Earth*
- Marcus Chown (*New Scientist* magazine): novel, *Double Planet*
- Donald Clayton (Clemson U.): novel, *The Joshua Factor*
- John Cramer (U. of Washington): novel, *Twistor*
- Paul Davies (U. of Adelaide): novel, *Fireball*
- Robert Forward (Hughes Res. Labs): novels such as *Dragon's Egg* and *Starquake*
- John Gribbin (*New Scientist* magazine): novel, *Double Planet*
- William Hartmann (U. of Arizona): a short story in *The Planets* (edited by Byron Preiss)
- Fred Hoyle (Cambridge U.): novels such as *The Black Cloud* and *October the First is Too Late*
- Eric Kotani (pseudonym of a senior NASA scientist): novels such as *Supernova*
- Thomas McDonough (Caltech): novels such as *The Missing Matter*
- William Rossow (Goddard Inst.): novels including *Lear's Daughters*
- Carl Sagan (Cornell U.): novel, *Contact*
- Stanley Schmidt (taught astronomy, now SF editor/author): novels, such as *Newton and the Quasi-Apple*
- Charles Sheffield (Earth Satellite Corp.): stories and novels such as *Between the Strokes of Night*
- J. Craig Wheeler (U. of Texas): novel, *The Krone Experiment*

3. Other Science Fiction Writers with Strong Background in Science

- Poul Anderson: many stories and novels, including *Tau Zero*
- Arthur Clarke: many stories and novels, such as *The Hammer of God* and *2010*
- Hal Clement (pseudonym for Harry Stubbs, a high school science teacher): stories and novels such as *Mission of Gravity*
- Larry Niven: many stories and novels, including *World Out of Time* and *World of Ptaavs*
- John Varley: many stories and novels, including *The Ophiuchi Hotline*.

An Astronomy Course for Preservice Elementary Teachers

Linda M. French
Wheelock College/Center for Astrophysics
Boston, MA
USA 02215

E-mail: linda@annie.wellesley.edu

Prospective early childhood and elementary teachers take few science courses and most receive little or no explicit instruction in what constitutes a good science curriculum for the children they will teach in the future. At Wheelock College, a liberal arts college with a tradition of preparing students to work with children, a team of scientists, mathematicians, and educators is designing a program to integrate science and math content courses, methods courses, and the student teaching experience for preservice elementary teachers. Our goal is to make clearer the connections and the distinctions between good science teaching at the college level and in the elementary classroom.

Science courses taken to fulfill distribution requirements for teachers must meet several different (and often conflicting) criteria:

1. The science content should be challenging, timely, and intellectually rewarding. These students will be the first science teachers of young children, and they deserve science courses which are intellectually rich and stimulating.
2. The methods of pedagogy should closely model those which we think desirable for teachers to follow in their own classrooms. Research has shown that all of us teach as we were taught. If elementary teachers have learned about science by memorizing facts and definitions out of books for multiple choice tests, it is likely that they will teach their students in the same way. If we wish them to teach in a different manner, we ourselves must model that way of teaching.
3. The topics and organizing themes chosen for such courses should be wide-ranging and fundamental. Because a survey course in astronomy may be a prospective teacher's only course in the physical sciences, fundamental physical principles should be a key part of the curriculum. Examples of relevant big ideas include the nature of light, density as a diagnostic property of matter, and the law of conservation of energy.

The design of a science course to meet such diverse criteria necessarily become a process of compromise; of striking a delicate balance. In keeping with the Professional Development Standards described in the working draft of the NAS Science Education Standards, students are challenged to analyze their own learning and to reflect on their understanding. Students in many courses keep journals, compile portfolios, and meet for extra class sessions to reflect on what they have learned and what has been confusing for them. Interesting, ongoing discussions have developed around questions such as "How could you teach about density to third graders?" and "Is this topic appropriate for elementary children?" Students appreciate our hands-on labs and report that these will help them in their teaching, but many have never before thought about the differences between the learning process for adults and for children.

At Wheelock, astronomy is offered as an independent course, and important astronomical themes are woven into a two-semester, interdisciplinary science course entitled "Introduction to the Natural Sciences". The content of the astronomy course varies from year to year, but the nature of light and the astronomer's use of light to study the universe are persistent themes. Although Wheelock is in the middle of Boston and has no observatory, we frequently visit other observatories and students are required to attend at least one Friday night observing session. Students carry out two observing projects. Last semester all students observed and drew the moon over a two-week period; for the other project students had the choice of observing the time and position of sunrise or sunset, or observing and drawing Mars at one-week intervals. Students begin these projects with apprehension and some self-doubt, but many report the projects to be the most rewarding part of the course.

The astronomy curriculum is designed around ideas of fundamental significance which can, in some way, be demonstrated with manipulatives. In learning about Galileo's early telescopic observations, students plot the orbits of the Galilean satellites using the slide set from the GEMS activity The Moons of Jupiter. The law of reflection is "discovered" through a series of exercises with small mirrors; students make sketches of the path light has taken from one mirror to another and then describe the path in their own words. Image formation is explored through construction of the Project STAR telescope from pinhole tube to completed telescope. Topics of great intrinsic interest to students , such as the expansion ot the universe and black holes, are presented in an informal lecture-discussion. Students are required to read before class and take notes on points they find particularly challenging or interesting. In-class discussions and occasional debates keep students actively engaged. In their class evaluations, most students report a greater sense of accomplishment and personal achievement in this class than in their previous science classes. The loss in "coverage" of material due to the increased emphasis on hands-on activities and in-class discussion time is the necessary price of greater student involvement.

This work was supported by an NSF Teacher Preparation grant to Wheelock College and by an NSF Collaborative for Excellence in Teacher Preparation.

A Local Newspaper Column About Astronomy and the Night Sky

David B. Friend
Department of Physics and Astronomy
University of Montana
Missoula, MT
USA 59812

Tel.: (406) 243-2073
E-mail: pc_dbf@lewis.umt.edu

Abstract: A good way to communicate astronomy to the general public is to convince your local newspaper to print an astronomy column. (Every newspaper has an astrology column, so you can argue for equal time for astronomy!) I write a semiweekly to monthly column (during the summer months) called "The Big Sky", in the Missoulian of Missoula, Montana. Each article always has two parts: a description of how to find something interesting in the sky, along with some astronomical background information about the object. Good examples would be a lunar eclipse or a meteor shower.

Most newspapers in smaller cities carry no stories about astronomy, except those that they get from the national wire services. These same newspapers always have a daily horoscope, however. Astronomers who find this situation deplorable should consider approaching their local newspaper about publishing a column about astronomy. Many newspapers welcome columns from local experts about many subjects. Perhaps your newspaper would run a monthly astronomy column. This would serve several purposes: 1) It would provide a welcome contrast to the astrology columns, 2) It would give readers some useful information about the sky and astronomical objects, and 3) It would provide you, your department, and your college or university with some good public exposure.

For the last three years I have written a column about astronomy (called "The Big Sky") in the Missoulian of Missoula, Montana. The column started as a semiweekly one, but this year it has changed to a monthly column which is longer and which contains a star chart. Each column always has two parts to it: a description of how to find something interesting in the sky that week (this is greatly facilitated by the star charts that have been included this year), and some astronomical background about the object. Topics I have covered include: conjunctions of planets and bright stars, the brightnesses of the planets, meteor showers, the Milky Way galaxy, the Andromeda galaxy, the phases of the Moon, the rings of Saturn, the ecliptic and the seasons, the constellations of the zodiac, lunar and solar eclipses, the impact of Shoemaker-Levy 9 on Jupiter, the distances of the stars, and the colors and temperatures of the stars.

To better illustrate the nature of my column, what follows is a typical example (published on July 29th, 1993).

If you spend much time looking at the sky late at night, you have almost certainly seen what is commonly called a shooting star: a bright light suddenly streaking across the heavens. Have you ever wondered what these are? They are certainly not stars, though they do come from beyond the Earth. They are bits of rock and dust that have broken off asteroids or comets. These bits are typically the size of sand grains or pebbles, so they are much too small for us to see them out in space. However, if they fall into the Earth's atmosphere, friction causes them to heat up and glow, and this is what leaves the glowing streak in the sky.

Most of these meteors, as astronomers call them, are pieces of comets. Comets are huge balls of ice and carbon-rich dirt that exist mostly in the outer solar system. Some of them (like Halley's comet) have orbits that bring them very close to the Sun every few decades. When they do, the intense sunlight vaporizes and ionizes some of the ice and it leaves an enormous glowing tail stretching out away from the Sun. Many people think comets and meteors are the same thing, but they behave very differently. Comets are orbiting the Sun like planets, so they don't shoot rapidly through the sky like a meteor. They move slowly against the background stars just like planets do.

So how do bits of comet make their way into our atmosphere? As sunlight vaporizes the ice in a comet, some of the dust and dirt is also released, and this is the source of our meteors. The dust accumulates in the comet's orbit, eventually spreading out to fill the entire orbit. Some of the comets have orbits that nearly intersect the Earth's orbit, and when we pass by one of these we pass through a cloud of cometary dust. At these times we see a meteor shower. They typically last a few days, and during these times we can see many more meteors than usual. Two of the best meteor showers are the Geminids in December and the Perseids in August. The Perseid shower normally produces about one visible meteor per minute; this year, they will peak the night of August 11th.

Just last year the comet that produces the Perseid shower was seen for the first time this century. It's called Swift-Tuttle, and it takes about 130 years to orbit the Sun. The comet itself passed by our orbit last December 31st, but the Earth was over on the other side of the Sun. In August of 1991 and again in 1992, the Perseid shower was much more intense than usual, probably because the cloud of cometary dust is denser near the comet itself. For that reason, this year's shower is predicted to be better than ever. During the last two years, the shower briefly became a meteor storm, in which dozens of meteors could be seen per minute. Such a storm is likely to occur before dark in North America, but this prediction could be off by a few hours. The best time to look for the Perseids this year will be on the night of the 11th, between 11:00 (when it first gets really dark) and 1:00 (when the Moon rises). The meteors will appear to come from low in the northeast.

Constellations in the Classroom

Mary Graham
Grijalva Elementary School
Tucson Unified School District
Tucson, Arizona
USA 85746

and

Mary Muratore
Acacia Elementary School
Vail School District
Vail, Arizona
USA 85641

As part of our two current programs, **Project ARTIST** (Astronomy-Related Teacher In-Service Training) and **Project ACCESS!** (All Children Can Explore the Solar System!), scientists, educators, and classroom teachers are developing a series of thematic units relating to astronomy (see accompanying abstract by Larry Lebofsky, Thea Cañizo, and Nancy Lebofsky). One of these units is *Stars and Constellations*.

At the Astronomical Society of the Pacific meeting in College Park, Maryland, teacher/ facilitators from **Project ARTIST** and **Project ACCESS!** presented materials developed by them and by other ARTIST and ACCESS staff. They shared activities for introducing stars and constellations to elementary and middle school students.

Examples of the activities presented included:

- **Constellation viewers:** Toilet paper tubes and potato chip cans were used to create constellation viewers that can facilitate the identification of actual constellations and asterisms in the classroom before the students go out to view the "real" constellations in the night sky.
- **Picture books on myths and legends of the sky:** There is a wealth of astronomy-related literature which teachers can use to introduce astronomy topics, review and reinforce concepts, and integrate social studies, language, and art into science themes. Since sky myths and legends of the constellations are present in all cultures, students can learn stories from around the world and can appreciate the rich diversity of the different ways in which humans have viewed the heavens and the patterns we see in the skies. Some examples are:
    - Birdseye, T. *A Song of Stars*
    - Cohlene, T. *Quillworker, A Cheyenne Legend*
    - Goble, P. *Her Seven Brothers*
    - Krupp, E.C. *The Big Dipper and You*
    - Lee, J.M. *Legend of the Milky Way*
    - Monroe, J. and Williamson, R. *They Dance in the Sky*
    - Staal, J. *The New Patterns in the Sky, Myths and Legends of the Stars*
    - Vautier, G. *The Way of the Stars, Greek Legends of the Constellations*
    - Winter, J. *Follow the Drinking Gourd*
- **Constellation transformations:** Using actual star patterns, students are challenged to create their own interpretation of that pattern. For example, when

given a pattern of Ursa Major, children have transformed it into horses, giraffes, a dancing cow, and even the face of a joker from a deck of cards. Cygnus the Swan has become a crawdad, a dolphin, bow and arrow, and a sting ray. Once the drawing is done, teachers can extend the activity into the language arts by having students write stories and poetry about the new constellation they created. This activity is especially effective for promoting language acquisition in the emergent reader and the English as a Second Language student.

- **Birth and death of a star:** Students first review many of the cycles of nature – the water cycle, the life cycle of the butterfly or frog, the rock cycle, etc. They then learn that stars also are born, pass through stages of development and then die, thus providing the material for another cycle to begin. Some reproducible pages from *Adventures in Astronomy, Ranger Rick's Nature Scope*, are used in this activity. Students work in groups to create a book, mobile, poster, or any other product which shows the life cycle of the different types of stars.
- **Big Dipper perspective:** In this activity students build a three-dimensional model of the Big Dipper asterism using thread or string and beads hanging from a piece of cardboard. From the Earth the constellations and asterisms form patterns which appear two-dimensional, and all the stars seem to be equidistant from the viewer. In reality the stars' distances from the viewer can vary by hundreds or thousands of light years. This activity was included in the packet for educators distributed for National Science and Technology Week by the National Science Foundation.

For more information on these programs and activities, please contact: Dr. Larry A. Lebofsky, Lunar and Planetary Laboratory, University of Arizona, Tucson, AZ, USA 85721; (520) 621-6947; (520) 621-4933 (FAX); E-mail: lebofsky@lpl.arizona.edu

Making Planetariums and Science Museums More Accessible for People With Disabilities

Noreen Grice, Education Coordinator
Charles Hayden Planetarium
Boston Museum of Science
Boston, MA
USA 02114-1099

Tel.: (617) 589-0273
FAX: (617) 589-0454
E-mail: grice@A1.mos.org

- Planetariums and Science Museums have an opportunity and obligation to make their facilities more accessible for people with special needs. Visitors with visual, hearing or mobility impairments may need only minor accommodations to make their visit more enjoyable.
- In this paper, I will describe some of the modifications we have made in the Planetarium, Omni and Exhibits halls of the Boston Museum of Science.

The Charles Hayden Planetarium

- Braille tactile materials are available to accompany planetarium shows. The pictures are made on a Versapoint-40 Braille printer.
- Williams-Sonoma Assistive Listening Devices are available for hearing-impaired visitors who need volume amplification. Some of these units can be used with hearing aids.
- Scripts are available for all planetarium shows. Red map lights in the armrests allow visitors, if they choose, to read the scripts during the show.
- American Sign Language interpretation is offered on the second Saturday of alternating months in the Planetarium. A three-sided wooden booth with a red-bulbed gooseneck lamp is provided for the interpreter to use.
- Some aisle seat armrests can be lifted to allow easier access for people with crutches, canes, or wheelchairs.
- Special audience response units are available to visitors with mobility impairments.

The Mugar Omni Theater

- Infrared headsets are available for hearing-impaired visitors who need volume amplification.
- Infrared headsets are offered for visually-impaired visitors to hear a descriptive narration track for most films.
- American Sign Language interpretation is offered on the second Saturday during alternating months.
- Scripts for Omni films are available in the Omni Theater.

Boston Museum of Science Exhibit Halls

- Braille Floor Plans are available at the Information Desk.
- Wheelchairs may be borrowed for use in the Main Lobby.

- Recorded guided tours and cassette players, for visually and mobility impaired visitors, may be borrowed at the Information Desk.
- Some exhibit areas (New England Habitats, Human Body Discovery Space, Seeing the Unseen, Arthur D. Little Discovery Space and Computer Discovery Space) offer many hands-on components.
- Blue stickers identify services for disabled visitors: Braille materials, Assistive Listening Devices and wheel chair accessible locations.

What You Can Do

- Be sure that your Theater or Science Center is in compliance with the Americans With Disabilities Act (ADA).
- Contact your local Commission for the Blind and Commission for the Deaf for suggestions on modifications.
- Invite people with disabilities to your facility to get feedback on what you offer and what you might include in the future.
- **Be pro-active rather than reactive in making an all-inclusive environment!**

Treatment of Creationism in an Introductory Astronomy Class

Donald E. Hall
Department of Physics and Astronomy
California State University
6000 J Street
Sacramento, CA
USA 95819-6041

Study of the origin of the Solar System offers an opportunity to consider important issues in what constitutes good science, without as many emotional barriers to rationality as in the case of biological evolution. This paper considers the rationale for including a discussion of creationism in this context, outlines some material used in lectures, and suggests attitudes and frameworks that may help creationism-oriented students become more open to scientific evidence and analysis.

I. Rationale: Why give lecture time to a topic outside the usual bounds of astronomy? Is it an appropriate topic even if time is available?

A. Student backgrounds and beliefs: An appreciable fraction of typical college students have had significant religious indoctrination, often including belief in some version of creationism. These students are better served by addressing the issue than by pretending that it doesn't exist. Other students with secular outlook are nevertheless likely at some point to have opportunity to discuss these issues with believers. It would be a valuable contribution to their education if they were aware of these issues and prepared to talk intelligently about their reasons for preferring a scientific view. Good education presents these students with a model of intelligent discourse, in preference to merely dismissing or ridiculing unscientific beliefs.

B. Further purposes served: Our understanding of any concept is often enhanced by presenting a contrasting concept; so here our reasons for believing in Earth's origin in a Solar Nebula may be better appreciated by contrasting them with arguments used by creationists. Our friends the biologists are generally left with the main responsibility to deal with the general issue of creation and evolution. But because of the extra emotional charge involved there, many students may find it easier to confront questions of ultimate origins rationally in the less personal context of Astronomy.

II. Content: Some possible points of interest to be included.

A. Appreciation for multiple meanings of important terms: "Evolution" does not have a single, simple meaning. It may refer to cosmological, galactic, stellar, planetary, chemical, biological, or cultural evolution. The nature and existence of each kind of evolution deserves independent consideration of its supporting evidence. "Creationism" too has no single, simple meaning. Some creationists ask only that human life be the product of recent divine fiat and are content with what science says about everything else. [Key question: Why then is our DNA so closely related to that of the other animals?] Others insist that all living forms are the product of a recent creation, but do not care if the inanimate Earth is old. [Key question: How do you explain young fossils in old rocks?] Others believe that both life forms and surface sedimentary rocks are young, but don't challenge 4.6 billion years for the body of the Earth. [Then what about igneous rocks with very old radioactive dates overlying fossil-bearing sediments?] Others believe the entire Earth or even Solar System are included in the

Genesis 1 story and are only thousands of years old. At the extreme, a significant number sincerely believe that the entire Universe is young; if some galaxies appear to be millions of lightyears away, they say that is only because the light we receive was created in mid-flight.

B. The choices are complex, and have public-policy consequences. Many religious people, and some institutional churches, regard it as entirely satisfactory theology to say that evolution may be God's chosen way of accomplishing creation. So scientists are often allied with church people in opposing the introduction of "scientific creationism" in school curricula. Items for discussion: Is "scientific creationism" a self-contradictory term? Who gets to decide what is or is not properly labeled as science?

C. Scientific methods for dating Earth's history raise several interesting points: The basic method of radioactive isotope dating should be explained, along with reasons for going from Earth to Moon to meteorites to get the full 4.6 billion year figure. As a way of avoiding many technical details, a very powerful approach is to scan a table of isotope abundances, asking whether those with half-lives in the range from 10,000 to 100,000,000 yr are found in nature; their absence is strong evidence for a very old Earth. Slowing the Earth's spin by tidal friction comes into play, with fossil records of eras with 400 or more days per year providing independent confirmation of Earth's great age, making it much more difficult for creationists who claim radioisotope ages are based on false assumptions.

D. The doctrine of "appearance of age" is used by creationists to superficially reconcile their beliefs with observed evidence. The classic case is a newly-created Adam in the Garden of Eden, seemingly 20 years old even though created only 5 minutes ago, for which it can be argued that there would be no point in a benevolent God creating a helpless infant. The reasonableness of the doctrine should be tested by asking why God should have carefully arranged isotope ratios in a newly-created rock to fool us into thinking it's a billion years old, if that has no recognizable connection to its function as a rock. At the extreme, how many students will be ready to believe after thinking about it that Supernova 1987A never really happened and that all we have seen is a lightshow created in midflight? This doctrine goes to the heart of the scientific enterprise, severely challenging our motivation for ever bothering to make any observations.

III. Framework: In what terms may we find a reasonable relationship between science and religion?

A. Classification of possibilities: Following J.B. Miller, I like to contrast several attitudes. (1) Dualism claims "no conflict" in the trivial sense that neither really has anything to say to the other. (2) Imperialism comes in two brands, (2a) Religious Fundamentalism [religion can and does reveal scientific truths, and its insight is always superior to scientific "evidence" – but then why bother to do science at all?] and (2b) Scientific Positivism, or in extreme form "scientism". (3) Interactionism says that both may have meaningful contributions but that science has primary responsibility for interpreting natural phenomena and religion should not place prior limits on what science may find.

The Educational Activities of the Astronomical Society of the Pacific

Robert J. Havlen
Astronomical Society of the Pacific
390 Ashton Avenue
San Francisco, CA
USA 94112

E-mail: rhavlen@stars.sfsu.edu

The Astronomical Society of the Pacific (ASP) is an international scientific and educational organization, founded in 1889, that brings together professional astronomers, educators at all levels, active amateur astronomers, and interested lay people. Its members today live in all 50 states and over 60 countries; thus its name is mostly a reminder of its origins in California over 100 years ago. The Society issues a refereed technical journal, publishes a series of conference proceedings, and holds technical symposia for its professional members. But it has an equally active program of educational projects and dissemination activities. These include:

- **Summer Workshops On Teaching Astronomy in Grades 3-12:** Held each summer at a different university around the country, these workshops, generally attended by 150-200 teachers, provide participants with hands-on activities, exemplary teaching resources, nontechnical introductions to current developments in astronomy, and a chance to network with teachers from other areas.
- **Newsletter On Teaching Astronomy:** The *Universe in the Classroom*, a free quarterly newsletter (supported with the assistance of several other astronomical societies) goes to more than 13,000 teachers in grades 3-12 around the US and Canada, focusing on information and activities that can be put to immediate use in the classroom. The newsletter is translated into several other languages and further distributed by educational institutions around the world.
- **Information Packets:** The Society produces and distributes information and resource packets on such topics as: *An Introduction to Black Holes, Debunking Astrology, Astronomy Education, Astronomy as a Hobby, Software for Astronomy, A Basic Astronomy Library,* etc. These materials are available in single copies in quantity, or for reproduction in non-profit educational publications and kits.
- **Project ASTRO:** With the support of the NSF, the ASP has organized a pilot program to bring visiting astronomers (professional and amateur) into 4th-9th grade classrooms. The astronomers don't just give a one-time talk, but (after attending a training workshop, and receiving a wide range of hands-on, age-appropriate activities) set up a genuine partnership with a teacher – involving regular visits, after-school activities, etc. Expansion to other sites around the country will begin in 1996.
- **The Resource Notebook For Teaching Astronomy:** One outgrowth of the Project ASTRO pilot is an extensive loose-leaf notebook of astronomy activities, resource lists, teaching suggestions, background material, and observing guides. *The Universe at your Fingertips* is available through the ASP catalog.
- **Brennan Award:** The Society recently initiated an award---called the Thomas J. Brennan Award – to recognize excellence in the teaching of astronomy at the high school level in North America. The winners have demonstrated exceptional commitment to classroom or planetarium education, as well as the training of other teachers.

Science On-Line: Partnership Approach for the Creation of Internet-based Classroom Resources

Isabel Hawkins
Center for EUV Astrophysics
U.C. Berkeley
2150 Kittredge Street
Berkeley, CA
USA 94720-5030

Tel.: (510) 643-5662
FAX: (510) 643-5660
E-mail: isabelh@cea.berkeley.edu

Robyn Battle
(U.C. Berkeley Graduate School of Education and Center for EUV Astrophysics)

Carol Christian, Roger Malina
(U.C. Berkeley, Center for EUV Astrophysics)

With funding from NASA's Astrophysics Division, the Center for Extreme Ultraviolet Astrophysics (CEA) at U.C. Berkeley carried out an eight-month pilot project, Science On-Line, which established a collaboration among formal and informal science centers, each of them having unique and complementary assets. The SOL project has investigated methods by which collaborations among teachers, other professional educators, and scientists can help the access of scientific data by teachers and students, taking advantage of a more technologically-driven, research-based model of learning. This investigation is developing case studies for effective ways of supporting teachers in the contextually relevant use of Internet technology in the classroom environment.

The introduction of technical innovations into the classroom has the potential to produce a dramatic redefinition of traditional teacher-student instruction (Park and Hannafin, 1993). In addition, the advent of platform independent World Wide Web (WWW) browsers makes possible the creation of more relevant curriculum that support models of learning which more closely reflect practices found in real workplace environments (Becker, 1972; Lave and Wenger, 1991; Collins, Brown, and Newman, 1989). Similarly, Park and Hannafin (1993) suggest that students and teachers benefit by learning how to use technological tools found in scientific, academic, and business workplaces. Use of technology in the classroom can foster an environment that more closely reflects the processes scientists use in doing research (Linn, diSessa, Pea and Songer, 1994). For instance, scientists rely on technological tools to model, analyze, and ultimately store data. Linn, diSessa, Pea and Songer (1994) suggest that technological tools be introduced to students from the earliest years because of their ability to facilitate scientific modeling, scientific collaborations, and electronic communications in the classroom.

Our investigation addresses the hypothesis that the transition of scientific data and research practices from the workplace to the classroom can be facilitated by the joint creation of curriculum by teams of education experts, scientists, and teachers. Our strategy for evaluating our partnership approach engaged the expertise available in scientific research institutions, centers of informal science learning, and schools. Our project, "Science On-Line (SOL)," was founded on the collaboration among U.C. Berkeley's Center for Extreme Ultraviolet Astrophysics, Lawrence Hall of

Science and U.C. Museum of Paleontology, with San Francisco's Exploratorium, and Chicago's Adler Planetarium for the purpose of developing on-line Earth and Space Science lesson plans for grades 6 through 12. The SOL Project serves as the basis for the UCB/CEA's NASA-funded Science Information Infrastructure Program. The Science Information Infrastructure Education Program includes the following additional partners: Smithsonian Astrophysical Observatory, National Air and Space Museum, Boston Museum of Science, New York Hall of Science, and Science Museum of Virginia.

The SOL lessons and information on the project methodology can be found on the World Wide Web at the URL: http://www.cea.berkeley.edu/Education. In addition, the SOL resources will be accessible through all the partners' WWW Home Pages, such as the Exploratorium's Home Page at the URL: http://www. exploratorium.edu.

As part of the SOL project, twelve teachers from Oakland (CA), Davis (CA), and Chicago (IL) area schools received training in the use of Internet tools and were supported by scientists and professional facilitators in the creation of six on-line Earth and Space Science lesson plans accessible through the WWW via Mosaic-type browsers. The lessons deal with various topics such as comparing Martian and Earth weather, using Internet images of solar system objects to calculate their rotation rates, and accessing Internet-based data to investigate earthquakes. An important aspect of the SOL project has been a lesson plan pilot-testing component which was carried out by four additional teachers from the San Francisco Bay Area with their own students. The lesson plans were subsequently modified in response to suggested improvements, increasing their effectiveness in different types of classroom settings.

An important part of the SOL program was an evaluation process that served both formative and summative purposes. The formative evaluation process channeled feedback from SOL participants (maintaining the anonymity of the source) to SOL teachers, organizers, and facilitators. This process facilitated responsiveness to changes or problems that arose during the course of the program. The summative aspect of the evaluation provided descriptive material and data that allowed parties to assess the success of the program as well as provided valuable insights into the SOL process for future SOL-like programs. There was also an evaluation of the SOL Internet-based lesson plans. The Exploratorium enlisted the efforts of four teachers for the purpose of testing the six lesson plans providing a formal evaluation to both the lesson developers and the SOL project organizers. Evaluations concerning the impact, organization and success of Science On-Line were carried out throughout the lifetime of the project. Data were collected through a series of methods, specifically, informal and formal discussions with participants, questionnaires and feedback forms, informal and formal reports from both teacher participants and organizer/facilitator formal and informal reports. Several classroom pilot testing sessions have been videotaped for analysis and will be presented in a "case study" format in subsequent publications.

A number of common difficulties and successes emerged for the teacher-developers during the lesson plan development stage. Common difficulties developer/facilitator teams encountered were: learning HTML/formatting in HTML; limiting lesson plan content and topics to meet deadline; finding images and materials on the Internet; connectivity difficulties or slow connectivity; lack of time; and, not enough robust connections to scientific community. Common successes include: finding the Internet meaningful in education; learning significantly enough to give advice to other teachers in their school; feeling that the lesson plan development task was relevant; feeling that their students were interested in the lesson they were developing; and, wanting to continue with development work.

The pilot testing phase results suggest that although the development process produced lessons and methodology relevant to those teachers who did the developing of the lessons, testing revealed that some of the developers' goals lacked relevancy in other teachers' practices in their own classrooms. The lessons did offer to user-teachers valuable Internet sources, Internet images, tools, content, format, and lesson ideas. However, results from this pilot project further suggest that a key to greater success for these lessons would be more user adaptability. That is, to produce lessons relevant to each teacher, as well as to involve user-teachers in active learning on the Internet, there must be more avenues for a teacher at the receiving end to create his/her own ideas in a manner similar to that of the developers.

One suggestion for increasing "user adaptability" is the creation of multiple level resource units. These resource units would include integrated sections varying in degree of formalization. Each section would be presented as a "coherent idea" subunit. Park and Hannafin (1993) emphasize the need for educational technology to be organized by contextually and conceptually cohesive themes. For instance, one section might include a page dedicated to illustrating and highlighting the use and understanding of the featured lesson Internet "tool" (interactive science-based simulations and tools, comparative image set, etc.). The context for using this Internet based tool would remain less specific than the lesson context. This will allow user-teachers to adapt the tool for use in their classrooms without the need to extract it from the more formalized lesson. The more formalized lesson would serve as one example of how to implement this tool. Teachers would be encouraged and instructed on how to participate in the development process from their school sites or at home. This might involve comprehensive Internet posted instructions for teachers explaining how to create their own lesson with the use of Mosaic and the WWW. This design would enable every user to participate in the self-directed, active process of adapting the resource unit into relevant lesson activities for the classroom. The resources presented in the SOL WWW server include examples of such resource units along with the developed lessons and activities for the classroom.

Acknowledgement: This work has been funded by NASA Astrophysics Division Grant NAGW-4174 to UCB/CEA.

References

Becker, H.S. *American Behavioral Scientist* 16 (Sept. - Oct 1972): 85-105. Sage Publications.

Collins, A., Brown, J.S. and Newman, S. (1989). Cognitive apprenticeship: teaching the crafts of reading, writing, and mathematics. In L.B. Resnick (Ed.), *Knowing, Learning, and Instruction*, (453-494). Hillsdale, NJ: Erlbaum.

Lave, J. and Wenger, E. (1991) *Situated Learning: Legitimate Peripheral Participation*. Cambridge: Cambridge University Press.

Linn, M., diSessa, A., Pea, R., and Songer, N. (1994) Can Research on Science Learning and Instruction Inform Standards for Science Education? *Journal of Science Education and Technology*.

Park, I. and Hannafin, M. (1993) Empirically-Based Guidelines for Design of Interactive Multimedia. *Educational Technology Research and Development* 41(3), 63-85. Association for Educational Communications and Technology.

Education Programs of the American Astronomical Society

Mary Kay Hemenway
AAS Education Officer
Astronomy Department
University of Texas
Austin, TX
USA 78712-1083

E-mail: aas@astro.as.utexas.edu

Abstract: The AAS engages in educational outreach at many levels. These include the Shapley Visiting Lectureship program to colleges, the Research Experience for Undergraduates program for college students, the Bok Awards for high school students at the International Science and Engineering Fair, the career brochure for high-school and college students, the AAS Teacher Resource Agent program and the Teachers' Days at AAS meetings for elementary/secondary school teachers. In addition, AAS members are supported through the Working Group on Astronomy Education and special sessions at AAS meetings.

Support for the REU and AASTRA programs by NSF and the Teachers' Day by NASA is acknowledged.

The American Astronomical Society (AAS) is the major organization of professional astronomers in the United States, Canada, and Mexico. The basic objective of the AAS is to promote the advancement of astronomy and closely related branches of science. The membership of approximately 6500 includes astronomers, physicists, mathematicians, historians of science, science teachers, geologists, and engineers whose interests lie within the broad spectrum of subject matter now comprising contemporary astronomy. The Society produces and/or participates in a broad range of programs to involve students and teachers, as well as the general public, in the excitement of modern astronomy.

The Harlow Shapley Visiting Lectureship program provides two-day visits by professional astronomers to college campuses, bringing the excitement of modern astronomy to the students and general public. A key feature of every visit is a free public lecture for the community. Lecturers also make presentations to a variety of classes at the college, visit with the college administrators, and often make outreach visits to local schools or planetariums. The level of the talks vary depending upon the needs of the host institution. Forty to eighty colleges are visited throughout Canada and the US every year. Over 80 astronomers are designated Shapley Lecturers each year. Program costs are provided by the Harlow Shapley Endowment fund of the American Astronomical Society. Host institutions are requested to make a financial contribution in support of the visit. For further information contact: A.G. Davis Philip, Shapley Program Director, 1125 Oxford Place, Schenectady, NY, USA 12308, E-mail: shapley@gar.union.edu.

Since 1994, the AAS Teacher Resource Agent (AASTRA) program has provided summer institutes and follow-up activities for 75 teachers per year. These K-12 teachers are prepared to become astronomy resource agents in their geographic regions. The objectives of the program are to enhance the teaching of astronomy in elementary and secondary schools, to provide astronomy workshops to elementary and secondary school teachers using hands-on astronomy activities, to encourage the professionalism of the participants, and to increase interactions among professional astronomers, science educators, and elementary/secondary school teachers. The summer institutes are conducted at Northern Arizona University, Loyola University of Chicago, and University of Maryland at College Park. All agents become associate members of the AAS as part of their participation. The AASTRA program is supported by the National Science Foundation as a teacher enhancement project. For further information, contact the AAS Education Office.

Mentoring of undergraduates by astronomers is provided in the AAS Research Experience for Undergraduates (REU) program. The student receives a stipend for either academic year or summer and travel funds for research. The mentor and student receive travel funds to attend an AAS meeting where they present the results of their research. Since Spring 1992, 35 awards have been made to mentors at 31 different institutions. Awards are made on the basis of peer review of short research proposals to the AAS/REU committee. The AAS/REU program is supported by the National Science Foundation. For further information contact the AAS Executive Office, 2000 Florida Avenue, NW Suite 400, Washington, DC 20009.

AAS/ REU Colleges

Colgate University
College of Charleston
Fisk University
Georgia Inst. of Technology
Gettysburg College
Millikin University
North Carolina State University
Pennsylvania State University
Pomona College
Rensselaer Polytechnic Inst.
Sam Houston State University
Sonoma State University
Southwest Research Institute
St. Cloud State University
St. John Fisher College
Swarthmore College
Trinity University
Tufts University
University of Kansas
University of Montana
University of Nebraska
University of Oklahoma
University of Pittsburgh
University. of Nebraska at Kearney
Vanderbilt University
Vassar College
Wellesley College
Wesleyan University
Western Carolina University
Wittenberg University
Yale University

The AAS and ASP jointly provide the Priscilla and Bart Bok Awards at the annual International Science and Engineering Fair for high school science fair projects in astronomy. Winners of the first and second place Bok Awards receive certificates, subscriptions to Mercury magazine, and cash awards. For further information, contact Science Service, 1719 N Street, NW, Washington, DC 20036, 202-785-2255.

The AAS Education Office produces and distributes a career brochure on astronomy. "A New Universe to Explore: Careers in Astronomy" is available upon request from the AAS Education Office, Astronomy Department, University of Texas, Austin, TX 78712-1083, aas@astro.as.utexas.edu. Individual copies are complimentary. Contact the office for prices for multiple copies.

The AAS sponsors 1.5 day meetings for teachers at some national meetings. Twenty to sixty teachers from the local region participate. The teachers receive substitute teachers' pay for one day, resource materials, and an opportunity for both special sessions of their own while attending parts of the regular meeting. Since 1991, the program has been supported by the NASA Astrophysics Division under the IDEA program. In 1995, the meeting was held in cooperation with the ASP. In addition, the Division of Planetary Sciences of the AAS sponsors similar meetings for teachers at their annual meetings.

Sites of AAS Teachers' Programs

Austin (1988)
Kansas City (1988)
Boston (1989)
Washington (1990 & 1994)
Philadelphia (1991)
Atlanta (1992)
Phoenix (1993)
Tucson (1995)

Sites of AAS Division of Planetary Sciences Teachers' Programs

Providence (1989)
Charlottesville (1990)
Palo Alto (1991)
Munich (1992)
Boulder (1993)
Bethesda (1994)

The Working Group on Astronomy Education was formed to provide interested astronomers with a forum for raising questions and gathering information. The Group sponsors education sessions at AAS meetings and publishes an electronic newsletter. For information, contact Steve Shawl at shawl@kuphsx.phsx.ukans.edu.

The Cosmology Distinction Course in New South Wales, Australia

Robert Hollow[1,2], Graeme L. White[1]
[1]Physics Department, Faculty of Science and Technology, UWS Nepean, PO Box 10 Kingswood NSW 2747 Australia
[2]Blue Mountains Grammar School,Private Mail Bag 6 Wentworth Falls NSW 2782 Australia

The Cosmology Distinction Course is an innovative subject introduced in 1994 for gifted students completing their Higher School Certificate (HSC) in New South Wales, Australia. It seeks to provide a stimulating and challenging one year course that goes beyond the normal confines of traditional matriculation subjects. Taught via distance-education mode, it includes a residential component and is provided free of charge to students from around the State. Organised into nine modules it covers historical, social and scientific aspects of Cosmology, emphasising the observational evidence and the development of the main models. Assessment is via assignments, exams and a major project. Now in its second year of operation, it currently has fourteen students enrolled.

Rationale Behind and Nature of the Cosmology Distinction Course

Distinction courses aim to encourage excellence and provide extended academic opportunities for highly gifted students. the Cosmology Distinction course provides students with a perception of the structure of the universe on the broadest scale. Students are asked to develop an understanding of the historical, social, theoretical and observational aspects of cosmology whilst being introduced to current research topics and techniques. They are expected to develop skills in evaluating data and interpreting and understanding theoretical models.

The Cosmology Distinction Course (Hollow et al. 1994) is a one-year course that can count towards a Year 12 student's Tertiary Entrance Ranking (TER) for university entrance and towards their Higher School Certificate (HSC). Along with the other Distinction Courses (Philosophy and Comparative Literature) students may have it considered for advanced standing or credit towards their undergranduate degree at the discretion of individual universities. Students completing the HSC in NSW typically study 11 to 13 units drawn from all three Key Learning Areas of English; Humanities; and Science, Mathematics and Technology. A 2-Unit course, including a Distinction Course, is generally assumed to occupy about 20% of school time per week or 120 hours per year plus a corresponding amount of time in home study.

All courses within NSW schools are developed by the Board of Studies. The Cosmology Distinction Course Committee, chaired by Dr. W.B. McAdam, an astrophysicist from University of Sydney, oversees the content, assessment and review of the course whilst students deal with the Course Coordinator, D. McKinnon from Charles Sturt University, Bathurst. The Committee comprises practising research astronomers from universities in NSW, the Board Inspector of Science and two experienced high school Physics teachers. Development began in 1992, the course first ran in 1994 with six students from across the state. Fourteen students are enrolled in 1995.

Board courses are "Outcomes" based. Objectives and Outcomes are classified into three areas: Knowledge and Understanding; Skills; and Values and Attitudes. Distinction courses differ, however, in several respects. They are taught by distance-learning mode. All course materials are provided free of charge by the

Board via Charles Sturt University who are contracted as the course provider. This material is provided in loose-format enabling students to order it as they see appropriate and to allow updates, when needed, at minimal cost.

Students cannot automatically elect to do a distinction course but must meet entrance criteria. They must sit the HSC exam in a subject at least one year ahead of their age cohort and be placed in the top 5% of the total candidature. Final selection is also dependent upon a recommendation by their school and a review by the Board of Studies. There are no prerequisites for any of the current Distinction Courses. This is both a strength and a potential weakness. The positive aspect is that it allows students in the "Humanities" to be exposed to "cutting-edge" Science presented in a non-traditional format. A drawback, however, is that the course cannot assume a common level of mathematical or scientific skill or knowledge and so must cover all relevant ideas internally.

Course Structure and Content

The course is divided into nine modules which are designed to take three weeks each to complete. Module headings are given below:

1. The Scale of the Known Universe
2. The Contents of the Universe
3. Development of Cosmological Ideas
4. Space and Time
5. Expansion and Observed Redshifts
6. Further Key Observations
7. What are the Models?
8. The Big Bang
9. Where to Now?

The text (Harrison 1981) was selected to provide an overview of the topic, supplemented by a more current work (Silk 1989) for recent developments. The guided commentary approach was adopted for writing course material. Readings are drawn from a range of sources from magazines such as *Sky and Telescope* and extracts from books. Self-assessment questions for the student to work through together with suggested answers are provided in separate book. Students are provided with e-mail access to allow feedback and prompt support.

Two residentials are included in the course. The first includes visits to the major astronomical observatories in NSW such as the Australia Telescope National Facility at Narrabri and the optical telescopes at Siding Spring Observatory including the Anglo-Australian Telescope, The four day tour also incorporates lectures and practical work. The second residential is held in Sydney over a weekend at one of the universities with a comprehensive astronomical library. Students research their major project, generally a literature review of an area of particular interest to them. Lectures and discussions as well as socialising complete the program.

Course Assessment

There are three main components of assessment for the Cosmology Course: assignments, exams and a Major Project.

Six assignments are completed by students, three of which each contribute 10% to the course mark whilst the other three are solely for feedback. Tasks range from mathematical problems to short written responses and essays.

There are two exams. The first, based on modules 1-3, is worth 10% and lasts one hour whilst the second of two hours is based on the rest of the course and is worth 30%.

The last assessment task is the Major Project (around 7000-12000 words) wihch comprises 30% of the final mark.

References

Harrison, E.R. 1981 *Cosmology, the Science of the Universe* (Cambridge: Cambridge University Press).

Hollow, R.P. et al. 1994 the Cosmology Distinction Course in NSW *Proceedings of the Astronomical Society of Australia* **11**(1) 39-43.

Silk, J. 1989 *The Big Bang* (New York: Freeman)

Teaching Astronomical Concepts Through IUE Data Analysis

Catherine L. Imhoff
Science Programs, Computer Sciences Corporation
10,000-A Aerospace Road,
Lanham-Seabrook, MD
USA 20706

E-mail: imhoff@iuegtc.gsfc.nasa.gov

and

Sarah Clemmitt
Mathematics, Science, and Computer Science Magnet
Montgomery Blair High School,
313 Wayne Avenue
Silver Spring, MD
USA 20910

E-mail: sclemmit@people.mbhs.edu

We are piloting a project to provide an active learning experience for high school students by involving them in small research activities using data from the International Ultraviolet Explorer (IUE) satellite. Three activities have been developed and used in this spring's astronomy class at Montgomery Blair High School. They are: (1) an examination of the Hubble expansion of the universe; (2) determining the density of gas in several planetary nebulae; and (3) demonstrating the decay in surface activity with time, due to magnetic spindown, in stars like our Sun.

During several classes, we discussed the IUE satellite and its instrumentation, how to use the IUE analysis software, how spectroscopy and the laws of radiation are used to determine information about astronomical objects, the origin of planetary nebulae, and the origin of the far-ultraviolet emission lines in cool stars. The group also took a field trip to Goddard Space Flight Center to the IUE and Space Telescope operations areas.

One of our goals was to give these students experience in dealing with "real" data, warts and all. IUE data filled this role very well. Another goal was to provide hands-on experience using spectroscopic analysis to do science. This gave the students a much better sense of "how we know", rather than just "what we know", in astronomy. We also used astronomy to help integrate the students' knowledge in the other sciences.

One difficulty we encountered was the lack of easily available, inexpensive spectroscopic analysis software. For our pilot, we used IDL and the IUE Data Analysis Center software used by many IUE researchers. However IDL is normally too expensive for the public schools, even when heavily discounted. We plan to address this issue in further work.

During the summer, we plan to further develop the lessons for use by other classes. We will then distribute the lessons, including the data, analysis activities, and background material over the Internet and by disk.

This work is supported by a grant to the Montgomery County Public Schools from NASA, under the Initiative to Develop Education through Astronomy (IDEA) program. Our thanks to Research Systems Inc., which loaned free copies of IDL to us for our pilot program, and to Randy Thompson, manager of the IUE Data Analysis Center, for his help in porting the software to the school's PCs.

NOAO K-12 Educational Outreach Activities

Suzanne H. Jacoby
National Optical Astronomy Observatories
NOAO Education Officer
P.O. Box 26732
Tucson, AZ
USA 85726

E-mail: sjacoby@noao.edu

Abstract: The expanded educational outreach activities of NOAO are presented and the astronomy education community encouraged to comment.

I. Introduction

The efforts of NOAO in the area of K-12 Educational Outreach are expanding. Our overall educational outreach plan includes direct classroom involvement, support of educational technology, and, if funding permits, a new program of summer workshops for teachers which includes observing time on Kitt Peak.

II. Direct Classroom Involvement

The NOAO K-12 Educational Outreach program includes on-going direct classroom involvement, funded in part by a NASA IDEA Grant titled "Active Learning Exercises in Planetary and Solar Astronomy for K-3 Students". Through this effort, we are gaining experience with the abundance of hands-on-science activities for classrooms, including those from Project ASTRO sponsored by the Astronomical Society of the Pacific. Lessons learned from the current project will help us formulate an effective educational outreach program which we expect to include these components:

- coordination with teacher ahead of time
- multiple visits by the same astronomer or NOAO tiger team
- a review or summary session for closure
- a follow-up mechanism, e.g., Science-by-E-mail, where classes with Internet capability correspond with NOAO staff members

III. Summer Teacher Workshops

Working with local educators and university professors, NOAO is seeking funding for a series of summer workshops for middle and high school teachers. The summer program would include a research experience using the observing facilities of Kitt Peak National Observatory. Students will be allowed to compete for service observing time on the 36" to extend the research experience back to the classroom. The summer workshops would also include enriched background instruction in astronomical concepts and observational techniques, instruction in the use of image processing and educational technology tools, and some hands-on resources for teaching science.

IV. Educational Technology

NOAO offers its extensive computing facilities to the science education community as an electronic home base for the dissemination of quality resources. We offer to format, maintain, and electronically distribute through the World Wide Web of the Internet quality science education materials such as brochures, lesson plans, and other printed materials. We are especially interested in serving members of the science education community without resources to do this themselves.

Our Educational Resources WWW page currently features an FAQ list of questions about being an astronomer and on-line versions of the exemplary Survival Guides for Sharing Science with Children produced by the North Carolina Museum of Life and Science.

V. Community Input Encouraged

We welcome input from participants on how the National Optical Astronomy Observatories might best serve the science education community in these and other ways. The NOAO Educational Outreach World Wide Web pages can be accessed at URL http://www.noao.edu.

Initiative to Develop Education through Astronomy (IDEA)

Anne Kinney[1]
Laura Danly[1]
Carole Rest

[1]Space Telescope Science Institute
3700 San Martin Drive
Baltimore, MD
USA 21218

E-mail: kinney@stsci.edu
E-mail: danly@stsci.edu
E-mail: crest@stsci.edu

The IDEA research grants program was developed in 1991 by NASA Headquarters' Astrophysics Division to create more opportunities for students and the public to participate directly in the excitement and rewards of space astronomy research. Today, the program is implemented by the Space Telescope Science Institute (STScI) on behalf of NASA. The IDEA program encourages research astronomers to use their talents and enthusiasm to undertake projects that promote greater mathematical, technological, and scientific literacy. A parallel goal of the IDEA grants is to introduce professional astronomers to the world of educational and outreach activities. Collaboration between partners in the professional education community, as well as links to active learning and education reform, are emphasized. Funding is available in two grant categories, up to $6,000 and $6,001 to $20,000. Funds may be used to support salary, travel, materials, stipends, etc. Funding for equipment is discouraged but allowable in exceptional circumstances.

In 1994, 49 proposals were selected for funding totaling $475,387. These proposals, from 26 states, cover a myriad range of activities reaching K-12 and college students along with the general public. Funded categories include Curriculum/Product Development ($107K), Exhibit Development ($40K), Internet Usage/Development ($32K), Multi-cultural Programs/Outreach ($68K), Public Outreach ($43K), Student Outreach ($44K), Student Research Opportunities ($35K), Teacher Resources and Training ($90K), Wilderness Outreach ($12K), and even support for this Symposium ($5K). A catalog of abstracts of proposals which received funding in 1994 can be found on the STScI World Wide Web site, or can be requested from the sources below.

To apply for an IDEA grant you must be a professional in astronomy. Strong preference is given to proposals that include the active participants of NASA-supported investigators, though astronomers at any institution are eligible. Preference is also given to proposals that include partners in the professional education community as co-creators, participants and evaluators. All proposals must have an astronomy focus and IDEA grant PIs are expected to be active participants in the educational endeavor. Each proposal should have the potential for multiplying its impact beyond its direct effect.

The 1995 IDEA announcement is available and can be obtained electronically via e-mail. The deadline for 1995 is October 31. If you would like to be included in the distribution for next year's announcement, please send email to IDEA@stsci.edu or call Carole Rest at (410) 338-4590. A copy of the announcement is also available on the World Wide Web at the following URL: http://www.stsci.edu/EPA/education.html.

The FOSTER Program: Teacher Enrichment Through Participation in NASA's Airborne Astronomy Program

D. Koch[1], C. Gillespie, Jr.[1], G. Hull[1], E. DeVore[2]

[1]NASA Ames Research Center
Moffett Field, CA
USA 94035

E-mail: koch@chandon.arc.nasa.gov

[2]SETI Institute
2035 Landings Dr.
Mountain View, CA
USA 94043

As part of NASA's effort to engage its community of research scientists, managers, engineers and support staff in an exemplary form of educational outreach, NASA's airborne astronomy program offers a unique opportunity for K-12 science teacher enrichment called **FOSTER** (Flight Opportunities for Science Teacher EnRichment). The program is based on a combination of summer workshops, curriculum supplement materials, training in Internet skills, building of partnerships between scientists and the schools, school visits by NASA aerospace educational specialists and ultimately the flying of the teachers on a research flight on NASA's C-141 Kuiper Airborne Observatory (KAO).

The KAO is unique in that it is a microcosm of the scientific method wherein the flight hardware, scientists and mission personnel are all working together at a singular place and time. Teachers share in the excitement, hardships, challenges, and discoveries. Further, work on board the KAO involves cooperation, teamwork and problem solving — critical skills for students in the 21st century — while in the process of doing cutting-edge scientific research. The teachers are then able to convey the elements of the scientific method to their students from a first-hand meaningful experience, rather than the students just reading about science in a textbook.

In 1995 the program expanded to include teachers from the eleven western states served by NASA Ames Research Center's Educational Programs Office as well as teachers from communities around the country where the scientists who fly on the observatory reside. Since the program started in 1992, it has served 79 teachers in fourteen states. The teachers are required to apply in teams of two, coming from different schools and at different grade levels. Generally, one of the schools is a feeder into the other. The team approach has proven to be beneficial to both the school districts by building bridges and to the teachers by having a compatriot in the program with whom they can generate support for science enrichment within their districts. Some of the teams have come from the communities of the scientists who use the KAO for their research. This has enabled the scientists to form partnerships with the schools in their communities and thereby share their research at a personal level.

FOSTER teachers attend a summer workshop at Ames Research Center where they learn about the many diverse programs being conducted by NASA and the airborne programs conducted by Ames Research Center. They are introduced to contemporary astrophysics with a "Grand Tour of the Universe". They then work with filters and gratings and build a spectroscope to understand colors and the electromagnetic spectrum. They build a planisphere; work with star charts and telescopes; tour and do observing at Lick Observatory; and receive hands-on training on the Internet. Teachers also contribute to the workshop by bringing and sharing lessons and activities related to astronomy. NASA scientists resident at Ames who fly on the KAO and who conduct laboratory research in the infrared describe their instruments and research programs to the teachers, very often providing interesting demonstrations related to their work that can be used in the classroom. Finally, the teachers are given an orientation to what they can expect to happen and how they can participate when they return for their research flight during the school year. Many teachers conduct experiments, during their flight, that their students have devised.

By integrating the entire **FOSTER** program into the curriculum the students ultimately learn first hand through the teacher's participation about the excitement of science, the scientific method as practiced, the team work involved, the relevance of science to their daily lives and the importance of a firm foundation in math and science in today's technologically oriented world. Through teacher workshops and inservice presentations, the **FOSTER** teachers share the resources and experiences with many hundreds of others within their schools, in their school districts, at the state level and at the national level such as at National Science Teachers Association regional and national meetings.

The current **FOSTER** program is based on NASA's Kuiper Airborne Observatory, a C-141 which has been flying research missions for twenty years. In a typical year, about eighty research flights are conducted. About half of these are appropriate for teachers to fly on. In the near future, the KAO will be retired and replaced by SOFIA - a Boeing 747 aircraft with a 2.5 meter telescope. It is planned for this aircraft to fly 120 or more research missions per year. It is anticipated that the **FOSTER** program would then be expanded to support one hundred or more teachers per year from around the nation, the limitation being the funding for the program. At this point NASA has fully funded the teachers' expenses for both the workshop and the return to NASA for the flight.

FOSTER is a cooperative program between NASA Ames Research Center and the SETI Institute and is funded by the Astrophysics Division of NASA Headquarters.

Planetaria and the Space Telescope Science Institute

Rob Landis[1], Martin Ratcliffe[2], Steve Fentress[3], Anne Kinney[4], Laura Danly[4]

[1]Space Telescope Science Institute
3700 San Martin Drive
Baltimore, MD
USA 21218

Tel.: (410) 338-4700
FAX: (410) 338-4767
E-mail: landis@stsci.edu.

[2]Buhl Planetarium
[3]Strasenburgh Planetarium
[4]STScI

Abstract: Planetaria are amongst the space program's most enthusiastic supporters. Throughout its history, the Hubble Space Telescope (HST) has enjoyed positive support in the planetarium community. Planetaria can be the most efficient and reliable means in which to communicate science to the public. Worldwide, planetarium attendance is roughly 4 million annually; it is in the interest of NASA institutions to nurture and maintain links to the planetarium realm. This poster documents past and ongoing support the Space Telescope Science Institute (STScI) provides planetaria and solicits comments for a more systematic means to support the planetarium community at large.

STScI's Support for Planetaria: Our support for planetaria runs the gamut and ranges from informal to more formal relationships. An example of an informal relationship would include supplying slides and images upon request, consultation on current planetarium productions (could be HST results, planetary missions, general astronomy), and contacts made at planetarium conferences. A more formal relationship is exemplified by the recently concluded collaboration with Buhl Planetarium of the Carnegie Science Center in Pittsburgh. Together, Buhl and STScI produced a world-class planetarium show entitled, *Through the Eyes of Hubble*. The program premiered in Pittsburgh on 17 March 1995 and is available to planetaria world-wide.

Prior to the Buhl/STScI collaboration, STScI supported several planetaria with details and materials on the first servicing mission to HST and the collision of Comet Shoemaker-Levy 9 into Jupiter. Comments from Strasenburgh and the Adler appear within this poster. Generally, the public is interested in topical events in space exploration as well as "real" astronomy. If motivated astronomy educators and planetarians properly prepare themselves and are nimble enough to capture current events and the public interest in them, a science center/planetarium can make an excellent investment in its reputation, and, sometimes, in their bank accounts.

Recent Informal Support: There are three major planetarium facilities "local" to the Space Telescope Science Institute. These include: Fels Planetarium, Philadelphia; Einstein Planetarium, NASM, Washington, DC; and the Davis Planetarium of the Maryland Science Center, Baltimore, MD. All have produced various programs which have involved STScI.

Fels Planetarium, The Franklin Institute: Fels is currently running *The Other Side of the Universe*, a planetarium show about the Hubble Space Telescope. The producer and manager at Fels made three visits to STScI in the production of their program. The unique feature of the program is that there is a live portion (~1-5 minutes in length) in which a staff member presents and discusses a current HST result. This feature adds a certain timelessness to the program. STScI continues to provide Hubble updates to Fels Planetarium as the latest results are released.

Einstein Planetarium, the Smithsonian National Air and Space Museum: Last year we assisted NASM in providing speakers for their "Exploring Space with Astronomers" lecture series. The lectures have been among the most popular series ever presented. The public has a hunger and a seemingly unquenched thirst for current development in astronomy and NASA's plans for the future. The speakers, most of whom have used HST in their research, had a unique opportunity to share their research with a public audience. Patrons, in turn, felt a little more connected to the space program and those involved in it.

Davis Planetarium, Maryland Science Center: Davis Planetarium has produced a recent show entitled *Pluto Express*. Davis consulted STScI re: the technical and astronomical expertise for the program. Pluto is a cold, icy, distant world that has guarded its secrets well. To date, HST has obtained the best views of the Pluto/Charon system. The program is an opportunity for astronomy educators to motivate and excite the public on a mission before it's launched, rather than discussing a mission after it has happened.

***Through the Eyes of Hubble:* A Joint Educational Outreach Project between the Buhl Planetarium and STScI**

At the 1994 Mid-Atlantic Planetarium Society (MAPS) meeting in Portland, ME, representatives from the Buhl Planetarium initiated contact with STScI to embark on a project to produce a world-class planetarium show (for worldwide distribution) centered on the serviced Hubble Space Telescope.

The astronomical accomplishments of HST following the servicing mission are the focus of this unique collaboration. Using the multiple video projection and Digistar capabilities of the Buhl Planetarium as inspiration, the planetarium project team branched out into small groups to tackle the task of transferring the Space Telescope story to the planetarium dome. The primary goal of this joint effort is to:

1. produce an astronomically articulate, accurate, and timely program detailing the space Telescope story;
2. share (and educate) the wonders of HST science to a broad audience (via the planetarium medium);
3. distribute the show via the existing network of professional planetarium associations.

Domestically, the program seems to maintain a broad appeal amongst major planetaria associated with science centers, universities, colleges, and high schools. Internationally, the program has been translated into German and is being translated into Japanese. To date, nearly 40 planetaria around the world have acquired *Through the Eyes of Hubble*.

The Internet: The public outreach office continues to post current Hubble results via the Internet. Our World Wide Web page may be accessed at: http://www.stsci.edu/ . The Internet was key to the success of many planetaria covering the collision of Comet Shoemaker-Levy 9 into Jupiter. We recently provided a link to ASP's listing of planetaria on our education page. We also alert planetarians to a variety of upcoming and ongoing astronomical events via regular postings to the dome-L listserver and the sci.astro.planetarium newsgroup. Undoubtedly, the Internet is an invaluable resource – not only for planetaria to acquire current astronomical happenings – but for STScI as well to get information to astronomy education quickly and efficiently.

Future Endeavors: STScI is exploring ways in which to better serve the planetarium community at large. The forthcoming issue of the IPS professional journal, *The Planetarian*, will contain a simple questionnaire as to how STScI might better serve planetaria world-wide. Planetaria typically have a need and hunger for current scientifically interesting astronomical images. Coincident to that need are cogent and accurate interpretations describing the image. Depending on current budget trends, STScI hopes to fulfill those needs.

It is in the best interest of NASA-type facilities to forge productive relationships with planetaria. Young people – our future scientists and engineers – need to be inspired. Planetaria are cathedrals of inspiration for young minds to explore and learn more on their own. A small investment of time and energy now will pay off untold dividends 20 and 30 years into the future – not only for NASA, but for our nation as well. These cathedrals of inspiration also serve as a cultural center for technological, scientific, and astronomical literacy.

Acknowledgements: The first author is grateful to staff of the Buhl Planetarium; namely, James Hughes and John French for their unwavering commitment to the joint Buhl/STScI collaboration; Steve Fentress at Strasenburgh Planetarium – coverage of *Voyager 2's* encounter at Neptune has been used as a model by numerous science centers and planetaria world-wide to cover fastbreaking science news (i.e., the first HST servicing mission and C/SL9's collision with Jupiter); and all others who may have been unintentionally omitted. Any errors in this document are the sole responsibility of the first author. All comments and correspondence should be addressed to Rob Landis, Space Telescope Science Institute, 3700 San Martin Drive, Baltimore, MD USA 21218; e-mail: landis@stsci.edu. Space Telescope Science Institute is operated for NASA by AURA. This work is supported by NASA contract number NAS5-26555.

The VEGA Extension of Project ASTRO

Neil L. Lark
Physics Department
University of the Pacific
Stockton, CA
USA 95211

Tel.: (209) 946-2221
E-mail: nlark@uop.edu

The Basic Idea: Build on our successful experiences with Project ASTRO in 1993-94, to create a local extension of the Project for 1994-95, called Valley Education Group for Astronomy.

Maintain Style and Goals of the Original Project ASTRO:

- select active amateur and professional astronomers
- select interested teachers of grades 4-9
- train teams of 1 astronomer and 1 or 2 teachers to work together in the classroom
- coordinate activities with regular science curriculum
- do mostly hands-on activities from the Resource Manual
- plan 4 to 6 visits by the astronomer during the year

Essential Components:

Astronomers: mostly members of the Stockton Astronomical Society, a strong active amateur club.

Teachers: from schools public and private, large and small, urban and rural, with a rich ethnic mix of students

Leaders of the Four Evening and Saturday Workshops: three local teacher/astronomer teams from the original 1993-94 Project ASTRO

Local Coordination and Facilities: Neil Lark, University of the Pacific

Official Sponsorship and Materials: Project ASTRO, Astronomical Society of the Pacific, Andrew Fraknoi, Director, Jessica Richter, Coordinator

Scope of the VEGA Extension, 1994-95

- 14 new teacher/astronomer teams
- 3 teams continuing from 1993-94 Project ASTRO
- 15 astronomers active in classrooms
- 5 other astronomers available as night assistants
- 28 teachers altogether, at 17 different schools

Evaluation of the Project: Of the 45 team members, 29 responded to a two-page questionnaire. Most of the 16 others were interviewed in person or by phone. About half of the teams, involving about 15 classes, were very clearly successful. They gave their partnerships "A" and "B" grades, and they reported three or more class visits. They reported success in using the "hands-on" activities from the Project ASTRO Resource Manual.

Despite much unfavorable weather, most teams had successful family star parties involving indoor activities as well as outdoor viewing through telescopes. Most star parties included more than one class, sometimes the whole school. Only two or three of the teams gave themselves a grade poorer than "C".

Several factors limited the success of other teams, with no single factor being very common:

- conflicting schedules
- differing expectations
- poor communication
- illness in family
- divorce

Where Do We Go From Here?

All of the responding participants think the local project should be continued next year. Over 80% of the respondents want to participate again themselves.

The main limitation is the small number of astronomers who can arrange to visit classrooms during regular school hours. It is unlikely that the size of the local project can grow much, but it can be continued indefinitely on the same scale.

PROJECT ARTIST AND PROJECT ACCESS!: Integrating Astronomy And Planetary Sciences Into The Elementary And Middle School Curriculum

Larry A. Lebofsky[1], Thea Cañizo[2], Nancy R. Lebofsky[3]

[1]Lunar and Planetary Laboratory
University of Arizona
Tucson, Arizona
USA 85721-0092

Tel.: (520) 621-6947
FAX: (520) 621-4933
E-mail: lebofsky@lpl.arizona.edu

[2]Vail Middle School Tucson Unified School District
Tucson, Arizona
USA 85711

[3]Steward Observatory,
University of Arizona
Tucson, Arizona 85721-0065

Scientists from the University of Arizona and teachers from southern Arizona are collaborating on ways to make space science (astronomy and planetary science) accessible and interesting to children from diverse backgrounds at both the elementary and middle school levels. Programs which offer instruction in space science through hands-on experiences, integrated curriculum, and translated materials are described below.

In elementary school, science is a subject that is often neglected by teachers for a variety of reasons. In particular, the demands of the basic "reading, writing, and arithmetic" can consume the entire school day leaving little or no time for science. Therefore, in order to teach science, science must be integrated into the existing school curriculum.

Space science can be used to introduce students to the natural world which is a part of their lives. Even children in an urban environment are aware of such phenomena as day and night, shadows, and the seasons. It is a science that transcends cultures, has been prominent in the news in recent years, and can generate excitement in young minds as no other science can. Space science also provides a useful tool for understanding other sciences and mathematics, and for developing problem solving skills which are important in our technological world.

Over the past five years we have been conducting a series of workshops for elementary and middle school teachers from southern Arizona. Our workshops have included one-day workshops for bilingual teachers, two-day aerospace workshops, and one-, two-, and four-week workshops in space sciences. Our two current programs are Project ARTIST (Astronomy-Related Teacher In-Service Training) and Project ACCESS! (All Children Can Explore the Solar System!).

Project ARTIST: is a four-year program of workshops and materials development funded by the National Science Foundation (NSF) Teacher Enhancement program. ARTIST provides four-week workshops for teachers of grades K-8 from schools throughout Arizona. The workshops stress hands-on activities and experiments, while providing instruction in space science content and concepts. ARTIST also encourages the use of space science materials across the curriculum. One more summer workshop will accommodate 25 teachers in 1995 under current funding. We are presently seeking additional funds from the NSF for the expansion of Project ARTIST to locations in several other states.

Project ACCESS!: is an outgrowth of ARTIST. ACCESS provides one-week workshops for K-5 teachers. It differs from ARTIST in that the activities are divided into thematic units that the teachers can incorporate directly into their curriculum. Many of the activities developed for ACCESS by staff and facilitator/teachers are now being incorporated into the final year of ARTIST.

Staff members have been assisted in both the planning and presentation of the workshops by approximately 20 facilitators, i.e., classroom teachers who have attended previous ARTIST workshops and have been using space science materials in their classes. Amateur astronomer David Levy was added to the staff in 1993. Of the over 130 ARTIST and ACCESS participants and facilitators, more than 100 are women, over 25 are Hispanic, more than 50 teach in schools with large minority populations, over 25 teach in bilingual or limited English proficiency classrooms, and four teach on reservations. Therefore, the student population affected by these teachers' new skills and knowledge is very broadbased, including students from groups traditionally underrepresented in the sciences.

Projects ARTIST and ACCESS seem to work well because of their partnership approach. Research scientists provide content and concepts, and facilitators (as well as the participants) put activities into the context of the classroom. Primary, intermediate, and middle school teachers bring a diversity of experiences and methodologies to the program, enhanced by the special insights of bilingual, ESL, gifted, and special ed teachers. The staff provides unique experiences in observational astronomy and planetary surfaces. Mutual respect by each component for the professionalism of every other component allows everyone involved to learn from each other.

At the Astronomical Society of the Pacific meeting in College Park, Maryland, a variety of materials from the workshops, activities developed by the workshop participants, and materials produced by the students of the workshop participants were presented. This presentation stressed the integration of science with language arts, fine arts, math, and social studies.

UNIVERSE: The Infinite Frontier

Joel M. Levine
Orange Coast College
Laguna Niguel, CA, USA 92677

Tel.: (714) 432-5888
Internet: joel=levine%mathscience%occ@banyan.cccd.edu

A Brief Description:

UNIVERSE: The Infinite Frontier is a one-semester college-level telecourse in introductory astronomy [High school/abridged adaptation in development – September 1995]. The telecourse covers four major areas: Exploring the Sky, The Stars, the Universe of Galaxies, and Planets in Pérspective. It explores a broad range of astronomy topics, concepts, and principles, from the motions of the visible sky to dark matter, from our own planet to the stars and galaxies. The telecourse examines evidence for the big bang and continuing evolution of the universe and tracks the formation, life, and death of stars. Throughout the telecourse, special emphasis is placed on the scientific evidence that astronomers use to support their conclusions. The course also depicts how astronomers have come to know what they know about the universe and shows how astronomers are still seeking answers to some of the most fundamental questions about the universe. The programs include interviews with leading experts; original computer graphics; footage from the Jet Propulsion Laboratory, NASA, and the European Space Agency; and images from leading observatories throughout the world, and the Hubble Space Telescope.

Course Themes: From its inception, the goal of this astronomy course was to provide students with the most reliable knowledge about basic astronomy and an understanding of how that knowledge was gained. With that in mind, the national advisory committee developed eight major themes for UNIVERSE: The Infinite Frontier.

1. The universe is dynamic and continually evolving.
2. The universality of physical laws discovered on Earth allows us to analyze and draw conclusions about celestial phenomena that can be studied only a great distances.
3. Scientific conclusions must be based on an exacting comparison of hypotheses to evidence obtained from observations and experimental data.
4. The universe, because it has objects and environments that cannot be duplicated on Earth, is a unique laboratory for testing scientific hypotheses.
5. Because astronomy is a human endeavor, it is subject to both the limitations and the enhancements of personal relationships, biases, inspiration, and creativity.
6. Most astronomical knowledge accumulates incrementally, with each new piece of knowledge providing a potential foundation for further understanding.
7. Observational and computational technologies play critical roles in shaping our understanding of the universe.
8. In addition to scientific value, astronomy has practical and philosophical value because humans are participants in, as well as observers of, of the universe.

Guided by these course themes, the academic advisors and course creators developed specific instructional objectives for each of the twenty-six lessons. These learning objectives are presented in the print material at the beginning of each lesson and provide the framework for the structure and content of each lesson.

Components of the Telecourse: A multimedia course, UNIVERSE: The Infinite Frontier includes the following elements: video programs, textbook, telecourse student guide, laboratory component (optional) and faculty manual.

Video Programs: The telecourse's 26 half-hour broadcast-quality video programs were designed and produced by academic and media production experts with the cooperation of KOCE-TV, an award-winning public television station.

Video Program Titles:

- The Scale of the Cosmos
- The Sky
- Cycles of the Sky
- The Origins of Modern Astronomy
- Newton, Einstein, and Gravity
- The Tools of Astronomy
- Atoms and Starlight
- The Sun
- Stellar Properties
- Stellar Formation
- The Lives of Stars
- The Deaths of Stars
- Neutron Stars/Black Holes
- The Milky Way
- Galaxies
- Peculiar Galaxies
- The Big Bang
- The Fate of the Universe
- The Origin of the Solar System
- Planet Earth
- The Moon and Mercury
- Venus and Mars
- Jupiter and Saturn
- Uranus, Neptune, and Pluto
- Meteorites, Asteroids, and Comets
- Life on Other Worlds

Textbook: The textbook for UNIVERSE: The Infinite Frontier is Horizons: Exploring the Universe by Michael A. Seeds, published by Wadsworth Publishing Company.

Telecourse Student Guide: The telecourse student guide assists students by coordinating all elements of the course and includes review activities and self-tests for each lesson. The telecourse student guide, written by Stephen P. Lattanzio, Joel M. Levine, and Valerie Lynch Lee, and reviewed by David Pierce, correlates and integrates the textbook, the video programs, and the objectives of the course.

Laboratory Component: For institutions that want to offer UNIVERSE: The Infinite Frontier with a laboratory component, the academic advisors and course developers have selected individual modules from the Astronomy through Practical Investigation series, published by L.S.W. Publications, Inc.

Faculty Manual/Test Bank: The faculty manual contains specific information for the instructor who manages the course. It offers information about the course; distance learning students; additional activities; student support; the laboratory component; astronomy resource guide; and a test bank.

For questions regarding the academic content of the course, please contact: Joel M. Levine Orange Coast College, Tel.: 714-432-5888, Internet: joel=levine%mathscience%occ@banyan.cccd.edu

For information or review copies of videos: Coastline Community College Attn: Coast Telecourses 11460 Warner Avenue, Fountain, CA USA 92708-2597.

Abridged Adaptation: Deborah Rush, Marketing; INTELECOM, 150E. Colorado Blvd., #300 Pasadena, CA USA 91105-1937.

The Remote Access Astronomy Project

Shea A. Lovan
Remote Access Astronomy Project
Department of Physics
University of California
Santa Barbara, CA
USA 93106

E-mail: shea@rot.ucsb.edu

and

Philip M. Lubin
Physics Department
University of California
Santa Barbara, CA
USA 93106

E-mail: RAAP@rot.physics.ucsb.edu

Abstract

The Remote Access Astronomy Project is a unique telescope and data distribution system that has the potential to change the way astronomical, earth science, and physics concepts are taught to high school and undergraduate students. The project uses high resolution images and image processing techniques that appeal to the natural curiosity people have about space and astronomy. In addition, particularly at the secondary level, it serves as a forum for low cost and rapid distribution of curriculum materials among teachers and as an educational network between high schools and universities.

By using a combination of high performance microcomputers, high resolution graphics, and high speed communications technology, the project can break down the traditional classroom boundaries and allow students and teachers access to a much richer environment. The system is currently in use at the undergraduate level at UCSB, a local junior college, and several California high schools. The project is sponsored by the University of California, The National Science Foundation's Center for Particle Astrophysics, Rockwell Corporation International, and the National Aeronautics and Space Administration.

Rationale behind the project

The Remote Access Astronomy Project at the University of California, Santa Barbara originally developed out of a desire to improve astronomy education at the undergraduate level. Students were disappointed when they observed celestial objects with the small portable telescopes used in the astronomy labs. These objects appeared only slightly better than with the naked eye, and nothing like the spectacular photographs taken with the large telescopes found in astronomy text books. By mounting a sensitive CCD camera on a 14-inch telescope, students can study stars, galaxies, nebulae, and other interesting items in a more meaningful way. Moreover, by taking advantage of computer networks and operating the telescope remotely, students need not gather on the roof of the physics building late at night to do their observing.

Using the communications technology

Within the last decade, computer networks and computer bulletin board services have become very popular. The technology already existed to extend access to our computer controlled telescope to high schools via a modem link. In addition, a tremendous volume of images has become available through the NASA space missions (Voyager I and II, Magellan, Viking), Earth-orbiting satellites (IRAS, COBE, Hubble Space Telescope), and other sources. At the university we have access to a wealth of data not generally accessible by high schools, but which could be used effectively by creative teachers in a variety of science courses. Since July 1990, we have maintained the ASTRO-RAAP bulletin board to allow access to these resources and provide a forum for discussion. In December 1994, we opened a home page on the World Wide Web to make the database more accessible.

Integrating the RAAP

The strength of a program such as ours lies in its ability to expand the resources normally available to high school students and teachers, specifically the use of our telescope and our experience with astronomy. To make incorporating astronomy into a classroom as painless as possible we have developed a core of image processing exercises that are available for download. These activities use digital image processing to examine different phenomena; from the rotation rate of our sun to the Doppler shift of the cosmic microwave background. In addition to the exercises, we have made our telescope available to students who are interested in pursuing a line of personal research. Thus far, students have conducted searches for dark matter and black holes and measured the light curves of variable stars and recent supernovae. Such research, done by individual students or groups, becomes a possibility once a classroom starts to exploit the potential of our program.

The Skymath Project: The teaching of middle school mathematics using the real-world problem of the weather

Beverly T. Lynds
P.I. Project Skymath Unidata
UCAR

E-mail: blynds@unidata.ucar.edu

The University Corporation for Atmospheric Research has received funding from the National Science Foundation to prepare a middle school mathematics module based on the use of real-time weather and climate data. The project, called Skymath (http://unidata.ucar.edu/ staff/blynds/Skymath.html), is led by a group of mathematicians, math educators, scientists, technology experts, and teachers.

The goals of the pilot project are to demonstrate that middle-school mathematical studies can be enriched with certain NCTM (National Council of Teachers of Mathematics) Curriculum Standards by incorporating current environmental and real-time weather data in ways that embrace both the dynamic and the uncertain natures of these data, and that teaching methods associated with the acquisition and use of these data by students are naturally consistent with the NCTM Professional and Assessment Standards.

The Skymath module is focused on diverse students - their experiences, interests, and understandings. The plan uses science as a context in which to teach significant mathematics, to bring to light and to learn the power of mathematics. It is a combination of math-focused, structured experiences with more open-ended exploratory experiences. The mathematical content, teaching and learning will incorporate highest national standards.

In an article published in the Mathematical Sciences Education Board's "On the Shoulders of Giants", Ian Stewart discusses mathematical concepts related to "change." Noting that every natural phenomenon is a manifestation of change, Stewart stresses that mathematics is the most effective tool for understanding patterns of change. He notes that to master the concept of change, one must be able to "represent changes in a comprehensible form, to understand the fundamental types of change, to recognize these types of changes when they occur, and to apply these techniques to the outside world."

The Skymath Design Team identified the mathematics of change as the most natural to address in a module that utilizes current weather data. The module articulates a mathematical study of the changing weather, focusing on temperature. It leads students to develop methods of representing change; how temperature changes with time and with location. Students are challenged to measure, represent, and analyze these changes.

Activities include developing symbol sets, preparing graphs for median-based analysis, predicting magnitudes of changes, and learning about the correctness of their work by seeing what really happens!

Thinking about how the real world changes with time and recording real-time data to accumulate a time series are activities that will evoke student ownership of the data and will provide experiences that retrospective data cannot. The combination of local data gathered by students with data conveyed electronically from around the world - depicting today's weather - will stimulate curiosity about natural variability and hence about the quantitative nature of change in spatial as well as temporal dimensions.

The tasks of the first Skymath module will flow from off-line experiences -

getting familiar with temperature measurements – to on-line collection of data, i.e., from small data sets to larger, computer-based data sets. The focus is on learning through questions and problems that motivate students by incorporating real data and authentic tasks, and through culminating projects which stimulate students to conduct their work with an audience in mind. Students will exchange information with other schools across the country that are participating in the Skymath program.

The Skymath educational strategies and materials will be prepared by the Education Development Center, Inc. (EDC) of Newton, MA, a nonprofit research and development organization, in a flexible curriculum module that includes guidance and tools for exploration plus a collection of resources and activities that use weather and real-time data to teach math concepts.

The end-product will be a loose-leaf handbook, available in on-line or hard-copy form. It will contain a series of activities accompanied by: suggestions for sequencing, assessment strategies for evaluating what students learn and understand, help – for students and teachers – with technologies needed in the classroom, a narrative describing successes and obstacles that have been encountered in the use of Skymath concepts, and an invitation to assist in further module development.

Skymath will utilize software and data dissemination methods developed in the University of Michigan's (http://groundhog.sprl.umich.edu) Blue Skies program, an endeavor developed under a National Science Foundation grant. Blue Skies software offers an extraordinarily simple interface so that users with a minimal computer experience can easily and quickly obtain needed information. Access is provided to hundreds of real-time weather and environmental images by using the client-server protocol developed for the University of Minnesota.

The module will be tested in collaboration with the University of Colorado's (http://www.cs.colorado.edu/homes/usha/public_html/kgs.html) Kids as Global Scientists (KGS) program, which is exploring the potential for and the value of telecommunication networks within the science classroom. KGS already is engaged in detailed studies of student learning in projects that entail explorations and interpretations of professional-quality weather data. These are pioneering studies in respect to middle school students' understanding of weather phenomena (but with scientific rather than mathematical emphasis) and KGS is developing new models of collaborative learning. Through the Internet, students become "global scientists," expanding their understanding of local phenomena through exchanges with peers worldwide and through access to professional observing networks. The Skymath module will complement other parts of KGS, providing a math-focused element that enriches materials used by the teams of participating middle school math and science teachers and helps create an integrated science/math curriculum.

Development sites for the first Skymath module will be located in Colorado and Boston, Massachusetts. Activities and materials will be developed in a series of design, classroom testing, and revision cycles. A third party evaluation will be conducted by Insites, Inc. of Ft. Collins, Colorado, using qualitative and quantitative data. The evaluation will focus on student learning, teacher experience, the role of technology, and the value of using real-time data.

The Skymath handbook is planned to be completed in 1997. It will be made available in hard copy and on the Internet. All feedback will be welcomed and should be sent to the Program Director, Beverly Lynds It is hoped that this module and its accompanying technology will be introduced to preservice teachers during their science/math studies.

The Light Pollution Monitoring Program

Ardis Maciolek
LPM Coordinator
Grosse Pointe North High School
707 Vernier
Grosse Pointe, MI
USA 48236

E-mail: maciolek@cirrus.sprl.umich.edu

The Light Pollution Monitoring program is a new venture to involve students and interested others in a study of light pollution at local and regional levels. You are invited to become a participant in this program!

Introduction: Light pollution is not only a problem for astronomers, but also for the average citizen. Lights that are improperly shielded or located contribute to light pollution. So do lights that are excessively bright. Light pollution adversely affects public safety, energy conservation and economics.

A number of excellent articles have been written about light pollution. One clearinghouse is the International Dark Sky Association (IDA). The IDA serves in both educational and political arenas to promote a better understanding of the issues involving light pollution.

- The address is: International Dark Sky Association, Inc., 3545 N. Stewart Ave., Tucson, AZ, USA 85716, or E-mail: Dr. David Crawford at dcrawford@noao.edu.
- There is also a page about the IDA on the World Wide Web at: http://fa-www.harvard.edu/ ~graff/nelpag/ida/IDA1.html

The Light Pollution Monitoring Program will enable students, teachers and interested others to study the variables that affect light pollution locally and to compare levels of light pollution on a regional basis. Students will engage in a process that involves data collection, communication with colleagues, and scientific analysis. They will be contributing more knowledge towards a better understanding of this little-known and often misunderstood environmental problem.

The Light Pollution Viewer: The instrument used to measure a light pollution level is made from an index card, a toilet paper tube, a sheet of black construction paper and a roll of tape. It can be cheaply and easily constructed and replicated. Calibration tests have been done (and are still in progress) to ensure that the transmission qualities of different brands of tape are considered.

The Light Pollution Monitoring Program: Participants will once a month measure their light pollution level and report it to a central location, where data will be processed and made accessible in the forms of text and graphics. Light pollution values for each reporting station will be archived, so that studies can be made to investigate cyclic or progressive levels of light pollution.

When to Monitor: Since the Moon is a natural source of light pollution, measurements cannot be made if it is in the sky. There is a window of approximately two weeks per month termed "dark sky" when the Moon is not visible in the evening sky. Clouds at different altitudes and thicknesses will intensify and reflect the effects of lighting, so readings must be done only on clear nights. Twilight must be over. Summarizing these points, readings should only be taken on clear, moonless nights after twilight ends.

Reporting Light Pollution Values: All data must be reported on the standard form provided upon registration. For those with full service Internet access: Report values to the Blue Skies Project at: telnet: groundhog.sprl.umich.edu. Enter your observations on the K-12 schools observations menu. To use this option you first need to register and be assigned a password. Register by sending an e-mail request to: maciolek@cirrus.sprl.umich.edu. For those with limited Internet access: E-mail your results to: maciolek@cirrus.sprl.umich.edu. Use the header "LPV" for your message. For those with little or no computer communications capability, mail your results to: Ardis Maciolek, LPM Coordinator, Grosse Pointe North High School, 707 Vernier, Grosse Pointe, MI 48236.

Analyzing and Viewing Data: Data will be archived and displayed in the form of charts and maps, similar to how temperature values are displayed on weather maps. If enough data are available, it will be possible to construct colour zone maps showing variations in sky darkness. These maps and text will be accessible through the Blue Skies computer program (mentioned earlier) or can be requested by mail.

A discussion group for those interested in or involved with this program is being developed as part of the Blue Skies program.

Suggestions for teachers or clubs: For schools or clubs: Average your results and report one value for the site per night. Calculate and report an average time, as well.

1. Create localized light pollution maps of your neighborhood.

2. Investigate the amount of change of light pollution throughout one night, then plot curves of LPV vs. Time.

3. Identify local "bright" and "dark" areas and use this information appropriately. For bright areas: What's the source? Is there too much light? Waste of money? Glare issues? For dark areas: A good place to observe the sky! Is it too dark (unsafe)? All of the above initiate opportunities for discussions, speeches, letter writing, e-mail, etc.

4. Investigate annual cyclic changes in light pollution. Is increased lighting associated with seasonal changes? Recreational or manufacturing activities?

5. Investigate year-to-year changes in light pollution. Can you see urban sprawl creeping across your sky?

6. Analyze regional patterns of light pollution.

Your data and participation are welcomed beginning in September 1995. Clear skies!!

PROJECT CLEA: Three Years of Developing Computerized Exercises for the Introductory Astronomy Laboratory

Laurence.A. Marschall
Department of Physics
Gettysburg College
Gettysburg, PA
USA 17325

FAX: (717) 337-6666
Internet: marschal@gettysburg.edu

and

G.A. Snyder
M.B. Hayden
P.R. Cooper
M.K. Luehrmann
R.F. Good (Gettysburg College)

Abstract: In an effort to bring modern research techniques into the introductory astronomy laboratory, we have been developing a series of modules containing software, student manuals, and teachers' guides. These modules incorporate simulations of actual observations and astronomical phenomena using Mac and PC based software, extensive reliance on digital images and spectra obtained at professional observatories, and realistic tools for the analysis of data. We present here descriptions of our work to date, including exercises on photometry, spectral classification, the Hubble redshift-distance relation, the large-scale structure of the universe, the mass of Jupiter, the rotation of Mercury, and the transfer of radiation out of the sun. We also discuss current software under development and plans for the future. CLEA (Contemporary Laboratory Experiences in Astronomy) has been supported by grants from the National Science Foundation and from Gettysburg College.

Introduction:

Astronomy is by its nature an observational rather than experimental science (like physics, chemistry or biology), because the objects it studies are so remote and the time-scales of events rather long. Even granting this distinction, astronomy classes, especially at the introductory level, have been conspicuously lacking in hands-on lab experiences that actively engage students. Most laboratories we know of, except for occasional night labs at the telescope, give students practice in copying tables, graphing data, or measuring photographs, but not in doing astronomy the way it's done by astronomers.

Yet because so many observations in astronomy are made using computer-controlled telescopes, and because so much analysis of astronomical data is done using computers, it is possible to create very realistic simulations of real observing experiences. With this in mind, we at Gettysburg College, with support from the NSF, have established Project CLEA (Contemporary Laboratory Experiences in Astronomy), which has, over the last 3 years, developed 7 modular exercises, based around computer simulations with accompanying workbooks and technical manuals, that illustrate important measurements and measuring techniques in modern astronomy. We describe the currently available material in the following section.

Exercises Developed By Project CLEA

- The Revolution of the Moons of Jupiter: Using a high resolution imaging telescope, students observe Jupiter and its satellites at regular intervals over a period of several weeks, and, by graphing projected radius versus time, they determine the mass of Jupiter from Kepler's 3rd law.

- The Rotation of Mercury Using Doppler Radar: Using a simulated radio telescope, students bounce a radar signal off Mercury, and determine its equatorial rotational period from the profile of the Doppler-shifted echo.
- Photoelectric Photometry of the Pleiades: Using a pulse counting aperture photometer attached to a 0.4-, 0.9-, or 4-meter telescope, students measure the UBV magnitudes of a sample of stars in the Pleiades, plot an H-R diagram, and determine the distance and age of the cluster.
- The Classification of Stellar Spectra: Using a photon counting spectrometer students measure the spectra of unknown stars in the field. They can match these spectra against digital spectra of representative standard stars, using a variety of graphic tools. Two dozen unknown spectra are also provided in a drill-and-practice option.
- The Hubble Redshift-Distance Relation: Using a photon-counting spectrometer students measure the spectra of unknown galaxies in 5 clusters of increasing distance, determining the velocity of the galaxies from the shift in the Calcium H and K lines. They plot galaxy velocity versus distance to determine the Hubble parameter H_0 and, from that, the expansion age of the universe.
- The Large-Scale Structure of the Universe: Using equipment similar to that in the Hubble Redshift lab, students plot up a wedge diagram showing the 3 dimensional spatial distribution of 218 galaxies in the neighborhood of the Milky Way. This exercise is best done as a collaborative project by an entire lab section.
- The Flow of Energy out of the Sun: Students shoot photons out of the sun, studying the mechanism of radiative diffusion in the solar interior and the formation of spectral absorption lines in the solar atmosphere.

Current Distribution And Future Plans

CLEA's software and manuals are distributed widely by regular mail (write Project CLEA, Gettysburg College, Gettysburg, PA, USA 17325; tel.: (717) 337-6028), by anonymous ftp: (io.cc.gettysburg.edu) and on the Web (http://www.gettysburg.edu/project/physics/clea/ CLEAhome.html). A recent survey of our mailing list indicates users in all 50 states and over 40 foreign countries, reaching at least 10,000 students each year. About a quarter of our users are high school teachers, and most of the rest teach introductory astronomy classes for non-majors.

In the remaining year of the current grant we are working on exercises on variable stars, asteroid astrometry, and pulsars. We hope to receive additional funding to continue the project through 1998, at which time we will have 12 to 16 exercises available, sufficient for a 1 or 2 semester course. We welcome comments and suggestions from users, both faculty and students, to help improve our effort to bring modern astronomy from the mountaintop into the classroom.

References:

L.A. Marschall, "Virtual Professional Astronomy", *Sky and Telescope*, 90, No. 2, p 92. (August, 1995)

Hands On Astrophysics: Variable Stars In Math/Science Lab

Janet A. Mattei
American Association of Variable Star Observers (AAVSO)
25 Birch Street
Cambridge, Massachusetts
USA 02138-1205

E-mail: aavso@aavso.org

and

Michael Saladyga
Lynn M. Anderson

and

John R. Percy
Erindale College
University of Toronto
Mississauga, Ontario
Canada L5L 1C6

E-mail: jpercy@credit.erin.utoronto.ca

Variable stars (variables in short) are stars that change in brightness. They are interesting, numerous (over 30,000 known and another 15,000 suspected), and important because they reveal much of what we know about the properties, processes, and evolution of stars.

The American Association of Variable Star Observers (AAVSO): The discovery and measurement of the brightness of variable stars is simple in principle, and astronomy is helped significantly by the efforts of thousands of skilled amateur astronomers who measure variable stars in a systematic way. The AAVSO is a non-profit worldwide scientific and educational organization. The AAVSO is the largest organization of variable star observers in the world, with members in 43 countries. The purpose of the AAVSO is to coordinate variable star observing (done primarily by amateur astronomers), evaluate the accuracy of these observations, compile, digitize, and publish them and make them available to researchers and educators. Over 7.5 million observations of variable stars have been compiled since the AAVSO was founded in 1911. The scientific activities of the AAVSO are coordinated by the Director, who is a professional astronomer. The current Director is Dr. Janet A. Mattei.

Over 300,000 observations are submitted to the AAVSO by observers around the world each year. These observations are digitized and archived by the AAVSO in the AAVSO International Database. The AAVSO Director and her technical staff at AAVSO Headquarters (25 Birch Street, Cambridge, MA, USA 02138-1205; Tel.: (617) 354-0484; FAX: (617) 354-0665; E-mail: aavso@aavso.org) receive more than 200 requests annually from researchers and educators for AAVSO data and services. The number of requests has increased by a factor of 20 in the last two decades.

AAVSO services are sought by astronomers: (a) to receive real-time, up-to-date information on unusual stellar behavior; (b) to assist scheduling and execution of variable star observing programs using earth-based telescopes and instruments aboard satellites; (c) to request simultaneous optical coverage of stars being observed during earth-based or satellite observing programs, and immediate notification of their activity; (d) to correlate optical data with data obtained using other techniques (photometry, spectroscopy, polarimetry), and at other wave-lengths from gamma-rays to radio waves; (e) to carry out collaborative long-term data analysis of the behavior of (particularly) large amplitude variable stars; and (f) to assist educators in setting up observing projects, and students in carrying them out.

Hands-On-Astrophysics – Variable Stars and Science/Math Education: Variable star observation is simple, but the analysis and interpretation of the data involves a wide range of scientific and mathematical skills. Students can experience the excitement of doing real science with actual variable star data. Hands-On-Astrophysics: Variable Stars in Math/Science Lab. – being developed by the AAVSO and funded by the National Science Foundation – is an astronomy educational project which emphasizes student involvement and scientific discovery.

The project can be used at the high school level for lab exercises in physical science, mathematics, and computer sciences; for project work for science fairs or individual research projects; for summer courses and camps; and for science clubs. It can also be used, with some modification, at the college or university level, in introductory astronomy courses.

The Hands-On-Astrophysics project materials include: a manual with background reading and activities for the student, and concepts and application to curricula and examples of typical results for the teacher; 900,000 AAVSO observations extending 25 years on 70 variables in 5 seasonal northern hemisphere constellations; computer programs for data analysis; star charts for the 70 variables; a set of slides of the 5 constellations taken over a year; instructional videotapes on variable stars, variable star observing, and backyard astronomy; and a newsletter. All participants of the project will be linked via Internet.

Project activities include:

- identifying and measuring the brightness of variable stars in the real sky and on the slides;
- accessing and analyzing data in the database provided using the AAVSO computer programs or programs written by students
- plotting observations and interpreting the resulting graphs
- identifying different types of variable stars from their behavior
- analyzing the long term behavior (regularity, periodicity, etc.) of variable stars in the database.

Thus, by carrying out all aspects of the research process, students can develop and integrate a wide variety of skills in astronomy, physical sciences, mathematics, and computer science. The scientific and mathematical skills developed include: background reading and literature searching on the individual variable stars; understanding measurement and errors; analysis and interpretation of graphs; application of numerical techniques and statistics; data processing, analysis and graphing; interpretation of trends and changes; presentation of written and oral reports. In addition, through observing, the students develop an appreciation of the night sky and the skills of making variable star observations.

We expect that the Hands-On-Astrophysics project material will be available in 1997.

We gratefully acknowledge the education grant ESI-9154091 from the National Science Foundation to develop this project.

PARTNERSHIP IN ASTRONOMY Linking Students and Teachers with a Community of Dedicated Amateur Astronomers

Janet A. Mattei
American Association of Variable Star Observers (AAVSO)
25 Birch Street
Cambridge, Massachusetts
USA 02138-1205

E-mail: aavso@aavso.org

and

Mario Motta
Michael Saladyga
Lynn M. Anderson

Partnership In Astronomy (PIA) is a pilot program of the AAVSO developed through funds from NASA Astrophysics Grant Supplement for Education. It links a community of dedicated amateur astronomers with middle and high school science teachers in the Boston, Massachusetts, area.

Our goal is to complement and enrich established units of study in middle and high schools, and to enhance the instructional expertise of participating teachers. We targeted our project particularly to grade 5 Math/Science, which includes a "Sky Observations" unit, and to grade 9, which studies earth sciences.

Through their years of observing experience, AAVSO members, mostly amateur astronomers, have an extensive knowledge of astronomy and the night sky, and some have even built their own observatories and telescopes. Furthermore, they are often the "ambassadors of astronomy" in their local communities and schools, giving talks or hosting star parties for students and neighbors which are meant to informally educate as well as inspire enthusiasm and interest in the science of astronomy.

Through activities involving students, teachers, and parents such as presentations, slide shows, star parties, trips to the science museum and planetarium, and workshops in grinding mirrors and constructing telescopes, the participating local amateur astronomers have shared their expertise to enhance science education in local classrooms; built appreciation and interest in science and astronomy; and provided mentoring along with active participation in hands-on observing projects for students and teachers.

We have held workshops for participating local middle and high school science teachers to inform them of the project and its activities, train them to carry out variable star observing programs, and provide them with information on astronomy resource and outreach materials.

We have also developed a lending astronomical library for teachers and students. This library, containing astronomical books at all levels, is open to area middle and high schools teachers and may be borrowed by teachers participating in the project.

The participating amateur astronomers – a music teacher, an optician, an engineer, and a cardiologist – developed a booklet of hands-on observing projects especially designed to be carried out as a family experience. These projects include: measuring positions of the moon, planets, and stars with a simple tool (sky bow);

observing and charting the positions of the planets; measuring the motion of the earth; observing and charting the position of the moon; and observing a cepheid variable (delta Cephei), recording its brightness variation, finding its period, and estimating its distance.

One school in particular - Lynnfield Middle School - enthusiastically supported the Partnership In Astronomy program, from the principal to the teachers, parents, and the students themselves. The entire school participated in a wide variety of astronomy-related activities, and astronomy was integrated into such other subject areas as art, English, history, and mathematics.

In a year-round workshop on telescope making held by one of us (Mario Motta, who has built 16-inch and 35-inch telescopes), Lynnfield students ground mirrors and completed the construction of eight telescopes that they and other students at their school will use to observe for years to come.

The Partnership in Astronomy project has been rewarding for everyone involved. We gratefully acknowledge the funding provided through NASA grants NAGW-3228 and NAGW-1493 to develop this project.

Cosmic Disaster in 2000!

John Mosley
Griffith Observatory
2800 East Observatory Road,
Los Angeles, CA
USA 90027

Tel.: (213) 664-1181
Home FAX: (818) 708-7314
FAX: (213) 664-4323
E-mail: jmosley@cello.gina.calstate.edu

The planets align in May, 2000, and the world will end – or so many people will say. Anyone involved in educating the public about astronomy will be involved. It will be the Jupiter Effect and Nostradamus all over again, but on a global scale. These planetary alignments will be, for many people, the most significant astronomical event of the next decade. We must prepare.

Doom at the End of the Millennium

We will hear from all quarters – psychics, astrologers, pyramid experts, prophesiers of the apocalypse, and good old Nostradamus – in books, in videos, and on TV that civilization will end at the end of the millennium. Some will cite the planetary alignments in May 2000 (and others), and as astronomers we will be asked to explain, defend, and even reassure. Recall the intensity of fears during previous similar scares; a lot of people were genuinely frightened. It will be an interesting few years.

The Planetary Alignments

In contrast to the Jupiter Effect and Nostradamus' 1988 prediction of doom, when there were no planetary alignments, in May 2000 there is a series of alignments, and astrologers and others will know of the events and bring them to everyone's attention.

On May 5 at 8:08 UT the five naked eye planets plus the sun and moon will fit within a span of 26° in celestial longitude – their minimum separation. (It will be difficult to see Mercury, Jupiter, and Saturn, which will be close to the sun). This is one day after new moon. Such a compact grouping of these objects occurs very roughly once a century but is not periodic.

On May 17 the five naked eye planets and the sun (but not the moon) will span 19½° – their minimum separation – at 10:30 UT. At that same moment, Venus and Jupiter are separated by only 42 arcseconds (!) and the moon is 21 hours before full and also in approximate alignment by being 170° opposite the sun. Only Mars might be visible just after sunset.

Dealing with the Media

As an astronomer at the Griffith Observatory in Los Angeles, I lived through the total destruction of Los Angeles by planetary alignments twice – in 1982 (the 'Jupiter Effect') and again in 1988 (Nostrodamus). Both were major media events. Like it or not, we will deal with the public and with the media at yet another end of the world at the end of the millennium.

Here are suggestions (from personal experience) on how to talk to the news media and to talk-show hosts, especially in a confrontational setting such as a debate, about planetary alignments:

- Be brief. There is no time to fool around. Answer in 20 seconds and don't use big words. Rehearse what you will say.
- Work from your agenda. Have a mental list of the points you want to make and be sure to get some in.
- TV is a visual medium. Bring and use visuals, including hand-held props. Use surprise.
- Know your facts. Do your homework.
- Use anecdotes. Tell stories about people, and use drama and humor.
- Use direct and forceful language. Don't be ambiguous or qualify unnecessarily.
- Avoid surprises! Anticipate the questions you will be asked and the arguments that will be used against you and be ready with good responses.
- Be sympathetic, not arrogant. Don't be dismissive (you will lose sympathy).
- Show that you understand your opponents' arguments. Exude wisdom and confidence.
- Rehearse. Rehearse. Rehearse.

At some point, you may choose to send out a press release prepared on your stationery and take the initiative.

Suggested Readings:

"5-5-2000," Wade E. Allan, *The Planetarian*, pp. 15-6, Sept. 1994.

"Compact Planetary Groupings," Jean Meeus, *Skeptical Inquirer*, pp. 290-92, Spring 1988.

"Great Earthquake Hoax," Edward Upton, *Griffith Observer*, pp. 2-8, Jan. 1982.

"Hooked on Horoscopes," Adrian Furnham, *New Scientist*, pp. 33-36, Jan. 26, 1991.

"How to Debate Astrology," John Mosley, *Astronomy*, p. 8, Feb. 1990.

"On the Trail of the Jupiter Effect," L.G. Thompson, *Sky and Telescope*, p. 220, Sept. 1981.

"Science in the Bear-pit," Tony Jones, *New Scientist*, pp. 70-71, Oct. 7, 1989.

"The 'End of the World' at the Griffith Observatory," Robert S. Richardson, *Griffith Observer*, pp. 62-5, May 1962.

"Your Astrology Defense Kit," Andrew Fraknoi, *Sky and Telescope*, pp. 146-50, Aug. 1989.

Any number of commercially available and shareware astronomy programs will show the alignments. See the August 1995 *Sky and Telescope* for recommended software. My thanks to Jean Meeus for working out details of the May 2000 alignments.

Astronomy Education and the AAPT

George S, Mumford
Department of Physics and Astronomy
Tufts University
Medford, MA
USA 02155

Tel.: (617) 627-3029
FAX: (617) 627-3878

Abstract: The American Association of Physics Teachers (AAPT) provides several resources for astronomy instructors. Its principal publications the *American Journal of Physics* and *The Physics Teacher* often carry relevant articles; national meetings feature hands-on workshops and papers on astronomical research and educational topics. In addition the Association has sponsored the development of computer software and published reprint books of interest to those teaching astronomy.

Introduction: The AAPT is one venue that brings together high-school instructors as well as teaching staff from junior and senior colleges to discuss areas of mutual interest and concern. The organization has some 11 000 members; holds three national meetings annually - one is joint with the American Physical Society - and promotes one or more regional meetings each year. Its major publications provide a resource for the dissemination of information related to the astronomy curriculum.

Committee On Astronomy Education: The major concerns of this group are to plan the various activities, either alone or jointly with other committees, that take place at national meetings, to serve as a liaison with other groups involved in astronomy education; and to provide a focus for AAPT's astronomical interests. The nine voting members represent high schools as well as junior and four-year colleges and universities. All committee meetings are open to participation by any interested person.

At national meetings, this committee generally sponsors half to full-day workshops devoted to such activities as image processing, remote telescope operation, or the latest computerized laboratory activities. Current astronomical research is described in half-hour talks by invited speakers. Both invited and contributed papers on a wide-range of topics of interest to educators are presented in other sessions.

Together with other committees such as those on Computers in Physics Education, Graduate Education in Physics, or Science Education for the Public, joint and plenary sessions have been sponsored. While the Astronomy Education committee may serve as a focus for astronomical interests much activity takes place in other spheres and under other auspices.

Publications:

- *The Physics Teacher* features articles on curriculum development, laboratory techniques and a variety of other topics of interest to teachers. The column *AstroNotes* gives suggestions for observational projects, classroom demonstrations and pencil and paper activities. Manuscripts may be submitted to The Editor, The Physics Teacher, Dept. of Physics, SUNY, Stony Brook, NY 11794-3800.

- The *American Journal of Physics* is devoted to the instructional and cultural aspects of the physical sciences. Its contents from time to time include current research, a variety of historical and educational topics, novel approaches and demonstrations for the laboratory or classroom, book reviews, and so forth.

Among the most valuable papers published are the so-called resource letters. These are intended as guides for teachers and others to some of the literature and teaching aids that may help to improve course content in a specific field. A recent one on radio pulsars (Weisberg, 1993) provides a guide to the literature on this topic. It cites papers at the elementary, intermediate, and advanced levels and lists historically important papers, including relevant ones published prior to the discovery of pulsars as well as those devoted to current research. Further, there is a discussion of pulsars' use in studying other phenomena from the interstellar medium to relativistic theories of gravitation. This paper has just been published as a reprint book that includes reprints of major papers cited and is available for twenty dollars ($20) from the AAPT at One Physics Ellipse, College Park, MD 20740-3845.

Other recent titles include Time and Frequency Measurement (Hackman and Sullivan, 1995), Experimental Tests of the Discrete Space-Time Symmetries (Commins, 1993) and the Anthropic Principle (Balashov, 1991). For a listing of the first one hundred resource letters see the article by French and Stuewer (1995).

References

Balashov, Y. V. 1991, *Am. J. Phys.*, **59**, 1069.
Commins, E. D. 1993, *Am. J. Phys.*, **61**, 13.
French, A. P., and Stuewer, R. H. 1995, *Am. J. Phys.*, **63**, 303.
Hackman, C., and Sullivan, D. B. 1995, *Am. J. Phys.*, **63**, 306.
Weisberg, J. M. 1993, *Am. J. Phys.*, **61**, 13.

Solar System – Activities and Rationales

Lois A. Nokleby
Farmington Middle School
4200 208th Street West
Farmington, Minnesota
USA 55024

The set of activities focuses on student-centered activities. Research skills, collegial collaboration, and public presentation are central to the projects. Assessments are written, practical, and reflective in nature. A description of the activities follows.

The students determine scaled distance and planet size of one of the nine planets, make a model of that planet, and place the planet (on the same scale as the size) in the main mall of the school. They proceed to take a "promenade through the Solar System" noting distances and sizes as they go along. They are forewarned to be conscious of things that may be a surprise to them. They are assessed through written observations and reflections. It has proven to be highly successful in that students see, experience, and realize how the solar system really is.

The next set of activities centers around the planets and their characteristics. Each group of 2-3 students researches what "their" planet is like. They design a creature that would be suitably adapted to the set of characteristics and make an annotated drawing of the creature. The annotations include the parts and how each part "fits" on the planet. The group then writes a creative story including a description of their creature. The stories, but not the drawings, are passed on to another group of students and they draw the creature based on the description in the story. The resulting drawings are a measure of how well the creatures were described originally and how well the "new" group followed the given description. The stories are read to the whole class, the drawings are displayed, and comparisons between the two drawings are made and analyzed. The groups then build their creature, present it to the class and justify to the class why they made their creature as they did. The groups also make and present advertisements for the planets. Finally, each group makes their "expert" presentation, summarizing for the entire class the characteristics of their planet. The other class members take and organize notes with this data. A prepared, but incomplete, summary is given to the students; they complete the summary based on their notes, and complete a written test on the material. Most students enjoy the experiences and do notice trends in the planets' characteristics.

The solar system activities span a period of 4-5 weeks and have been refined many times. A strong characteristic is that the students do the work, are held accountable for it, and are assessed in a variety of ways. The activities are popular with the students, based on survey responses.

Education Programs Of The International Astronomical Union

John R. Percy
Erindale College
University of Toronto
3359 Mississauga Road North
Mississauga, Ontario
Canada L5L 1C6

Internet: percy@astro.utoronto.ca

The International Astronomical Union

The International Astronomical Union is a non-governmental union founded in 1922 to "promote and safeguard astronomy ... and to develop it through international co-operation". There are currently 60 adhering countries, and 7839 individual members. Members are nominated by the adhering countries. The normal qualification is an advanced degree in astronomy, and a few years of experience as a professional astronomer. There are no individual membership fees. The IAU is funded by the adhering countries, on the basis of the number of IAU members in that country, and its ability to pay. Almost all of the funds are used for the development of astronomy. The IAU is administratively "lean".

IAU Commission 46

The IAU is organized in 40 interest groups called commissions. Commission 46 (The Teaching of Astronomy) is the only one responsible for astronomy education. Its mandate is "to further the development and improvement of astronomy education at all levels, throughout the world". There are about 150 members, including representatives from each of the adhering countries, as well as members-at-large. The commission is guided by an Organizing Committee of individuals with responsibilities for specific projects.

In general, Commission 46:

1. Works with other scientific and educational organizations, and with other IAU commissions and working groups, to promote the development of astronomy and other sciences.
2. Through the national representatives, promotes astronomy education in the adhering countries, and informs the IAU of the state and needs of astronomy education in these countries.
3. Encourages all programs which promote astronomy and astronomy education in the developing countries.

IAU Commission 46 Meetings

Commission 46 holds business and education sessions at every IAU General Assembly. At the 1994 General Assembly in The Hague, Commission 46 organized a one-day session on "Current Developments in Astronomy Education". The proceedings will be published in the Transactions of the IAU.

Teachers' Workshops

Commission 46 also organizes one-day workshops for local schoolteachers at every IAU General Assembly, and at many regional meetings. In February 1995, in Cape Town, South Africa, John Percy and Patricia Whitelock organized a very successful workshop for 100 local teachers – the first such workshop to be held in conjunction with an IAU research conference.

IAU Colloquium #105: The Teaching of Astronomy

Of the over 300 IAU symposia and colloquia, this is the only one devoted to astronomy education. It was hosted by Williams College, in Williamstown MA. Over 160 astronomy educators, from over 30 countries, enjoyed five days of talks, posters, formal and informal discussion. The informal aspects of the meeting were especially important in forging new friendships and exchanges of ideas. The proceedings of the colloquium were edited by Jay Pasachoff and John Percy, and published by Cambridge University Press (1990); a paperback edition is available. It has been one of the bestselling IAU conference proceedings ever, and is still the best overview of astronomy education worldwide.

International Schools for Young Astronomers

ISYA's are intensive three-week schools attracting 20 to 50 students, mostly at the graduate level. They are held approximately once a year, in different parts of the developing world. About half of the students come from the host country; the rest come from neighbouring countries. A handful of faculty members give mini-courses on topics chosen by the host institution, in consultation with the ISYA secretaries (currently Don Wentzel and Michele Gerbaldi). The students also give short talks on the state of astronomy at their institution, and on their research projects. Informality, interactive discussion, and interchange of ideas are strongly encouraged.

The most recent ISYA's have been in China (1992), India (1994), Egypt (1994) and Brazil (1995).

Visiting Lecturers Program

The VLP sent a series of lecturers into two target countries-- Paraguay and Peru – for two to three months, to give courses, to work with local astronomers and students, and to generally help to improve the astronomical capacities of these countries. The target countries were selected in an open competition. The IAU covered travel costs, and the host country covered local expenses. It was expected that the host country would continue to support astronomy after the VLP and, keeping in mind the economic and political situations in the target countries, this has been achieved.

The VLP has now been replaced by a more flexible program – Teaching for Astronomical Development (TAD) - which is intended to help a small number of additional countries. The IAU would support primarily the cost of travel – for a visiting lecturer to give a course, for an advanced student to work in another country, or to establish and advance an appropriate research collaboration, for instance. The host institution would cover local costs, and establish a plan whereby astronomy could be maintained after the TAD program ends.

The Travelling Telescope

The Travelling Telescope is a Celestron-8 telescope with a photometer and other instruments; it is intended for use in countries without significant astronomical research facilities. The project was funded through the Canadian Committee for UNESCO, and was developed by Dieter Brueckner, of the University of Toronto. The photometer is currently being used for research in variable stars, using an existing telescope in Paraguay.

Newsletter on the Teaching of Astronomy

The Newsletter of IAU Commission 46 is published twice a year, and is sent to the members of the Commission, and to several hundred other addresses around the world. It publishes short announcements, reports and news notes on any topic related to astronomy education, at all levels, throughout the world. The editor is currently John Percy. Because of the cost of printing and mailing the newsletter, it is now being published in electronic form for all those with access to electronic mail. The paper edition is still available to key addresses without electronic mail. To receive the electronic edition, contact Armando Arellano Ferro (armando@ astroscu.unam.mx).

Once every three years, the Commission 46 national representatives provide a report on the state of astronomy education in their country; these are published in a special edition of the newsletter. They provide a fascinating look at astronomy education worldwide.

The Astronomy Village

Stephen M. Pompea
Pompea & Associates
1321 East Tenth Street
Tucson AZ
USA 85719-5808

E-mail: spompea@as.arizona.edu

The Astronomy Village was developed by the NASA Classroom of the Future Program at Wheeling Jesuit College. It is designed as a supplement to the 9th grade earth science or physical science curriculum, although it has been tested and used at a variety of grade levels. The basis of the software is that students research teams go the "The Astronomy Village" to work on an astronomical investigation for about three weeks. In the Astronomy Village, students actively construct inferences, relations, comparisons, questions, mental models, and the like. They actively solve problems and use image processing. This can be a culminating activity to their astronomy unit in their general science class. The NASA Astronomy Village can also be used as a resource for teachers and as a reference for students.

In this CD-ROM based multimedia program, student teams can pursue one of (10) ten research "investigations". In each investigation they are guided by a mentor, receive e-mail, hear a lecture in the Village auditorium, and make observations using ground or space-based telescopes in a virtual observatory. The students also process data using the NIH Image image processing program. They also keep a detailed logbook of their research activities. They can run simulations on stellar evolution and can manipulate 3-D visualization tools. At the end, they present their research to others in their class and answer questions in a press conference about their results.

The investigations emphasize open-ended research topics that are similar to front-line astronomy research questions. For example, the investigations revolve around searches for: Obscured supernovae using neutrino data; Earth-crossing asteroids; Planets around nearby stars; Undiscovered nearby stars: and Protoplanetary disks. Other investigations address long standing astronomical problems: Characterizing galaxy clusters; Comparing views towards the Milky Way with views away from it; Using Cepheids for distance determination; Characterizing how star forming regions look at different wavelengths; and Astronomical site selection.

Our goal was to create a multimedia product aligned with the national standards in mathematics and science education and consistent with sound pedagogical practice. We wanted to encourage students to "learn science in ways that reflect the modes of inquiry that scientists use to understand the natural world". The national standards also advocate covering less content but in greater depth. We are consistent with these standards; the Village experience is an intense introduction to the research and research models. The Village research problems engage students in a model of scientific inquiry similar to that used by astronomers. The steps in this model include: identifying a question or problem; conducting background research; conducting experiments and collecting data; analyzing data; interpreting their results and conclusions; and presenting their results to an audience of peers for comment. However, we have taken great care to not overwhelm the students.

Components of the Village: The village metaphor is used to frame the research activities in each path. The interface allows students to interact with the program, navigate to different "facilities" within the village, and make use of various tools and resources. Here is a brief outline of the Astronomy Village.

The Orientation Center : The Orientation Center contains a Village tour that introduces the Village and its tools and resources to students. The tour is presented as a combination of computer graphics, audio, and video clips. The Orientation Center contains "skills building" activities that guide students through introductory activities to show them how to use the various facilities, tools, and resources within the Village.

The Conference Room: The Conference Room contains the ten research questions from which teachers may assign projects or from which students may select projects.

The Observatory: While in the Observatory, students are able to use a simulated telescope control panel to "capture" images of various objects at a variety of wavelengths. Both local observing and remote observing at a variety of sites are simulated. Students can also browse through an archive of over 300 images.

The Library: The library contains an electronic catalog and search tools, the "COTF Times" newspaper (which contains articles illustrated with images), and a collection of NASA publications that focus on various aspects of astronomy (e.g., "Exploring the Universe with the Hubble Space Telescope"). There are over 150 articles in the Library.

The Computer Lab: Within the Computer Lab students are able to access and use three tools:

1. NIH Image (an image processing program)
2. Simulation programs, including a "Star Toolkit" which allows them to create stars of different masses and observe their life cycle
3. Simulated e-mail, and a telecommunications program that allows them to connect to the NASA Classroom of the Future's server and NASA SpaceLink.

The Hands-On Lab: The Hands-On Labs contains directions for conducting experiments on and off the computer that teach, illustrate, or reinforce the astronomical concepts. Students will be able to print the directions to these experiments. There are over 25 Hands-On Labs.

The Auditorium: The Auditorium provide students with twelve different 5-10 minute talks by astronomers related to the ten investigations. Each lecture is presented by means of audio recordings and illustrative computer graphics, video clips, and images.

The Cafeteria: The Cafeteria is used by students to "overhear" astronomers informally talking about why they chose astronomy as a career, what being an astronomer is like, what it means to be engaged in scientific inquiry, and their specific research projects. There is also humor in the conversations, the menus, and the wall decorations.

The Press Conference: At the end of each investigation, students get a chance to answer questions from the press. Sometimes the simplest questions are the hardest to answer!

Meta-tools: There are four "meta-tools" students may access from any location within the Village. These include the Student Log Book, Calculator, Village Map, and Help. A fifth meta-tool is the "path" diagram, which outlines the types of activities possible when a more structured approach is desirable.

Minimum System Requirements: Macintosh LC with 8 MB RAM, CD-ROM Drive; System 7.0 or higher, HyperCard 2.1; 13" Color Monitor (256 colors).

How to get the Astronomy Village software: The Astronomy Software Village is being made available as this goes to press. The cost is nominal. For more information on obtaining the software, contact: The Astronomy Village Project, NASA Classroom of the Future, Wheeling Jesuit College, Wheeling WV 26003 Phone (304) 243-2388; Fax (304) 243-2497.

Cosmic Explorers: An Astronomy Day Camp for Children

Alison Procter
Department of Astronomy
Yale University
P.O. Box 208101
New Haven, CT
USA 06520-8101

E-mail: procter@astro.yale.edu

and

Heather Johanson
Department of Astronomy
University of Western Ontario
London, Ontario
Canada N6A 3K7

Abstract: We created and ran a week-long astronomy day camp for children ages 9-12 through the YM-YWCA in London, Ontario in July 1993. We included as much 'hey, wow!' science as possible since this was a summer camp. To this end, we took two trips to Cronyn Observatory on campus at the University of Western Ontario to look at the sun and at the night sky. We also cooked up a comet and we designed spacecraft.

Camp Format

The camp was located at the London YM-YWCA, with one day trip to the observatory, and one night session which included the parents as well as the children. It ran from Monday to Friday, 9 am - 5 pm, and one evening session 8:30 - 10:30 pm. We included a half-hour break in both the morning and afternoon to go outside and play some active games. We also had an hour each day in the pool (a welcome relief for the leaders!) and an hour for lunch. This left us with about 5 hours a day for astronomy, which we filled with lectures, crafts, videos and games. We planned for 24 children, ages 9-12. We actually got about 20 children each of the 4 weeks, ages 7-15. The leaders were Heather Johanson and Alison Procter, UWO astronomy graduates, and Jerome Neufeld, grade 10 Oakridge High School student.

Activities

The highlight of the camp was the evening visit to Cronyn Observatory, where the children got to look through the 12 inch refractor at the moon, planets and other interesting objects. This outing was also a hit with the parents, who were invited for the evening. Some of our successful indoor activities were a scale model of the distances in the solar system on a string, a balloon model of the expanding universe, and making small telescopes from cardboard tubes and lenses. The following are two activities that also went over well.

The Tin Canetarium

Purpose: To familiarize the student with a constellation.

Materials: large tin can with one end removed; hammer; nails of assorted sizes; tracing paper; pencil

Method: Draw a constellation or group of constellations on the tracing paper so that they will fit on the bottom of the tin can. Place the tracing paper on the tin can upside down, so that when you look through the can, the constellation will look as it does on the sky. Using nails of different sizes to correspond to the magnitudes of the stars in the constellation, punch holes in the bottom of the can.

Historical Models of the Universe

Purpose: To discuss historical models of the earth/moon/sun system and to present today's.

Materials: world map; paragraphs on ancient models; paper; pencils, markers, crayons etc

Method: We had the children read a short paragraph on each of six ancient models (Egyptian, Indian, Chinese, Japanese, Greek and Norse) and discussed how the geography of the region affected their view of the universe. Then we had the children draw the model of their 'universe' as they saw it from their home. We stressed that they had to rely only on what they could see, not what they had been told (for example, the earth looks flat). We asked them to make up a myth about how the sun and moon move through the sky. As they were drawing, the leaders would go around and ask what the children were thinking about. This helped get some slow starters going, let students bounce ideas off each other, and kept others from getting off topic. The children then presented their models, and we discussed the modern ideas ofthe solar system.

Acknowledgments: Mark Whitcombe of the Sheldon Centre for Outdoor Education and the Algonquin Space Campus, and Dr. John Percy of the University of Toronto for their help in finding ideas and materials; Beth Martin of the London YM-YWCA for her support and for taking the chance on us; The Astronomy Department of the University of Western Ontario for the use of Cronyn Observatory and other resources.

Undergraduate Astronomical Spectrographs-A Modular Approach

Stephen J. Ratcliff
Dept. of Physics
Middlebury College
Middlebury, VT
USA 05753

E-mail: ratcliff@middlebury.edu

Abstract: Spectroscopy of celestial objects plays a vital role in modern astronomy, and therefore has an important role in astronomy education as well. The enormous range in brightness of objects of interest, from the Sun through the planets to naked-eye stars, supports a variety of observations suitable for advanced undergraduate lab exercises. However, it also calls for a variety of dispersions. By assembling spectrographs from individual optical components on an optical table, students learn much more about how a spectrograph works and are able to design the spectrograph to suit the observations. An optical fiber is used to couple the light from the telescope to the spectrograph.

Recognizing the value of spectrographic laboratory work in undergraduate astronomy (and physics) programs, we have been working on the development of instrumentation for astronomical spectrographic observations which could serve as the foundation of spectroscopic instrumentation for institutions starting essentially from scratch.

The two parameters which perhaps best describe the variety of spectroscopy experiments we envision are (i) the available intensity of the light source (ranging from the Sun at the bright end, to stars and nebulae at the faint end), and (ii) dispersion and/or resolution required (from spectral features at lower-than - "classification" dispersion, to the measurement of the < 0.01 nm shifts of solar absorption lines due to solar rotation). To accommodate the enormous ranges, we need in principle only supply a variety of the optical components (slits, gratings, lenses/mirrors, and mechanical holders) which make up a spectrograph, and ensure that the detection of the light is done as efficiently as possible. A corollary is that since the spectrograph will be assembled from individual components on an optical bench, and in any case may require a very large layout (if high dispersion is required), it will not be possible in general to attach the spectrograph to the telescope. Hence, a separate light-delivery system will be required. This may consist of an optical fiber cable, perhaps attached to a telescope; or it may consist of a heliostat and set of mirrors, for solar observations. For greatest observing efficiency, scanning-spectrometer mode should be avoided in favor of use of a panoramic detector (CCD array, video camera, or even photographic emulsions). This also simplifies the mechanical needs, as all optical components can remain fixed in location and orientation during the observations.

The required components, though numerous, are to a great degree interchangeable and, unless faint objects and/or high dispersion are part of the experiment, none costs more than a few hundred dollars (and in some cases could be machined in-house). Required components for the spectrograph include: a fixed or adjustable slit; a collimating lens or mirror; a dispersive element (most likely a reflection grating, though transmission grating or prism may be used); a camera lens or mirror; and mechanical positioners for these items as well as for the detector.

Note that a single mechanical positioner may be used with, for example, mirrors in a variety of focal lengths, or identically-sized gratings with different groove spacings. Hence, a selection of optical components can be used to assemble a wide variety of spectrographs, and the marginal cost of a "new" spectrograph may be only a few hundred dollars!

Following are descriptions of two spectrographs we have assembled in this manner, using a grating with 1800 lines/mm.

(A) A moderate-dispersion spectrograph: This was designed and assembled by Middlebury undergraduate Aaron Ambuske for use in an experiment to detect and measure the rotational speed of Jupiter by observing the Doppler-shifted solar spectrum. The optical fiber from a 0.41 m telescope illuminates an adjustable slit (used at a width of a few tens of microns) at the focus of 0.45 m focal length collimator (mirror). The camera consists of two components: a 1.0 m focal length mirror and a 35 mm focal length lens, which used in conjunction provide an effective camera focal length of approximately 2.0 m. In this configuration, the limb-to-limb Doppler shift in Jupiter's spectrum was expected, and found, to be several pixels (each 27 μmm) in our CCD detector.

(B) A spectrograph designed primarily for use in stellar spectroscopy: The collimator is a 305 mm focal length achromat (originally marketed as a finder telescope) and the camera is a Nikon 135 mm f/2 photographic lens. An additional feature of this spectrograph is the lack of an entrance slit; the end of the optical fiber (200 μm diameter) serves as the slit. With our CCD detector of dimension 28 mm in the dispersion direction, this configuration provides coverage of about 100 nm and a resolution of approximately .25 nm. In addition, the optical fiber cable consists of four separate fibers: one from the telescope, and three shorter ones which can be used to feed comparison lamp spectra (we usually use a Neon bulb) into the spectrograph. The four fiber ends are stacked in a column at the end of a 1/4 in diameter ferrule which is held in the focal plane of the collimator. This cable assembly was constructed in-house.

This project has been supported by Middlebury College and by the National Science Foundation (through grant DUE-9252109).

Light! Spectra! Action! Teaming up to create excellent astronomical experiences

Elizabeth E. Roettger, Amy L. Singel, Douglas K. Duncan[1]

[1]Manager of Astronomy Content
The Adler Planetarium
1300 South Lake Shore Drive
Chicago, IL
USA 60605

Tel.: (312)-322-0338
E-mail: eroettge@midway.uchicago.edu
http://astro.uchicago.edu/adler

Abstract: Adler Planetarium and Astronomy Museum creates "excellent audience experiences" through a process of prototyping and evaluation. We try to find good projects to replicate or adapt and good partners to reduce the effort, achieve higher quality, or multiply the impact of exhibits, shows, and programs. Adler recently created the exhibit Light! Spectra! Action! by starting with exhibits from the NSF and Exploratorium Science Exhibits Starter Set program. Exhibit text appropriate to astronomy was written, evaluated with test audiences, and revised. Demonstrations and photographs about color, light, and spectra (some based on previous activities) are part of the exhibit. Photographs were enthusiastically provided by David Malin/Anglo Australian Observatory, the Astronomical Society of the Pacific, The Space Telescope Science Institute, Roger Ressmeyer, and Kitt Peak National Observatory. A NASA IDEA Grant was used to enhance the exhibit and enable Kitt Peak to replicate parts of the exhibit for their Visitor Center.

The Main Idea: We're in the business of creating experiences for our visitors (we're no longer focusing on the exhibits as things to be produced, but as a tool to affect our visitors). We wish to create excellent audience experiences, but how do we go about it?

Start from something good: A good idea that works already is a good place to start. Using it, we produce new ideas, come up with a new combination of old ideas, use an existing process in a new way, improve something without changing the substance, or change something to use it for a slightly altered purpose or focus. Starting from something good saves time and keeps us from going too far down a dead-end path.

Work with good partners: It's becoming more apparent that few of us have the full range of knowledge and talents to create an excellent experience for the very diverse audiences we're trying to reach (e.g., AAAS 1989), and that fewer still can keep multiple points of view in mind at one time. This is true of institutions as well as individuals. Cooperation and partnerships are necessary to create the multifaceted quality experiences we desire.

Test your theories: Wonderful ideas don't always work out in practice. It's important to define what you're trying to achieve, and then test it in as unbiased a way as possible. It's best to do this early and often, so as to avoid wasting resources on ineffective ideas. In our museum, the first questions are often "can a visitor use this successfully?" and "does a visitor enjoy using this enough to choose to use it?", but we also must ask "is the visitor learning what we want them to learn?", "is the visitor having the positive emotional responses we wish?", "are the visitors interacting with each other in the way we wish?" (we want visitors to learn something, but also to like astronomy and feel its excitement, and to be able to experience the museum together with their family or partner).

Light! Spectra! Action!: An example in process

Start with something good: We started with exhibits originally developed by the Exploratorium in San Francisco. We applied and received a grant through the NSF's and Exploratorium's Science Exhibits Starter Set program, from which science museums can obtain copies of Exploratorium exhibits. We chose exhibits on light (Bruman 1991, Hipschman 1990, Hipschman 1987), as these seemed most relevant to astronomy.

The exhibits were primarily about physics; we're an astronomy museum. We decided to rewrite the labels (the instructions and other text that are part of an exhibit) to make the exhibits more relevant to astronomy. In some cases, we altered the exhibit itself.

We also had an activity (from Edna DeVore and the FOSTER program, date unknown) where participants look at astronomical images through colored filters. We incorporated this into the exhibit by using astronomical images and photographs to augment the exhibit, tie the exhibit components together, and illustrate the ties to astronomy. We put photographs of telescopes (particularly the interiors) with an exhibit where visitors play with curved mirrors, lenses, and prisms (affectionately referred to as "telescope guts"). We illustrated an exhibit that shows color separation with colored filters and a prism with five images of the Ring Nebula – one in full color, and four black-and-white images, each taken through one of four colored filters. We took an exhibit on spectra that uses gas discharge tubes and diffraction gratings to show the "fingerprints" or "bar codes" of light, and tied it to astronomy with an image of red hydrogen gas in a nebula.

Work with good partners: We successfully applied for an IDEA grant together with the Kitt Peak National Observatory's Visitor Center. After prototyping at Adler, some of the exhibits will be replicated for Kitt Peak. Further testing in that (rather different) environment will lead to more refinements and improvements. Meanwhile, Kitt Peak supplied some of the images and photographs, and we anticipate that they will bring additional scientific expertise to the project. We also anticipate that we will learn from each other, and that the partnership thus established will lead to further cooperation and advantages for both.

The Exploratorium has only recently begun replicating exhibits for other museums. The new exhibits are not always identical to the ones in the Exploratorium. We have requested some changes unanticipated by the Exploratorium. This, too, has been a cooperative process with learning on both sides. For example, we were trying to decide whether to make the sides of one exhibit component white or black. We called the Exploratorium and asked them to hold black material to the sides of the original exhibit. After doing this, they were able to advise us not to choose the black.

Test your theories: Our development process involves prototyping and evaluation (see for example Taylor 1991). In the theoretical linear process (much like the textbook description of the scientific process), we start with "front end" evaluation to determine audience interests and existing ideas (which we then build on or challenge), and to collect previous experiences and research which may apply. Then we move to a series of prototypes (cardboard and tape) and "formative" evaluation to create something that works, that can be used successfully by visitors, that they find interesting, and that create the effects in our visitors that we wish to create (learning, interest, emotional responses, interactions). Then we build the real thing and do a formal, "summative" evaluation to assess the effectiveness and give us information for future projects.

Of course it doesn't work like that in reality, but the theory is instructive. It is challenging to apply the process to something that already exists, and to use it in cooperation with partners, but it is well worth the trouble.

For Light! Spectra! Action!: We did some front-end evaluation before designing the exhibit. We found some of the visitors' existing ideas, collected relevant research, and tested several possible exhibit titles with visitors. This process continues as we adapt the exhibits and develop demonstrations and facilitation ("explainers" in action) to augment the exhibit components. Because we were starting with something existing, it was less crucial to do all of the front end evaluation before starting.

We wrote the exhibit text (labels for exhibit components and captions for the images), designed the exhibit layout to encourage reasonable circulation from component to component, received most of the exhibit components, and opened the exhibit. We have since added the images, and are working on one last exhibit component before installing it. We've found the parts that break or are misplaced most often, and are both improving these and using the information to help us understand how people use the exhibit. Educators and astronomers have worked together and separately to facilitate the exhibit and observe people using it. (It's important for the team to get first-hand experience with visitors.)

Evaluators have now done one form of evaluation on two of the exhibit components. (They asked people to use the exhibit, observed them doing so, and interviewed them afterwards.) This turned up both strengths and weaknesses. We are now trying to rework one exhibit component to respond to what we've learned (see Borun and Adams 1991, Kennedy 1990). Eventually, we will evaluate the other components and the exhibit as a whole (see Serrell 1993), using several different techniques (such as observing people who haven't been primed to use the exhibit). We will also do the same for demos and facilitation we develop.

Conclusions: We're changing the way we're doing things, and are adapting our processes to fit. We use established projects, good partners, and a series of prototyping and evaluation to create excellent audience experiences. It isn't fast and it isn't easy, but the results are worthwhile (and it's also quite satisfying). The process has parallels to the (real) scientific process - and the unbiased testing of theory is crucial.

References

AAAS 1989. *Science for All Americans*. (USA: Amer. Assoc. for the Advancement of Science), p. 155.

Borun, Minda, and Adams, Katherine A. 1991, "From hands on to minds on: Labelling interactive exhibits", in *Visitor Studies: Theory, Research, and Practice*, V. 4 , ed. A. Benefield, S. Bitgood, and H. Shettel (Alabama: Visitor Studies Association).

Bruman, R. et al. 1991. *Exploratorium Cookbook I*, Revised (San Francisco: The Exploratorium), sections 9 and18.

DeVore, Edna, undated handout. Flight Opportunities for Science Teacher EnRichment -- FOSTER Project. (SETI Institute, Mountain View, CA, and

NASA Ames Res. Ctr., Moffet Field, CA.)
Hipschman 1990. *Exploratorium Cookbook II*, Revised (San Francisco: The Exploratorium), sections 95 and 131.
Hipschman 1987. *Exploratorium Cookbook III* (San Francisco: The Exploratorium), sections 174 and 178.
Kennedy, Jeff 1990, *User Friendly: Hands-On Exhibits That Work* (Washington DC: ASTC), pp. 59-68.
Serrel, Beverly 1993, "The question of visitor styles", in *Visitor Studies: Theory, Research, and Practice*, V. 6 , ed. D. Thompson, S. Bitgood, A. Benefield, H. Shettel, and R. Williams, (Alabama: Visitor Studies Association), p. 48.
Taylor, Samuel, 1991. *Try It! Improving Exhibits Through Formative Evaluation*. ASTC, Washington DC.

Astronomy Distance Education Courses at the University of South Carolina

John L. Safko
Department of Physics and Astronomy
University of South Carolina
Columbia, SC
USA 29208

Tel.: (803) 777-6466
FAX: (803) 777-3065
E-mail: safko@psc.psc.sc.edu

I. Introduction

Telecommunications courses have been taught at the University of South Carolina since 1965. The enrollments have grown until there are currently over 10,000 students per term enrolled in distance education courses. The distance education astronomy courses, which enroll between 30-50 per term, are only a small part of the university's program. The Office of Distance Education provides coordination for these courses. They arrange for rooms or video tapes, copying and distribution of materials and texts, enrollment and communications with students, and proctoring of exams. They also provide a record keeping service and receive, distribute, collect and return assignments to students.

I adapted our graduate level course PHYS 781 "Astronomy for Teachers" to television using video series Project Universe from Coast Community College. Although the astronomy content of this course is at the level of an introductory course, it carries graduate credit to meet the South Carolina Department of Education requirements in effect in the 1970's. Since the content was so similar, we also offered the first undergraduate astronomy courses (ASTR 111 and ASTR 111A) at the same time. For the supporting print material, I used a modification of the study guide used in our on-campus self-paced courses.

Initially the courses were broadcast open circuit over the state-wide South Carolina Educational Network. Open circuit broadcasting has several disadvantages, ranging from the broadcast time being chosen by others to the necessity of having to present a program even during midcourse exam time. After three years the courses were distributed on videocassettes and additional, locally produced video segments were added. We have also added the second semester to the undergraduate video options, upgraded the Coast Video Tapes as new releases were produced, and incorporated The Astronomers series.

II. Philosophy

The Astronomy Distance Education courses follow the same philosophy used in our on-campus self-paced, mastery-oriented courses. The student is given clearly stated performance-based learning objectives. Some hands on experience with the concepts is included even if the student is only taking 3 credit hours. This means that contrary to most 4 semester credit hour introductory courses that have the "laboratory" component (termed "exercises" in our material) in the one hour segment, we have integrated the hands-on experience into the three credit hour portions as well as in the one hour courses. We are also working on integrating the constructivist approach into the exercises.

Since the K-12 teachers and many of our undergraduates take only one course in astronomy, I decided it is necessary for the first semester to cover more than just the solar system. The teachers' course and the first term undergraduate course both present a summary of astronomy from naked eye astronomy through a very brief introduction to the big bang theory. They also include material on extraterrestrial life and on pseudoscience.

III. Telecommunication Courses

There are five telecommunications courses currently offered to our distant learners. The undergraduate courses are ASTR J111 for 3 semester credit hours, ASTR J111A for 1 credit hour, ASTR J112 for 3 hours and ASTR J112A for 1 hour. The "J" designator shows that these are telecommunication courses. They parallel the on campus courses in material and in expected achievement. The three hour courses have written assignments, exercises and observations, a midterm, and a final exam. The one hour courses have written assignments, exercises and observations, and a final exam.

The graduate level course for K-12 teachers (PHYS J781) carries 3 semester credit hours and is accepted as a content course in several states. The course has written assignments, exercises and observations, a midterm, and a final exam in parallel with the first undergraduate course. There is also an additional exercise from the study guide and the students are expected to perform and critique 12 astronomy exercises from a book produced by our State Department of Education. The exams in all courses consist of 80% multiple choice and 20% essay questions. The multiple choice exam is graded as correct answers minus one-fourth the wrong answers.

The courses are offered three times per year (Fall, Spring and Summer). The total undergraduate enrollment varies between 20-50 per term while the graduate course enrollment is between 5 and 15. Most students are from the Southeast, although there have been students from as far away as Alaska, Canada, and Egypt.

IV. Correspondence Courses

The correspondence courses are designated by a "C" designator. The four undergraduate astronomy courses are identical with the telecommunications courses except they do not have the video component. Correspondence students have 12 months to complete a course with a 6 month extension upon request. There are typically between 10-15 students enrolled at any time. The completion rate is, as might be expected, lower than for the telecommunications courses. As in the Telecommunications Courses, most of the students are from the Southeast, although there have been students from as far away as Africa and Japan.

V. Resources for the Video Courses

The current text is Hartmann & Impey: *Astronomy: The Cosmic Journey*, 5th Edition, Wadsworth: Belmont CA. The study guide is John L. Safko: *Astronomy Study Guide and Exercises for Distance Education*, 4th Edition, Kendall/Hunt: Dubuque. The three hour courses use the Activities Kit produced by John Wiley for R.R. Robbins, W.H. Jeffreys and S.J. Shawl: Discovering Astronomy. The graduate course also uses *Astronomy Activities for the Classroom*, South Carolina Department of Education.

The videos are UNIVERSE: The Infinite Frontier which consists of twenty-six 1/2 hour programs, The Astronomers which consists of 6 programs of approximately one hour length, eight locally produced 20 minute programs from our on-campus courses, and two brief programs explaining how the video courses operate and how to prepare for the exams.

Students' Hands-On Physics (SHOP)

Jon M. Saken[1], Todd J. Henry

[1]Space Telescope Science Institute
3700 San Martin Drive
Baltimore, MD
USA 21218

E-mail: saken@stsci.edu

The "Students' Hands-On Physics" program (SHOP), now in its third year, is designed to expose students in one of Baltimore City's most disadvantaged schools to fundamental exercises in physics and astronomy. The program typically involves 150-200 students a semester, at 6th, 7th, and 8th grade levels, including special education classes. Over the course of a semester, a team of four to six volunteer instructors from the Space Telescope Science Institute and Johns Hopkins University conduct six sessions in four classes at West Baltimore Middle School, one of the largest middle schools in the nation. These sessions consist of bi-weekly labs and a follow-up session to reinforce the concepts covered in the experiment. These repeated visits to the same group of students and teachers is the greatest strength of the program. It provides a sense of continuity for the students and teachers, and allows a great deal of personal interaction between the volunteers and the students. Students have a chance to work with and get to know scientists personally.

Tieing the experiments together is a project to design a Habitat to live on another planet in the solar system. In this scenario, astronomers have discovered a new group of near-Earth asteroids that may impact the planet in 2029. The students are given the task of designing a Habitat on their choice of Mercury, the Moon, Mars, Europa, Titan, or Pluto to ensure that someone will survive to resettle the Earth. Each experiment teaches the students about something they will need for survival on their new world. This "theme" is used to relate the experiments to each other, demonstrate a practical application for each experiment, and encourage some creativity from the students.

In addition to teaching basic physical concepts, the experiments are designed to provide experience taking and analyzing scientific data, and give the students a chance to use equipment they might otherwise never have a chance to use. Almost every experiment requires the students to work cooperatively together to assemble the equipment and obtain the data. Some quantitative measurement is always involved. In the follow-up sessions the students have a chance to discuss the experiment, ask questions, and incorporate the principles into the Habitat by adding some elements, such as solar cells or UV protection, to their design.

The current set of experiments are: The Solar System, Light, The Greenhouse Effect, Water and Oxygen, Energy Sources, and Using Power. Each of the later experiments builds on some aspect of a previous experiment. In past years, experiments have included: Gravity, Buoyancy, Electricity, Magnetism, Parallax, Telescopes, and Osmosis and Diffusion.

The SHOP program is funded by the Maryland Space Grant Consortium, and starting in the fall of 1995 will be administered by the Office of Public Outreach at the Space Telescope Science Institute. For further information contact Flavio Mendez at STScI.

A High School Astronomy Course For A Wide Range of Student Abilities

Gary E. Sampson
Wauwatosa West High School
11400 West Center Street
Wauwatosa, Wisconsin
USA 53222

Tel.: (414) 778-6550
FAX: (414) 778-6546
E-mail: gsampson@omnifest.uwm.edu

Astronomy is, inherently, a high-interest subject. However, at the high school level there is a tendency to teach astronomy using higher-level abstractions and complex mathematics. This teaching approach thus eliminates a large number of students who have difficulties with abstractions and complex mathematics, thereby restricting the study of astronomy to a rather select group of students.

The astronomy course offered since 1976 by the Wauwatosa School District was developed to reach a wide range of students with differing abilities. The prerequisite for this one-semester elective course is the successful completion of one year of high school science. Also, since algebra is not a prerequisite, the "math phobic" students have been attracted to the course. The higher ability students enjoy the challenges posed by astronomy and often take this course as a supplement to their physics classes. Students who normally have difficulty with science suddenly discover that they can succeed in this high-interest subject, and their basic curiosity about astronomy is fulfilled.

As a means of avoiding higher-level abstractions which often become a part of astronomy teaching, this course focuses on hands-on activities. In particular, the Project STAR activities are used extensively throughout the semester. These activities include construction of celestial globes, use of sun-tracking hemispheres, building a scale model of the solar system, study of the seasons using celestial globes, building moon phase dials, construction and use of pinhole tubes and telescopes, development of three-dimensional constellation models, doing the brightness-distance activity, observing the structure of both the Milky Way and the universe. The course textbook is also the STAR text, The Universe in Your Hands.

Observational astronomy is enhanced by the plotting of constellations, use of star finders, regular observation in the planetarium, and direct observation whenever feasible. Five major constellation groups are studied: circumpolar, the summer triangle, autumnal constellations, the winter hexagon, and the spring diamond.

The astronomy course is divided into seven units as follows:

- **Astronomy and the Beginning of Science:** the origins of science through astronomy, interpreting sky phenomena, methods of astronomical inquiry, including the uses of technology.
- **Solar System Dynamics:** measuring size, distance and scale in space, modeling the solar system, and observing the planets.
- **Observing Earth Motions**: apparent daily motion of the sun, timekeeping systems, time zones, revolution of the Earth, the seasons, and precession.
- **Earth-Sun-Moon:** scale of earth-sun-moon system, lunar phases, solar and lunar eclipses.

- **The Sun and Stars in General:** measuring and observing magnitudes of stars, magnitude scales, physical properties of the sun, and solar activity.
- **Stellar Evolution:** spectral classification; the Hertzsprung-Russell Diagram; energy production in stars; and evolution of low mass, medium mass and more massive stars.
- **Galaxies and Cosmology:** observing the Milky Way and deep space objects, types of galaxies, large-scale structure of the universe (the Bubble Theory), and theories of cosmogeny.

Application of basic skills is another major thrust of the course. As new vocabulary terms are introduced, their use is integrated into laboratory, audiovisual, planetarium and observational activities. Scientific notation and unit conversion activities are used extensively throughout the course. The use and understanding of time zone changes is studied thoroughly in the unit on earth motions. Use of outside reference material is integral to each unit. Interpretation of charts, graphs and coordinate systems facilitate basic understandings of potentially complex material. Models are used extensively. Computer technology is used on a classroom basis, particularly to study constellation maps. There are provisions for students to do independent study, and even for a separate independent astronomy course.

Formative assessment includes regular quizzes on constellations and basic skills. Summative assessment is based on unit tests, a constellation unit test, a final examination, as well as written homework and laboratory procedures. The grading scale is deliberately widened at the C and D range to accommodate those students who truly work hard, but have difficulties in academic situations.

Astronomy has become a very popular course in the Wauwatosa high schools, but it is not widely taught in high schools elsewhere. Perhaps the present model could be used to provide an incentive for more widespread teaching of astronomy at the high school level.

Space Odyssey and Astro Adventures: A Comprehensive Astronomy Outreach Program

Dennis Schatz
Pacific Science Center and
University of Washington Space Grant Program
Seattle, WA
USA 98109-4895

Tel.: (206) 443-2867
E-mail: schatz@pacsci.org

Introduction: The mission of Pacific Science Center is to serve the greater Northwest region, especially Washington State. Over the years the Science Center has developed a comprehensive outreach program with the goal of reaching diverse audiences, especially rural underserved populations. Pacific Science Center served approximately 1.4 million people during the 1994/95 fiscal year. Of this number, over 25 percent (400,000) were served through our outreach programs.

The mission of the University of Washington Space Grant Program is to provide space science related resources to residents of Washington State. Space Grant provided funding for the development of the *Astro Adventures* curriculum, and for the teacher training associated with the curriculum.

Space Odyssey Van: *The Space Odyssey* van is typical of the seven outreach van programs Pacific Science Center offers to schools. Each year, these vans provide over 600 days of delivery, reaching more than 40 percent of the public elementary schools in the state.

Space Odyssey spends one or two days in each elementary school, often as the culmination of the school's study of astronomy. The day features:

1. an assembly which starts the day and prepares the students for what they will learn during the remainder of the day
2. an exhibit area set up in the library or multi-purpose room, which all students visit
3. classroom experiences for every class, held in our inflatable planetarium or in each classroom.

Astro Adventures + Teacher Training: With funding from the Space Grant Program, members of Pacific Science Center teacher education staff spend a day working with teachers from the elementary schools to train them in the use of the activity based *Astro Adventures* astronomy curriculum in preparation for a visit by the *Space Odyssey* van. This training provides the background information, teaching activities and strategies, and all teaching materials required to present a multi-week unit on astronomy that links directly — but can also be used independently — to the *Space Odyssey* outreach van experience.

A Revised Approach to a Self-Paced Introductory Astronomy Course

Stephen J. Shawl
University of Kansas
Lawrence, KS
USA 66045

E-mail: s-shawl@ukans.edu

My teaching of introductory astronomy is currently based on 9 premises:

1. "Learning is something done by, not to, the student."
2. Students should know in detail what is expected of them.
3. Learning will occur best if the student is actively involved.
4. Testing should be a vehicle for learning, not only for assigning grades (i.e. students can learn from their mistakes).
5. Within certain guidelines, students should be allowed some flexibility in what material they study.
6. Students may be able to learn fundamentals from other students better than from a professional, since students may not hesitate to ask "stupid" questions of their peers.
7. Lectures can play an important role in learning.
8. Comprehensive exams allow students to bring the pieces of the subject matter together.
9. "Students are a lot like people" and may procrastinate.

The method used to incorporate these premises is that of **self-pacing**, which has many **advantages** over the lock-step methods of most courses:

- Allows for the fact that students have different backgrounds and abilities.
- Allows for the fact that students learn at different rates.
- Emphasizes that the *student* does the learning and must take responsibility for learning.
- Allows students to choose to study different topics beyond a specified core.
- Provides continuous feedback.

The system also contains some distinct perceived **disadvantages**:

- Allows students to procrastinate.
- Provides more work for the faculty when compared with the usual lecture method.
- The continuous feedback can produce frustration on the part of students not passing quizzes right away.

Since self-paced methods have been used for some 25 years, **what's new here?**

- Self-paced classes are normally without lectures; this revised approach uses lectures that proceed at the rate a student should follow to earn an "A"

- Lectures are an integral part because they allow for demonstrations of physical concepts; they provide a venue for showing films, slides, computer software; and they provide an opportunity for additional explanation.
- For students who get behind or do not want to attend lecture, class materials are available in the library. Furthermore, students can purchase notes taken by a for-profit note taking service.
- The system provides a means by which students may decide what topics they want to investigate beyond a specified core.
- The class has "Section Exams" that test over multiple units of related material, such as the solar system, stars and stellar evolution, or galaxies and cosmology. The section exams provide an overview and review. Without such exams, after passing a small unit students tend to think they no longer need to know the material.
- The course now contains a variety of time constraints. The presence of deadlines has increased in each of the three semesters I have used self-pacing. For example, the first time taught, no constraints were present. This was a mistake! In the second semester there was a date after which no more units could be passed than were passed before the date. In other words, at least half the course had to be completed in the first half of the semester. The third semester had dates by which the section exams had to be completed; the dates were staggered for each grade. For the Fall 1995 semester, in addition to the above, attendance at a certain number of lectures will be required as suggested by students.
- An "A" student must take an oral final exam unless the average score on all five section exams is ≥85% (from zero to 50% of the "A" grades have met this percentage). These students are supplied with a short list of general questions that require integration of material. It takes about 30 minutes per student; for the 3-7 students who need the exam, this method provides me with excellent (and sometimes frustrating) feedback and with an opportunity to see prospective tutors explaining the material.

Grades are determined by the number of sections completed with satisfactory exam scores:

A. completion of 5 sections with average section exam scores ≥70%
B. completion of 4 sections with average section exam scores ≥70%
C. completion of 3 sections with average section exam scores ≥70%
D. completion of 3 sections with average section exam scores <70%
F. completion of less than 3 sections or average section exam scores <60%.

If a section exam score on one of the non-core sections is less than 60%, the section is failed and none of the work from that section will count; another section will need to be done. In this situation, the maximum grade obtainable in the course is effectively lowered.

The approach that has been described has **three main problems** that I have identified:

1. Procrastination is a problem for unmotivated students and especially freshmen. It has also been a problem for a few seniors.
2. Frustration, caused by the need to repeat quizzes, is wide spread.

3. At each grade level there can be a huge range in the knowledge obtained (one student can complete 3 units with 70% while another has 85%; both earn a "C").

Possible Solutions to these problems are:

1. Procrastination can be solved only with the imposition of deadlines, as discussed above. Requiring lecture attendance will, if nothing else, provide a reminder to each student that he/she is enrolled and should be studying astronomy on a continuing basis.
2. Frustration could be alleviated by making quizzes easier! Beyond that, the faculty and tutors must provide continuous encouragement and reminders that the purpose of the quizzes is to help them learn the material and prepare for the section exams.
3. The grading range problem has no obvious solution if grades are to be determined only by number of units completed. No algorithm has yet been developed to combine the number of units completed with the knowledge obtained within each section. Such a system might readily be incorporated if pluses and minuses were allowed in the grading system.

Interdisciplinary Astronomy in Planetarium Shows

Dale W. Smith
BGSU Planetarium
Department of Physics and Astronomy
Bowling Green State University
Bowling Green, Ohio
USA 43403

Tel.: 419-372-8666
FAX: 419-372-9938
E-mail: dsmith@newton.bgsu.edu

Abstract: Astronomy is one of the most interdisciplinary sciences. It has connections with most other sciences and with several fields beyond the sciences. Planetarium programs can utilize these ties to broaden their appeal and to show astronomy in its widest context. Several examples of interdisciplinary programs are briefly described.

Astronomy is a naturally interdisciplinary science. It has direct connections with all the other natural sciences, including physics, chemistry, geology, biology, and meteorology. It also has links with many areas beyond the natural sciences, such as archaeology, mythology, history, and geography, and it affects everyday life in a variety of ways. Many of us explore these links directly or indirectly in the courses we teach. Planetarium programs are a public avenue of astronomy education in which an interdisciplinary approach can be used. An interdisciplinary component in a planetarium show can help to (1) create shows of broader educational appeal, (2) display astronomy in its broadest context, and (3) demonstrate astronomy's relation to other topics and to everyday life. This approach can provide effective astronomy education and also draw an audience that might not come for the astronomy alone.

Many of the planetarium programs we have created at Bowling Green during the past decade have had a strong interdisciplinary emphasis. Many people in our audience now expect it and find their astronomy education experience is enhanced by its placement in an interdisciplinary context. In scripting these programs, I try to weave the themes together in a seamless way so the audience will see them as naturally connected. Some of our shows which have included other subject areas are:

It's About Time (1987): The story of how we keep time. Our clocks and calendars record all the prominent cycles of the sky. The program shows these cycles and tells how many early cultures such as the Babylonians, Egyptians, and Romans based their calendars on them, thereby creating the hours, days, weeks, months, and years we use today. (Includes history, geography, and effects on everyday life.)

I Paint The Sky (1989): A visual guide to the colors in the sky: daytime blues, sunset reds, vibrant rainbows, shimmering auroras, icy halos, deceptive mirages, and more. (Includes meteorology, physics, and effects on everyday life.)

Secret Of The Star (1989): A December-Christmas program. It shows the origin of many Christmas customs in ancient solstice festivals and demonstrates several theories, astronomical and other, of the star of Bethlehem. (Includes history and effects on everyday life.)

Sky Stones (1991): An interpretive trip to Stonehenge, the pyramids of Egypt, the temples of the Maya, and many other native American sites, all part of the legacy of ancient skywatchers who recorded the celestial cycles in their great stone buildings and monuments. (Includes archaeology and cultural geography.)

Water World (1992): The life story of the Earth: born by accretion from the solar nebula; divided into crust, mantle, and core; awash in oceans and an atmosphere outgassed by volcanoes; shaped by drifting continents and their volcanoes and by mountain ranges on land and sea; subject to ice ages and a human greenhouse; and destined to perish when the Sun becomes a red giant. (Includes geology and an appeal for the environment.)

New Worlds? (1992): A show on exploration of the Earth for the Columbus quincentenary. It includes the voyages of the ancient Greeks and Phoenicians, the Vikings, and Columbus; an evocative section on native Americans; Ptolemy's work as astronomer and geographer; Copernicus making the Sun central; and modern spacecraft missions. (Includes history and physical and cultural geography.)

The Seabird Show (1993): A celebration of seabirds and their environment. Includes an illustrated alphabet of seabirds, a section on migration and navigation, and a puffin's narrative of its life. Inspired by a trip to seabird colonies in Quebec and Labrador. Only a little astronomy here but a lot of fun. (Includes biology and an appeal for the environment.)

The Day The Earth Turned The Wrong Way (1993): A story-format program for young children in which the Earth protests environmental abuse by rotating backwards and telling how we abuse it. The other planets travel to Earth to show how they are not suitable for human life. (Includes an appeal for the environment.)

Unworldly Weather (1994): Weather on the Earth, on other planets and moons, and on the Sun, including hurricanes, tornados, thunderstorms, drought, and extreme hot and cold, as well as many types of clouds, rain, and snow. (Includes meteorology and effects on everyday life.)

Dinosaur Light (1996, in prep.): The light from stars and galaxies that we see now left there in the past. This show will weave together a description of selected astronomical objects and the events of human or Earth history occurring when the light now reaching us left those sources. (Includes history, geology, and biology.)

We have also run some purchased planetarium shows which include topics beyond astronomy. Of course, we also purchase and create shows whose topics are almost exclusively astronomical.

Naturally, the choice of topics and how to convey and relate them depends on the writer/educator's own interests and style. But whether you are in planetariums or in another avenue of astronomy education, I would encourage you to include an interdisciplinary element whenever you can. It will add interest for you and for your class or audience and it will contribute to more effective astronomy education.

Students Submit Proposal For Amateur Viewing Time On The Hubble Space Telescope

Sallie Teames,
Texas Society of Astronomy Educators
1033 Reed Street,
Hurst, TX
USA 76053

Tel.: (817) 268-3206
E-mail: steames@tenet.edu

July 1993: I attended the NSF-funded West Virginia University/NRAO Summer Institute "Investigative Research in Radio Astronomy" at the National Radio Astronomy Observatory at Green Bank, WV. For two weeks, we learned and practiced techniques of student investigative research to take back to our classrooms and use with our students. My research group's project was "Using HI Emissions to Study Galactic Motions and Structure." We used the 40 foot radio telescope at NRAO to detect the HI emissions.

March-April 1994: My seventh grade honors class used student investigative research techniques to research supernovae, pulsars, and the Hubble Space Telescope and its functions. They used the research results to submit a proposal to use amateur viewing time on the Hubble Space Telescope. Each student was responsible for an article, (magazine or journal), on supernova remnants to contribute to the proposal. Each student was responsible for keeping all information in an individual proposal notebook.

Some parts of the project were worked on collectively by the whole class. For other tasks, the class was divided into several committees responsible for different parts of the project. For example, the Pulsar Committee corresponded with the National Radio Astronomy Observatory, who sent them the current chart of pulsars, which was used in the proposal.

The students practiced research techniques in the school library, using the fax to request copies of articles from the public library, and using the computer encyclopedia resources as well as books. A few students searched the astrophysics journals stacks at a nearby university library with an active astronomy department (Texas Christian University). In this way, we were able to build an impressive bibliography for the proposal. Each student pieced together five pages from the *Uranometria 2000, the Northern Hemisphere*, to make a poster mosaic of the region of sky we were targeting. The class spent several days in the school computer lab, organizing and recording their findings.

The students worked in conjunction with two adult co-investigators and advisors from the Fort Worth Astronomical Society: Don Garland, planetarium director at the Fort Worth Museum of Science and History, and Wade Weaver, the leading local telescope dealer and avid amateur astronomer.

February 1995: We were notified that our proposal had made it to the final selection of four proposals before being turned down.

April 1995: We resubmitted the proposal with a new title: "To detect a supernova remnant associated with the young pulsar, PSR1737-30." Because we needed a more precise target, we changed our target to a faint nebula near the young pulsar, PSR 1737-30.

February 1996: We will be notified if our revised proposal for amateur observing time is accepted for the next cycle of observations.

Acknowledgements: 1) Dr. Mark McKinnon, astrophysicist and pulsar astronomer at the National Radio Astronomy Observatories at Green Bank, West Virginia, for a good deal of information that was unobtainable elsewhere. 2) Sue Ann Heatherly, educational director at NRAO for directing us to various sources of information. 3) Dr. Pat Obenauf, West Virginia University, for her instruction in student investigative research techniques at the NSF teacher workshop at NRAO, July 1993.

Addresses:

1. For information on student investigative research (constructivism): Dr. Pat Obenauf, West Virginia University, Dept. of Curriculum and Education, P.O. Box 6122, Morgantown, WV, 26506-6122, Tel. (304) 293-3442, ext 325.
2. For educational ideas in radio astronomy: Sue Ann Heatherly, Educational Director , National Radio Astronomy Observatories, P.O. Box 2, Green Bank, WV 24944-0002, Tel. (304) 456-2209.
3. For HST information, photos, posters, etc.: Sheryl Gundy, Pat Momberger, Space Telescope Science Institute, Public Outreach, 3700 San Martin Drive, Baltimore, MD 21218, Tel. (410) 338-4562.

To Submit Your HST Proposal:

Deadline for next cycle is April 30, 1996. Mail your requests for proposal applications and information booklets to: American Association of Variable Star Observers, 25 Birch Street, Cambridge, MA 02138, Tel. (617) 354-0484.

The American Astronomical Society's Teacher Resource Agents Program

John Trasco[1]

Kathy DeGioia-Eastwood
Northern Arizona University

David Slavsky, Loyola University

Mary Kay Hemenway, University of Texas at Austin

[1]University of Maryland
Astronomy Program
Room 1205
Space Science Building
College Park, MD
USA 20742-2421

E-mail: jtrasco@astro.umd.edu

The American Astronomical Society's Teacher Resource Agents (AASTRA) program prepares elementary/secondary school teachers to become Teacher Resource Agents for the American Astronomical Society (AAS). Each year, 25 teachers participate at each of three national sites: University of Maryland in College Park, Maryland, Northern Arizona University in Flagstaff, Arizona, and Loyola University of Chicago, Illinois.

The participants attend four-week programs at a site, reconvene at the summer meeting of the AAS, become associate members of the AAS, and agree to give at least two local workshops per year for the following two years. Participants are provided with stipends, travel support, room and board, materials, and AAS membership for two years.

The basic goal of the AAS Teacher Resource Program is to improve the teaching of science in our nation's schools. More specifically, the program seeks:

- to use astronomy as a vehicle to promote activity-based science teaching
- to enhance the teaching of astronomy in elementary and secondary schools
- to bring science education in line with the national topical goals as reflected by the Project 2061, the NSTA's SSC project, and the NRC Science Standards
- to provide astronomy workshops to elementary and secondary teachers using hands-on astronomy activities
- to encourage professionalism of the participants
- to increase interactions among professional astronomers, science educators, and elementary and secondary school teachers.

All sites use a common set of hands-on teaching materials. Site specific activities include the use of small and large telescopes, lectures by visiting astronomers, field trips to museums and planetaria, and computer-based astronomy laboratories. Activities appropriate for students in grades K-9 are presented by instructional teams composed of two master teachers and professional astronomers. A special emphasis is on materials for grades 3-8, the level where many future workshops will be presented. The 5E approach (engage, explore, explain, expand, and evaluate) is used as a model for elementary science instruction.

Materials common to all sites are several teacher guides from the Great Explorations in Math and Science (GEMS) series of Lawrence Hall of Science: "Color Analyzers," "Earth, Moon, and Stars," "Moons of Jupiter," "More than Magnifiers," and "Oobleck." Each participant also receives copies of the Peterson Field Guide "Stars and Planets," "Project SPICA: A Teacher Resource to Enhance Astronomy Education", Ranger Rick's "Astronomy Adventures," a "Sky Challenger" star wheel, a copy of the special Odyssey magazine "Women in Astronomy," the Optical Society of America optics kit, and Lynn Moroney's cassette tapes of American Indian and other ethnic star tales.

The AASTRA institutes are scheduled for completion in the summer of 1996. For more information or an application, contact:

Dr. Mary Kay Hemenway
American Astronomical Society Education Office
Astronomy Department
University of Texas at Austin
Austin, TX USA 78712-1083

E-mail: aas@astro.as.utexas.edu
Tel.: (512) 471-6503

Planet Twister

Kirsten Vanstone
Ontario Science Centre
770 Don Mills Rd.
North York, Ontario
Canada M3C 1T3

What is it? Planet Twister is an educational game developed at the Ontario Science Centre by Jenny Krestow, Devon Hamilton, Janet McLellan-Winship, and Kirsten Vanstone. Planet Twister has been used effectively in the classroom and on the exhibit floor, helping students and visitors learn about the order of the planets, orbit shapes, sizes, periods and distances. It can also be used to illustrate more advanced concepts such as Kepler's Laws. Since the only requirement is flexibility, Planet Twister can be used with people of many ages and backgrounds. Most of all, though, Planet Twister is fun!

Twister With a Twist: Planet Twister works under the same general rules as Twister. In Twister, the objective is to balance yourself on coloured spots placed randomly on a mat. The spot is chosen by a spinner. Planet Twister works the same way, but it is played on a mat with the orbits of the planets marked on it, and players must balance on the positions of the planets on a specific date.

Planet Twister works best with three or more people. The players spin in turns. If the spinner lands on a planet, place a sticker where that planet would be on the chosen date. The player must then balance on the sticker. If the spinner lands on the Sun, the player must balance on the middle of the board. The Asteroid Belt, marked by a dotted line, is wild, and the player may balance anywhere on it. The Oort Cloud is a miss-a-turn, as it would be too far away to reach from the board. Once a body part has been placed on a sticker, it cannot be lifted off again. Two players may not occupy the same spot. If a player spins to a spot already taken, they must spin again. Once a player falls, they are out of the game. The last person standing wins. Simple!

How to Make Planet Twister: Planet Twister is very easy to make. The board can be made by sticking tape in circles on the floor or painting orbits on a canvas or vinyl mat. It is not practical to make the orbits to scale, for the inner planets would be too squished together. It is possible to use one scale inside the asteroid belt, and then re-scale for the outer planets. A simple spinner can be made with cardboard and a butterfly pin: mark off twelve sections on the spinner, one for each planet, the Sun, the Asteroid Belt and the Oort Cloud. The positions of the planets may be obtained from an ephemeris or commercially available astronomical software. Try to use dates of historic, or astronomical significance.

Other Ideas: Planet Twister is a versatile tool in the hands of creative people. With the proper preparation, it can be used to illustrate many other topics, and so can be several games in one. Here are some suggestions and challenges to keep the game interesting.

To help introduce Kepler's Laws: Get the players to help make the game board. They can draw the elliptical orbits by putting a pin in the centre of the board to represent the Sun at one focus of the planet's orbit, and another pin farther out such that, when a pencil is run through the inside of a string stretched taut between it and the two pins, it draws the ellipse.

To show how spacecraft trajectories depend on planet positions: Use dates that show how the planets were aligned when various spacecraft were either launched, or at important points of their voyage. For example, use the dates when Galileo was at Venus, then at Earth, etc.

Have "Guest Appearances" by interesting comets or asteroids: Use dates that show Shoemaker-Levy 9 occupying the same position as Jupiter!

Where and how to use Planet Twister: Planet Twister is a great hands-on, feet-on, elbows-on activity which accommodates many levels of understanding and can be used in lots of different ways. In Ontario, the curriculum that deals with space science and astronomy is almost exclusively centred on the Solar System. Planet Twister is a fun way to introduce the planets to younger grades, and is effective when done in conjunction with other activities about the Solar System. It also works with older students to reinforce their knowledge of the solar system, and to extend that knowledge to more advanced and interesting concepts. We have found that even adults can learn a lot from playing a round or two!

The Ontario Science Centre: The Ontario Science Centre, located in Toronto, has just celebrated 25 years of making science interesting, accessible and fun. Visitors can explore exhibits about many different areas of science. "Hosts" in the Centre perform demonstrations presenting everything from Paper Making to Lasers. The Centre has a small planetarium, Starlab, in which live shows about the current sky are presented.

In addition to the exhibits and demonstrations, small, portable "exhibit extensions" are created that allow visitors to investigate topics in more detail with the help of a host. This is a less formal but extremely rewarding way of interacting with visitors. Planet Twister was developed as an exhibit extension by four hosts for International Astronomy Day, 1991.

An Exercise on the Scientific Method

Mary Lou West
Math/CS/Physics
Montclair State University
Upper Montclair, NJ
USA 07043

E-mail: west@apollo.montclair.edu

This activity is to illustrate the type of critical thinking used by scientists to sharpen up a hypothesis and test it against real data. Each student brings to class two preliminary hypotheses (what is correlated with what) in astronomy and why they think this might be so. Working in pairs, they explore the possibilities available with Voyager (Carina Software, about $120) running on Macintosh SE/30's. It is an interactive desktop planetarium program with many options, such as zooming, displaying constellations, displaying much information about a star or planet, observing from different times or different places on Earth or other planets, tracking movements, and searching for conjunctions.

After about fifteen minutes of "playing around" checking out qualitative correlations, each pair is asked to choose one "working hypothesis" to investigate deeply. They are encouraged to be alert to unexpected possibilities which might make them modify or abandon their preliminary hypotheses. Each group then designs a procedure to test one working hypothesis. Defining variables and clarifying methods is crucial here. Discussion with the instructor and guided questioning leads the students to refine their search method and to move from vague to precise thinking. They are encouraged to be as quantitative as possible. Focusing on one issue is a difficult task for some people who want to write down a lot of irrelevant data. Extracting a pattern from the data is sometimes easy in tabulated form but often much easier in graphical form, so graph paper and Excel (Microsoft) spreadsheets are available. A spreadsheet is especially useful for comparing various data sets or their differences. Students are encouraged to circulate and offer help to each other.

After a correlation has been found the students extrapolate (linearly, if possible) and make a prediction (value for another planet, time when something will repeat, ...) and then test it. Success sometimes leads to exuberant displays of emotion! They write their conclusion in the form "The hypothesis is (not) correct" and then state it in words. A short written report is due a week later.

Some of the preliminary hypotheses which have been suggested:

- The planets are found only in a small part of the sky
- The moon's orbit repeats itself.
- The inner planets are different from the outer planets.
- The planets move at different but steady speeds.
- Eclipses happen with periodicities.
- Eclipses are seen more frequently from locations near the equator.
- Venus crosses the sun periodically.
- Stars of a certain brightness are not distributed randomly.
- The brightness distribution itself is not uniform across the sky.
- The locations of galaxies are correlated with diffuse nebulae.
- Open clusters are correlated with globular clusters.

Such a vague hypothesis is sharpened up by discussion. For example, c) became "c1) larger mass planets rotate slower, and c2) planets farther from the sun have a lower density." The data for Mercury, Venus, and Mars support both hypotheses. However, the rotation prediction for Jupiter is totally wrong and so disproves c1). Jupiter is not far from the linear prediction of c2) but Uranus and Neptune are. Modifying the hypothesis to consider the outer planets as a group saves c2).

Another example showed that there are indeed many more faint stars than bright ones, and that the distribution is very similar except in one region (Sagittarius) which has an over-abundance of all types of stars. These students were very relieved to find out that they had stumbled onto the center of the Milky Way.

Conclusions: This activity works well because students know enough about planets and stars to feel comfortable making astronomical hypotheses. They learn first-hand that it is not easy to operationalize some variables. They learn to value clear definitions and critical thinking. and they learn to be alert to unexpected possibilities (serendipity). They learn to appreciate how powerful a tool graphing is. They learn that some people are better at pattern recognition than others and that some people tolerate frustration better than others, and are willing to play around more before they design an experiment. They learn to value both imagination and reason, but most importantly they learn that science can be great fun.

The Nepean Astronomy Centre and Its Educational Programs

Graeme L. White[1]
Paul A. Jones[1]
A.W. Hons[1]
K. Watson[2]
T.M. Young[2,3]
M.A. Honywood[2,3]
R.P. Hollow[1,3,4]
J.R. Field[3]
C.M. Bauer[1]
A.J. Jarman[5]

[1] Physics Department, Faculty of Science and Technology, UWS Nepean
[2] Faculty of Education, UWS Nepean
[3] Linkwest, UWS Nepean
[4] Blue Mountains Grammar School, Wentworth Falls, NSW 2782, Australia
[5] Western Sydney Amateur Astronomy Group, P.O. Box 400, Kingswood, NSW 2747, Australia

The Nepean Astronomy Centre (White et al. 1994) is on the Werrington North campus of the University of Western Sydney (UWS) Nepean, about 30 miles from Sydney city. The Centre is prominent within the local community and consists of a lecture theatre for 50 students, office space for administration and student activities, and a research 24" Ritchey-Chrétien telescope. There is a second small dome housing a 8" Celestron and adjacent observation areas for a 17.5" Dobsonian, several 8" Newtonians and a 4" refracting telescope. An adjacent area is earmarked for a building to house our small Spitz planetarium projector.

Research programs at the Honours, Master's and PhD level are in the areas of astrometry of double stars (including speckle interferometry), variability of southern stars and QSOs, and instrumentation. Other programs are undertaken in collaboration with the Australia Telescope National Facility.

The Astronomy Centre is the focus for the Liberal Arts Astronomy elective programs (White et al. 1995b) and undergraduate programs in Physics (optics and instrumentation).

An important role of the Centre is to promote the University within the local school student community of over 250 000, and to advertise the science faculty within that community. An outline of the teaching programs is given in table 1; they consist of 10 minute hands-on activities for groups of 3 - 6 students. In addition, we run programs for teacher training for the Higher School Certificate (HSC) Astronomy Option (Bhathal 1993) and the new HSC Cosmology Distinction Course (Hollow et al. 1994, Hollow 1995 and Hollow and White – this meeting).

The Astronomy Centre is also used for public-access with evening group visits for the local community. A typical program consists of a lecture and telescope work. There is also an adult continuing education program (White et al. 1995a). Some of these activities run as entrepreneurial activities, creating employment for postgraduate students as lecturing and support staff. The Centre is also the meeting venue for the new Western Sydney Amateur Astronomy Group.

Acknowledgments: The authors acknowledge grants from DEET, UWS Nepean, the Faculty of Science and Technology, the Physics Department and a UWS Nepean Seed Grant.

Table 1. Education Programs of the Nepean Astronomy Centre

Primary Schools Programs; Day Programs

Making a Sundial	Demonstrating Planetary Orbits
Measuring Brightness with Distance	Making a Refracting Telescope
Testing Alternative Energy (Solar Collector)	Receiving Radio Waves
The Intensity of the Sun	Experiments with Prisms and Light
Measuring the Temperature of Infra-red light	Making a Model of a Black Hole
Examining Sun Spots	Proving the Rotation of the Earth

Year 7 - 10; Day Programs

Construct a Model Rocket	Rocket Stages
Explore Gyroscopic Action	The Moon - what is it like?
Weight and Mass on Various Planets	

Year 7 - 10; Evening Programs

Build a Simple Telescope	Using a Spectroscope
Observing Stars and Constellations	Colour of Stars

Science for Life Program - Year 11 - 12; Day Programs

Evolution of Science Fiction	Faster than Light travel
Science Fiction vs Real Life	Matter Transfer
What are Aliens like?	

Physics Astronomy Elective - Year 12; Overnight Visits

Stellar Evolution	Magnitude Determination
CCD Photometry	Hertzsprung-Russell Diagram
Limiting Magnitude Determination	

General Public; Evening Visits - One lecture each night plus telescope viewing

The Big Bang	The Solar System
The Great Race to the Moon	Tour of the Universe
Beginners Stars	Life in Space – SETI

References:

Bhathal, R., 1993, *The HSC Astronomy Option*. Kangaroo Press.

Hollow, R., 1994, The Cosmology Distinction Course for Gifted Children. *Physics Education*, 30, 129 - 134.

Hollow, R. et al., 1994, The Cosmology Distinction Course in NSW. *Proc. Astron. Soc. Austr.*, 11, 39 - 43.

White, G.L. et al., 1994, The New Teaching and Public Access Observatory at the University of Western Sydney, Nepean. *Proc. Astron. Soc. Austr.*, 11, 188 - 190.

White, G.L. et al., 1995a. The Liberal Arts Astronomy Program at UWS Nepean. For the AGM Astron. Soc. Austr, July 1995, and for publication in proceedings.

White, G.L., et al., 1995b. "The Night Sky and Beyond" Continuing Education West Program at UWS Nepean. For the AGM Astron. Soc. Austr, July 1995, and for publication in proceedings.

Follow The Drinking Gourd: Bibliography

Christina Wilder
Educational Consultant
377 Mandarin Drive, #208
Daly City, CA 94015-2746

Tel.: (415) 991-3966

Adult Astronomy Resources:

Beyond the Blue Horizon: Krupp, E.C., HarperCollins Publishers, 1991, New York, Chapter 14, pp. 226 - 240, includes many stories from around the world, plus an excellent circumpolar diagram on p. 238.

365 Starry Nights: Raymo, Chet, Prentice Hall Inc., 1982, New York. I am indebted to this book for clear illustrations, by the author.

Star Names: Their Lore and Meaning: Allen, Richard Hinckley, Dover Publications, Inc., 1963 reprint of 1899 original, New York.

Children's Literature and Background:

Eight Hands Round: A Patchwork Alphabet: Paul, Ann Whitford, HarperCollins, Publishers, 1991, New York. Source of "Variable Star" and "Underground Railroad" quilt patterns.

Faith Ringgold: Turner, R. M., Little, Brown and Company, 1993, Boston. Biography of Afro-American artist using quilts as an art form.

Follow the Drinking Gourd: Winter, Jeanette, Alfred A. Knopf, 1988, New York. An extremely attractive book, with lyrics for people who can sing or play an instrument.

Patchwork Math 2: Baycura, Debra, Scholastic, 1990, USA. A series of quilt patterns to be colored in multiplication/division facts. Nice correlation! Lots of astronomy here!

Sweet Clara and the Freedom Quilt: Hopkinson, Deborah, Alfred A. Knopf, 1993, New York. Another attractive book, beautifully illustrated.

Tar Beach: Ringgold, Faith, Crown Publishers, Inc., 1991, New York. An exceptional book with an inspirational message and quilt associations.

The Underground Railroad: Bial, Raymond, Houghton Mifflin, 1995, Boston. Photos and background on the Underground Railroad. Exceptional.

Magazines:

Sky & Telescope: Feb., 1995, pp. 36 - 38. Astronomy reference.

Teaching Children Mathematics: March, 1995, pp. 438- 443. Excellent resource for mathematics correlations in geometry and number concepts.

The Universe in Your Hand: teaching astronomy by using an astrolabe

Harold Williams, Planetarium Director
Montgomery College,
Takoma Avenue and Fenton Street
Takoma Park, Maryland
USA 20912-4197

Tel.: (301) 650-1463
FAX: (301) 650-1550 at college library
E-mail: haroldw@umd5.umd.edu

An astrolabe is first and foremost an elegant model of the universe - bring some of the majesty of the celestial realm down to something that you can hold in your hand. It is a powerful astronomical calculation device; in fact, it is the earliest personal computer. It is a clock capable of 4 minute resolution. It is a device for estimating when and where bright stars or the sun rise, transit, and set. It easily allows coordinates transformations to or from celestial coordinates to horizon coordinates. It allows you to read off altitude and azimuth. The adiale, usually on the back of the astrolabe, allows you to measure altitude. The word astrolabe is from Greek; it means star taker, finder, or thief.

History of Astrolabes

- **Hipparchus** 150 BC Rhodes and Alexandria invented the stereographic projection mathematics.140 AD Alexandria
- **Theon of Alexandria** (Hypatia father) First know book on the astrolabe 375 AD - manuscripts lost, but the table of contents is preserved in Yaquibi's History of the World - 800 AD
- **John Philoponus** also of Alexandria - 530 AD earliest book on the astrolabe where a complete copy of the book is known today.
- **Bishop Severus Sabokt of Nisibis in Syria** - 650 ADanother ancient astrolabe book.
- **Masha allah Spanish Jew** - 800 AD Arabic text none currently known to exist, but it was translated into Latin in 1276 AD this manuscript survives.
- **Geoffrey Chaucer** - 1387 *A Treatise on the Astrolabe* subtitled **Bread and Milk for Children**
- Earliest work in science education in a language that might be called English. Here is a quote from the introduction:

"Little Lewis my son, I have perceived by certain evidences your ability to learn sciences concerning numbers and proportions. I have also considered your anxious and special request to learn the Treatise of the Astrolabe. Then, forasmuch as a philosopher has said, "he wrappeth him in his friend, who accedes to the rightful prayers of his friend," therefore have I given you an astrolabe for our horizon, constructed for the latitude of Oxford. And with this little treatise, I propose to teach you some conclusions pertaining to the same instrument. I say some conclusions, for three reasons. The first is this: you can be sure that all the conclusions that have been found, or possibly might be found in so noble an instrument as an astrolabe, are not known perfectly to any mortal man in this region, as I supposed."

Antique Paper Astrolabes

- Georg Hartmann of Nuremberg (1489-1564) mass production with numbered lots – 1525 astrolabes and sundials. None of the paper ones actually survive.
- Peter Jordon 1535 kit book with hand colored components.
- Nicolas Bion (1652-1733) paper version in 1702 of the La Hire projection.

For more than a thousand years astrolabes were the premier astronomical measuring and calculation device, but in 1652 the pendulum clock was invented so astrolabes were no longer the most accurate clocks. In the "Age of Reason" the popularity of astrology declined precipitously, but with more astrology columns in newspapers than astronomy columns today we are in the "new age." Furthermore, the material-rich west divided the function of the astrolabe into many separate devices. Astrolabes were still commonly used in India until the 19th century for horoscopes.

Currently a few paper astrolabes are made by Glen Ellen Scientific of California, Adler Planetarium of Chicago, the National Maritime Museum of Greenwich England, and Janus Personal Astrolabes. The Janus Personal Astrolabes are probably the most accurate astrolabe ever made and one of the most affordable of the current paper astrolabes. My students buy these laminated in plastic astrolabes, which come with an excellent booklet, in the book store when they buy their text books.

Will it mean anything to AST101 students?

After students have learned about the celestial sphere by building a Learning Technologies Inc. "Celestial Sphere" and doing some exercises with it they are ready for the astrolabe. The astrolabe will enable them to do many things with ease and fairly good precision that the celestial globe will not. The astrolabe is the perfect interdisciplinary study object that combines astronomy, history, mathematics, and real hands-on measurement.

You must teach the coordinate systems on the celestial sphere first. Coordinates can not be taught effectively with flat pictures to most students excepting physics and mathematics majors. A one time classroom exercise with celestial sphere is only a little more effective. Coordinates viewed in a planetarium against a star field help, but still won't do it. Students must build their own celestial sphere and label it carefully with thought out exercises following the assembly. This is now affordable, cheap, thanks to Learning Technologies celestial spheres. Then they can use the Janus astrolabes which have the accuracy that a hand built celestial sphere lacks. Why bother with the coordinates at all if all you want your astronomy course to do is amaze your students with how neat and keen astronomy and astronomers are? Spend more class time with black holes, cosmic inflation, and SETI; but don't be surprised if your students can't think critically, have never done anything approaching real science, and believe in magic – since they will probably not even understand what causes the seasons. Must real science education involve more than reading the book and attending the lectures? Yes if you want to teach anything more than a little vocabulary that your students will soon forget. If real hands-on inquiry based science is the best way to learn even for smart people (How did you really learn most of the science that you remember and actually use?), how come we still give more credit for the lecture part of the course and grade (assess) as if the laboratory counted less than the lecture material? We lecture because we are afraid our students can't read or don't read; or maybe we haven't realized that the lecture method was invented in medieval universities before Gutenberg's printing press. Lectures are cheap – one professor, many students in a big hall. I feel in control in the lecture; after all I am the expert. We teach as we were taught, even if it was not very effective. It worked for me why can't it work for them? Did it?

How do I get an astrolabe?

Janus, 9 Bigelow Road, New Fairfield, CT 06812, Tel. (203) 746-6716, or FAX (203) 746-0815

Conceptual Astronomy: A Cognitive Approach for Teaching Science to Non-Majors

Michael Zeilik
Department of Physics and Astronomy
The University of New Mexico
Albuquerque, NM
U.S.A. 87131-1156

Internet: zeilik@chicoma.la.unm.edu

We have developed a process to alter radically the course structure and assessment of a large (200 to 300 students) introductory astronomy class for non-science majors. This process, based on principles of cognitive psychology (Carey, 1986), transformed a traditional lecture course into a conceptually-based one. As a major instructional strategy, we utilized heterogeneous cooperative groups on a weekly basis with activities specifically targeted toward astronomy's most difficult concepts and common misconceptions. We created alternative assessment tools such as concept mapping, which was also used to teach explicitly the structural knowledge of astronomy. In addition, we have developed conceptually-designed multiple choice quizzes and tests as our traditional means of assessment.

The large introductory astronomy class for non-science students is a typical lecture course for lower-division undergraduates. Some 200,000 students take this course in the US each year. Prior research has indicated that non-science major students have significant misconceptions in astronomy (and physics), but little sustained work has used new forms of instructional strategies or assessment vehicles to improve conceptual understanding.

Teachers of introductory science classes know that novice students typically miss the "big picture," no matter how well-honed the lectures or well-structured the course. This frustration illuminates the expert/novice divide, which "standard model" courses fail to bridge. By a "standard model course," we mean one that has most elements of a traditional large-lecture survey course in astronomy. Our work presents an alternative model to this one, a model in which the expert/novice divide impels a new paradigm based on the structural knowledge of astronomy – how concepts interrelate (Jonassen et al., 1993).

Lecturing is the usual mode of instruction in post-secondary education, especially in physical science and mathematics courses: 89% of US professors in these areas lecture. Instructors of large classes in the natural sciences typically talk over 90% of the class time. Although lecturing can have advantages as a delivery technique (such as imparting information not available in other forms), research shows that lectures do not promote high-order intellectual skills or change misconceptions. This consequence has been especially dramatic in physics instruction at many levels (Wells et al., 1995).

Cooperative learning groups promote such skills (Johnson et al., 1991) and form the core of our instructional strategy. The students make up their own groups – called focused discussion groups – of 3 to 4 people. They have well-defined but flexible social roles (leader, skeptic, reporter) and must deliver a consensus written report. Students can shift roles, and groups do not have to include the same people for every class. We use the groups with hands-on activities of difficult concepts, quizzes, and demonstrations.

Activities take about 30 minutes and are carefully focused on difficult core concepts. Many require interpretation of graphs that display key analytical relationships (such as the inverse-square law for light). A class discussion moderated by the instructor closes the activities. Quizzes have 20 multiple-choice questions. Students work on these individually for about 20 minutes and then form into their groups to discuss the questions for another 20 minutes. Each person can change answers in light of the discussion before handing in the answer sheet. After demonstrations, groups take about 5-10 minutes to answer the question, "What was the point of the demonstration?" A class discussion then follows, again moderated by the instructor.

We find that the groups are effective even with a diverse student population; that performance does NOT depend on gender or ethnicity; that they are practical with even as many as 200-300 students and one instructor (no TA help!); and that they inexpensive to set up and take no more resources (once developed) than teaching the class by the standard model.

The cooperative groups enhance performance on quizzes and tests (which students still take individually). Class averages on quizzes were 63% before groups were used, 73% after. On tests, the class averaged 70% before and 75% after. (The test and quiz questions were similar in all cases.) We also used a special subset of questions as a pre/post test of achievement of common (and important) misconceptions. For 15 questions in Fall 1994, the pretest mean was 5.60 (standard deviation 2.06); posttest, 10.57 (2.36). At a chance probability level $p < 0.001$, the effect size (the difference of the means scaled to the standard deviation) was 1.86. That means that 97% of the post scores were above the mean of the pretest scores. Our instructional techniques showed a large, positive impact on student's achievement with conceptually-based questions. Preliminary results from Spring 1995 show sustained effects in achievement.

Our results indicate that we can transform our teaching in introductory science classes to develop more powerful scientific habits of mind and understanding by using a systematic process to design and assess conceptually-based instruction. It is a concrete place to start for those faculty in higher education who want to develop their courses conceptually. We believe that our model can be applied to almost all disciplines as we focus more on "how" we teach rather than "what."

This research was supported in part by NSF grant DUE-9253983.

References

Carey, S., 1986, "Cognitive Science and Science Education," *American Psychologist*, vol. 41 (10), pp. 1123-1130.

Johnson, D. W., Johnson, R. T., and Smith, K. A., 1991, Cooperative Learning: Increasing College Faculty Productivity, ASHE-ERIC Higher Education Report No. 4, Washington, D.C.

Jonassen, D. H., Beissner, K., and Yacci, M., 1993, Structural Knowledge: Techniques for Representing, Conveying, and Acquiring Structural Knowledge, Lawrence Erlbaum Associates, Publishers, Hillsdale, N.J.

Wells, M., Hestenes, D., and Swackhamer, G., 1995, "A Modeling Method for High School Physics Instruction," *American Journal of Physics*, vol. 63 (7), pp. 606-619.

TELESCOPES IN EDUCATION

Lee Ann A. Hennig
Thomas Jefferson High School for Science
7714 Lookout Ct.
Alexandria, VA
USA 22306

Tel.: (703) 750-8380
FAX: (703)750-5010
E-mail: lahennig@pen.k12.VA.US

Over the course of several evenings in June 1995, student observers from Apple Valley Science and Technology Center in California, and Thomas Jefferson High School for Science and Technology in Virginia, participated in a joint effort to establish refined positions for the planet Pluto. Pluto's close passage to SAO 140853, a star whose coordinates are precisely known, will enable scientists to more accurately calculate Pluto's orbit. The accuracy of the distant planet's orbit is paramount in plans for spacecraft missions to Pluto and Charon.

Telescopes In Education provides opportunities for students to do research with a 24" telescope equipped with a ST6 CCD camera. In 1993-94, Apple Valley and Thomas Jefferson HSST served as Beta Test Sites for this innovative program. Using ***The Sky Software*** by Bisque, students can remotely control the telescope, image the object, download the image, and enhance the image – all of this from their workstation in the classroom.

Support for the **TIE Project** comes from funding grants out of NASA's Office of Space Science and Information Infrastructure Technology and Applications Office. **TIE** is a special adjuct of the Mt. Wilson Institute directed by Dr. Robert Jastrow. A wide variety of astronomical instruments are housed on the mountain as well as the historic 100" Hooker Telescope. Under the management of Dr. Gil Clark, the program has grown since 1993 to include many more school group users as well as amateur astronomers and international observers.

Our students were most excited to be able to be a part of this historic event. The Apple Valley students made their observations from 10 p.m. to 11 p.m. PDT and the Jefferson students had the 2 a.m. to 3 a.m. EDT observing watch. At some points in the sessions our students were conversing with each other over the phone lines merging in the Observatory on Mt. Wilson.

The students' digital images will supplement those made by Dr. William Owen of the Jet Propulsion Lab. Certainly observations by other professional and amateur astromonmers in this coordinated effort will enhance NASA's effort in planning its **FAST Flyby** to Pluto. Students interacting with each other, even collaborating in joint projects using a real research instrument offers a unique and memorable experience for students!

For further information on the **Telescopes In Education** Project contact: Gil Clark, TIE, Box 24, Mt. Wilson, CA, USA 91023, Tel.: (818) 793-3100, FAX: (818) 793-4570.

Appropriate Curricula Materials for K-12 Astronomy Topics

Darrel Hoff
Luther College
Decorah, IA
USA 52132

E-mail: hoffdarr@martin.luther.edu

There has been a great deal of interest in developing and promoting activity-based materials for K-12 mathematics and science topics and courses during the last three decades. Before I share with you my views on what materials I would recommend for use in American schools, it might be well to review the data that shows the effectiveness of activity-based teaching/learning.

While data to support the use of such an approach is not as commonly available for the employment of this teaching strategy at the secondary level, a great deal of data is available that supports the efficacy of this approach at the elementary level. The National Science Foundation (NSF) and other agencies funded the development of three major programs for elementary science. These programs were The Elementary Science Study (ESS), Science - A Process Approach (SAPA) and The Science Curriculum Improvement Study (SCIS). These programs were developed in the late 1960's and early 1970's for use in grades K-8 and covered the general areas of biological, physical, and earth and space science. They were promoted as alternatives to the traditional text/discussion approach to science instruction commonly used in schools at that time.

In the early 1980's two investigators, Ted Bredderman of the State University of New York at Albany[1] and Jim Shymansky of the University of Iowa[2], were funded by the NSF to complete a meta-analysis on the many comparative studies that had been done by a variety of investigators comparing activity-based science teaching with traditional didactic methods. In all over 13,000 students in approximately 1,000 classrooms were involved in the controlled studies. A variety of student performance areas were examined, including science processes, science content, language development, creativity, attitudes toward science, and logic development. While the two investigators worked independently on the meta-analysis, their results were very similar. Their combined results showed the clear superiority of an active learning classroom. In the following tables, both taken from Bredderman's work, the gain scores are all normalized to a maximum of 100 percent. The table gives data that shows that on all of the common measures students in the active classroom showed striking gains in their attitudes toward science, understanding of the processes of science and the like. What was particularly striking was that, in spite of the fact that activity-based teaching takes more time and as a result limits the amount of content or topic coverage, the scores on science content tests also showed significant gains.

---

[1]Ted Bredderman, "What research says", *Science and Children*, September, 1962

[2]James A. Shymansky, et al., "How effective were the hands-on science programs of yesterday?" *Science and Children*, November/December, 1982.

Average Performance of Students in Activity-Based Science Programs Compared with Students in Traditional Programs

Student Performance Area	Average Percentile Gain	Number of Studies
Science Processes	20	28
Creativity	16	5
Attitude	11	12
Perception	10	5
Logic Development	10	9
Language Development	9	11
Science Content	6	14
Math	5	7

Of particular interest in today's world with our concern for the under-represented students in our classrooms are the results of Bredderman's analysis of comparative results where economic status was considered. Students who were disadvantaged, academically, economically, or both, gained more from activity-based programs than did students who were not disadvantaged.

Average Percentile Gain of Students in Activity-Based Science Programs Compared to Students in Traditional Classrooms

Student's Econ. Status	Science Process Skills	Science Content	Logic Develop.	Language	Math Develop.
Disadvantaged	34* (9)**	30 (3)	12 (3)		
Average	14 (13)	1 (7)	13 (4)	9 (6)	8 (4)
Advantaged	17 (9)	-4 (5)	-5 (3)	8 (3)	4 (2)

* Percentile gain
** Number of studies including this test category

Yet, in spite of a general belief in the effectiveness of this approach among both teachers and administrators, surveys indicate that the use of active learning classroom practices are on the decrease rather than on the increase. While there are many factors that go into determining what should be done in the classroom, we would be remiss if those of us who are interested in K-12 science education didn't advocate the use of active teaching/learning materials. I have examined the commonly available materials for astronomy that are activity-based and suitable for the K-12 audience. The following list includes those material (most developed with NSF support) that meet the following tests:

(1) Do they meet the test of activity-based and are they age-appropriate?

(2) Do they model good science practices?

(3) Are they reasonably inexpensive and use commonly available materials.

Selected Bibliography of Materials Useful for K-12 Students

Solar Energy; Lenses and Mirrors; Color and Light – Delta Science Modules. Delta Education, PO Box 915, Hudson, NH 03051. These are derived from the old Elementary Science Study (ESS) curriculum and stress a process theme.

Astronomy Adventures, Ranger Rick's Nature Scope from National Wildlife Federation, 1400 Sixteenth St., NW, Washington, DC 20036-2266. Selected activities with an outdoor focus.

Earth, Moon, and Stars; Height of Meteors; Moons of Jupiter; Color Analyzers; Oobleck; More than Magnifiers; Hot Water and Warm Homes from Sunlight; Convection; Discovering Density; Experimenting with Model Rockets – Alan Gould, Cary Sneider and other authors, Great Explorations in Math and Science (GEMS) Lawrence Hall of Science, University of California, Berkeley, CA 94720. These cover a variety of topics for upper elementary grades, generally.

Astro Adventures, Dennis Schatz and Doug Cooper, Pacific Science Center, 200 Second Avenue North, Seattle, WA 98109-4895. Complete with a set of slides.

Planetarium Activities for Student Success. Twelve volumes by Alan Friedman, Cary Sneider, Alan Gould and other authors. Lawrence Hall of Science, University of California, Berkeley, CA 94720. This PASS series set the tone for interactive planetarium programs.

Project SPICA. (1994), Edited by N.B. Ball, et. al. – Kendall/Hunt Publishers, 4050 Westmark Drive, Dubuque, IA 52004-1840. Tested activities from the Support Program for Instructional Competency in Astronomy (SPICA). This NSF supported teacher resource agent program was conducted at the Harvard-Smithsonian Center for Astrophysics. Contains 43 activities suitable for generally middle school through high school.

Project STAR. (1993), Edited by H. Coyle, et. al. – Kendall/Hunt Publishers, 4050 Westmark Drive, Dubuque, IA 52004-1840. This publication resulted from a six-year, $2.4 million project funded by the NSF and at the Harvard-Smithsonian Center for Astrophysics. Contains 15 chapters and is activity-based. It is deliberately not a comprehensive look at "all" of the universe. Contains only black and white drawings and photographs.

APPENDICES

The following two Appendices - "National Astronomy Education Projects: A Catalog", and "Selected Resources for Teaching Astronomy" - are taken from the Astronomical Society of the Pacific's *Project ASTRO Resource Notebook*. This is an outstanding 800-page compendium of activities and other resource material which is invaluable for teaching astronomy at any level. Project ASTRO was a partnership between astronomers (both amateur and professional) and teachers. It was supported by a generous grant from the National Science Foundation, by the ASP itself, and by dozens of astronomers and educators who participated in the project, either directly or indirectly. Project ASTRO is now being expanded from its original site in California, to sites across the country. Suggestions and other feedback about the project can be addressed to Andrew Fraknoi.

The Project ASTRO Resource Notebook (and other excellent educational material) is available from: ASP, 390 Ashton Avenue, San Francisco CA 94112. We are grateful to Andrew Fraknoi, the author of these documents, for giving permission for them to be included here.

NATIONAL ASTRONOMY EDUCATION PROJECTS: A CATALOG

by Andrew Fraknoi
Foothill College &
Astronomical Society of the Pacific

This is an evolving list of projects and programs in astronomy education *to* which anyone from around the U.S. can apply or *from* which anyone can receive materials. It does not include the many worthwhile projects that are designed to serve only one city, one state, or one institution (although we recognize that such programs may nevertheless serve as models for the rest of the country). We very much welcome suggestions and additions for future versions of this list. Please contact the author at the above address or e-mail: fraknoi@admin.fhda.edu.

NOTE: Organizations that are involved with a number of projects are just listed with their names; see the key at the end of the list for their addresses and telephone numbers. We do not list publications such as astronomy magazines, college-level textbooks, or popular books, some of which are listed elsewhere in the notebook.

Table of Contents:

1. WORKSHOPS AND TRAINING FOR TEACHERS OF ASTRONOMY

American Association of Physics Teachers (AAPT): Has astronomy education sessions aimed at secondary and community college teachers at meetings. Has sponsored a Physics Teacher Resource Agent program to train physics teachers to help other teachers [*see address on last page*]

American Astronomical Society: A-ASTRA Program: Intensive summer workshops for elementary and secondary teachers to become astronomy teaching resource agents in their own communities; offered at several locations around the U.S. Contact: Mary Kay Hemenway, Dept. of Astronomy, University of Texas, Austin, TX 78712 (512-471-6503)

American Astronomical Society (AAS): 2-day "From Black Holes to Blackboards" workshops for teachers, in conjunction with the Society's meetings in different cities [*see AAS address on last page*]

Association of Astronomy Educators (AAE): Puts on astronomy education sessions at meetings of the National Science Teachers Association [*see address on last page*]

Astronomical Society of the Pacific (ASP): 2-day "Universe in the Classroom" workshops for about 200 teachers in grades 3-12, each summer, in conjunction with the Society's meetings [*see address on the last page*]

Astronomical Society of the Pacific (ASP): Project ASTRO: forms and trains ongoing partnerships between astronomers (professional & amateur) and local 4th-9th grade school teachers for class visits [*see address on last page*]

Center for Astrophysics (CfA): Project ARIES Summer Workshops for Elementary School Teachers (grew out of Project STAR and Project SPICA workshops for upper level teachers) [*see address on last page*]

National Science Teachers' Association (NSTA): NEWMAST and NEWEST Workshops, offered at various NASA centers around the country (astronomy is only one part of the workshops' focus) [*see address on last page*]

NSTA also holds large national and regional conventions for science teachers; most of them have astronomy lectures and programs

Project FOSTER: Holds summer workshops and flies K-12 teachers (from selected states) aboard NASA's Kuiper Airborne Observatory. Contact: Edna DeVore, SETI Institute, 2035 Landings Dr., Mountain View, CA 94043 (415-961-6633)

University of Colorado: Two-week summer workshops on

introductory astronomy for 2- and 4-year college instructors with limited background in astronomy. Contact: Stephen Little, Campus Box 389, University of Colorado, Boulder, CO 80309 (303-492-7627)

2. CURRICULUM AND INFORMATION MATERIALS

Remember this is only a listing of projects, not of all publications. See other resource lists in this Notebook for many recommended activity guides, magazines, and books.

Astronomical Society of the Pacific (ASP): *The Universe at Your Fingertips: An Astronomy Activity and Resource Notebook*, 800+ pages of activities and resources for teachers at all levels, especially grades 4-12. Available through the ASP Catalog.

Astronomical Society of the Pacific (ASP): Information Packets for beginners on astronomical topics, such as black holes, debunking astrology, selecting a first telescope, etc.

Center for Astrophysics (CfA): Project STAR & Project SPICA were NSF-supported programs that developed activity-based curriculum & workbooks for teaching astronomy in secondary schools. Materials can be purchased from Kendall Hunt Publishers, P.O. Box 1840, Dubuque, IA 52004. (The projects also left a legacy of some 200 trained "astronomy resource agent" teachers around the U.S.)

Jet Propulsion Lab: For those who do not live near a NASA center (see below), it is often possible to get NASA lithographs and booklets by writing to: Teaching Resource Center, CS-530, JPL, 4800 Oak Grove Dr., Pasadena, CA 91109. Write on school stationery and indicate what mission or missions you are interested in.

Lawrence Hall of Science (LHS): *Great Explorations in Math & Science* (GEMS), *Planetarium Activities for Student Success* (PASS) are two series of superb hands-on activity guides for teaching astronomy in grades 3-9. Available from the Hall's Eureka store (415-642-1016) or through the ASP catalog [*see addresses at the end*]

NASA: A colorful series of booklets, posters, prints & other materials on space astronomy is available through NASA teacher resource centers around the country. Contact a local NASA center or: Education Division, NASA Headquarters, Washington, DC 20546 for current list of centers. The list of available materials changes constantly. (See also NASA CORE in section 3, and Jet Propulsion Lab above. See also the addresses of organizations listings elsewhere in this notebook.)

National Radio Astronomy Observatory: To get some high school-level radio astronomy activities developed by teachers, write: Sue Ann Heatherly, NRAO, P.O. Box 2, Green Bank, WV 24944 (304-456-2209)

National Science Teachers Association (NSTA): Develops and distributes a range of books and activity collections for teaching astronomy in grades K-12. Ask for their catalog.

New Mexico State University.: Dr. Bernard McNamara is developing a series of writing exercises focused on cosmology. Contact him at: Dept. of Astronomy, Box 30001, Dept. 4500, New Mexico State U., Las Cruces, NM 88003 (505-646-2614) [email: bmcnamar@nmsu.edu]

Pacific Science Center: *AstroAdventures* Curriculum, a series of astronomy activities for grades 3-12, assembled by respected astronomy educator Dennis Schatz, and supported by a NASA Space Grant. Available through the Astronomical Society of the Pacific Catalog.

Project ARTIST: Develops a variety of K-9 activities (some in Spanish, some using multi-cultural ideas) and trains teachers how to use them during summer workshops at the U. of Arizona. Write for list of available materials to: Dr. Larry Lebofsky, Lunar & Planetary Lab, University of Arizona, Tucson, AZ 85721.

SETI Institute: *Life in the Universe* Curriculum Project is developing six interdisciplinary curriculum guides and accompanying materials focusing on a SETI theme at the middle school level. Contact at: 2035 Landings Dr., Mountain View, CA 94043 (415-961-5006)

University of Texas McDonald Observatory: Teacher Kit of poster, activities, and planetary fact sheet for $3. Contact at RLM 15.308, U. of Texas, Austin, TX 78712 (512-471-5285)

Young Astronauts Program: Produces simple activities and materials on space science for youngsters; has many local chapters. Contact at 1308 19th St., NW, Washington, DC 20036 (202-682-1984)

3. AUDIOVISUAL MATERIALS

American Association of Physics Teachers (AAPT): has a modest catalog of audiovisual materials for teaching physics, which includes some astronomy [*see address on last page*]

Astronomical Society of the Pacific (ASP): Produces and distributes slide sets on astronomical topics with extensive booklets of captions and background material, as well as many videotapes and CD-ROM's. Ask for their catalog. (E-mail: asp@stars.sfsu.edu).

Center for Astrophysics (CfA): The Private Universe Project is producing a series of videotapes on student misconceptions in science and strategies for promoting conceptual change. Contact: Nancy Finkelstein (617-496-7676)

Coast Telecourses: Developed 26 half-hour episodes of *Universe: The Infinite Frontier*, a new educational TV show. Contact at: 11460 Warner Ave., Fountain Valley, CA 92708 (714-241-6109)

Finley-Holiday Films: This commercial company has been

designated by both the Jet Propulsion Laboratory and the Space Telescope Science Institute to distribute slides and videos to educators at reduced cost. Contact them for a current catalog: 12607 E. Philadelphia St., P.O. Box 619, Whittier, CA 90608 (800-345-6707)

NASA CORE: Distributes a wide range of NASA audiovisual materials at low cost to teachers. Ask for their catalog & updates. Contact at: Lorain County JVS, 15181 Route 58 South, Oberlin, OH 44074 (216-774-1051, ext 293, 294)

Science Graphics: Small company that produces and distributes interesting slide sets in astronomy and earth science. Contact: Richard Norton, P.O. Box 7516, Bend, OR 97702 (503-389-5652)

Space Telescope Science Institute: Has had a strong program of supplying audiovisual materials for educators. Entire education program is under review and is expected to emerge in a new form in 1995. Contact: Anne Kinney or Laura Danly, STScI, 3700 San Martin Dr., Baltimore, MD 21218.

4. Computer Materials or Projects

American Astronomical Society Working Group on Astronomy Education: Provides an internet bulletin board for announcements and discussion. Contact: Steve Shawl at shawl@kuphsx.phsx.ukans.edu.

Astronomy Software List Project: John Mosley of the Griffith Observatory (and *Sky & Telescope's* primary software reviewer) will supply an up-to-date listing of astronomical software for $2.00. Make check out to John Mosley, 7303 Enfield Ave., Reseda, CA 91335. (It's a great resource.)

The Astronomy Village: CD-ROM, Mac-based supplementary curriculum for 9th grade students. Contact: Craig Blurton, NASA Classroom of the Future, Wheeling Jesuit College, 220 Washington Ave, Wheeling, WV 26003 (304-243-2388)

Exploration in Education (ExInEd): Project directed by Robert Brown at the Space Telescope Science Institute to produce inexpensive Macintosh discs of images (ranging from Hubble Space Telescope images to paintings of an asteroid hitting the Earth) and background material in HyperCard format. Discs are available from the Astronomical Society of the Pacific catalog.

Hands-On Universe Project: trains teachers to do image processing of student-acquired images. Contact: Elizabeth Arsem, Lawrence Berkeley Labs, U. of Calif., Berkeley, CA 94720

Image Processing for Teaching Project: (led by Richard Greenberg at the U. of Arizona) trains teachers to use image processing to stimulate student interest in science. A range of excellent materials is available from the Center for Image Processing, Suite 201, 5343 E. Pima St., Tucson, AZ 85712 (800-322-9884)

Joint Education Initiative: develops CD-ROMs & Internet projects for teachers in geologic and planetary sciences. Contact: Robert Ridky, 3433 A. V. Williams Bldg., U. of Maryland, College Park, MD 20742 (301-405-2324)

Lowell Observatory: Has an innovative interactive exhibit simulating a night at an observatory, with plans available to other institutions. Contact Bill Buckingham, Lowell Obs., 1400 East Mars Hill Rd., Flagstaff, AZ 86001 (602-774-3358).

MicroObservatory (Center for Astrophysics): Developing fully automated, CCD-equipped telescopes with Internet links for remote observing by students. (Ken Brecher at Boston University is also a leader of this project, but contact the Education Dept. at the Center for Astrophysics; address on last page.)

Modern Observational Astronomy Lab Project: Series of lab exercises using an 8-inch telescope and modern instrumentation; developed under the supervision of Daniel Caton at Appalachian State University. Available by anonymous ftp from ili.phys.appstate.edu or 152.10.26.23

NASA Spacelink: An educational bulletin board accessible via modem or internet. For instructions, contact: Spacelink, Code CL01, NASA Marshall SFC, Huntsville, AL 35812 (205-544—6360) [*See the sheet on Spacelink in this section of the Notebook.*]

Project CLEA: Produces innovative computer-based astronomy lab exercises for undergraduates on IBM and Mac. Contact at: Dept. of Physics, Gettysburg College, Gettysburg, PA 17325 (717-337-6028)

Remote Access Astronomy Project: computerized telescope and dial-in data distribution (using images from many telescopes) with image processing software & activities. Contact them c/o Dept. of Physics, University. of California, Santa Barbara, CA 93106 (805-893-8432) [e-mail: raap@rot.ucsb.edu]

Science Information Infrastructure Project: makes NASA remote sensing data and activities available on-line to museums and schools. Contact Carol Christian, Center for EUV Astrophysics, 2150 Kittredge St., #5030, Berkeley, CA 94720. (510-643-5658) [e-mail: carolc@cea.berkeley.edu]

Many organizations and projects are setting up a "home page" on the World Wide Web about their projects, with files that can be downloaded. For a review of resources on the internet, see the article by David Bruning in the June 1995 issue of *Astronomy* magazine.

5. Planetarium Education Activities

International Planetarium Society: holds conferences, publishes *The Planetarian* magazine, offers special publications and a directory. Contact them c/o Hansen Planetarium, 15 S. State St., Salt Lake City, UT 84111. (There are also active regional organizations of planetarium staff, whose work is described in *The Planetarian.*)

Lawrence Hall of Science (LHS) has offered a series of training workshops for those working with portable planetaria and has a series of activity books for such planetaria (see section 2)

Learning Technologies, Inc.: Small company that makes the *Starlab* inflatable/portable planetaria and trains teachers on how to use them. Also distributes excellent kits of material for high school and college astronomy activities. Contact at: 59 Walden St., Cambridge, MA 02140 (800-537-8703)

Project STARWALK: Offers materials and training for school planetarium activities. Contact: Bob Riddle, Southwest Science H.S., 6512 Wornall Rd., Kansas City, MO 64113 (816-871-0913)

A number of planetaria sell pre-packaged planetarium shows, including the Hayden Planetarium, New York City; the Strassenburgh Planetarium, Rochester, NY; and the Hansen Planetarium, Salt Lake City, Utah.

6. Programs Involving Amateur Astronomers

American Association of Variable Star Observers: Hands-on Astrophysics Project develops activities and materials for students involving real variable star data. Contact: Janet Mattei, AAVSO, 25 Birch St., Cambridge, MA 02138 (617-354-0484)

American Association of Variable Star Observers: Partnership in Astronomy project links amateur astronomers to school classes (operates in only one area at present) [*see address above*]

Astronomical League (umbrella group of amateur clubs in US): Offers handbook developed by amateurs for teaching astronomy to public. Contact: A.L. Sales Service, c/o Marion Bachtell, 1901 S. 10th St., Burlington, IA 52601. Also has an electronic bulletin board and Messier observing club. Contact John Wagoner, 1409 Sequoia, Plano, TX 75023 (214-422-1886)

Astronomical Society of the Pacific (ASP): Project ASTRO: forms and trains ongoing partnerships between astronomers (professional & amateur) and local 4th-9th grade school teachers for class visits [*see address on last page*]

Astronomy Day: Annual day when amateurs around the country bring telescopes to shopping centers, schools, and other sites to let the public view the sky. Co-sponsored by many organizations. Contact: Gary Tomlinson, Public Museum, 54 Jefferson SE, Grand Rapids, MI 49503 (616-456-3987)

7. Awards or Grants for Astronomy Education

American Astronomical Society (AAS): The Annenberg Foundation Prize of the AAS is given for a lifetime of contributions to astronomy education (offered since 1992). Contact them to get the name of the current Chair of the Annenberg Prize committee [*See address on last page*]

Astronomical Society of the Pacific (ASP): The Klumpke-Roberts Award is given each year for a lifetime of contributions to the public understanding of astronomy (offered since 1974). Contact the Society for nomination guidelines.

Astronomical Society of the Pacific (ASP): The Brennan Award is given each year to an outstanding high school teacher of astronomy (offered since 1993). Write for guidelines.

NASA Astrophysics Division (through the Space Telescope Science Institute): IDEA Grants Program funds small and medium size projects in astronomy education by astronomers. Contact: Project Scientist for Education, STScI, 3700 San Martin Dr., Baltimore, MD 21218 [e-mail: idea@stsci.edu]

National Science Foundation: Offers grants for projects in science education through various education and training programs. Write for brochures about current programs and grant guidelines to: NSF Directorate for Education & Human Resources, 4201 Wilson Blvd., Arlington, VA 22230 (703-306-1600).

Slipher Fund of the National Academy of Sciences: supports small grants for innovative astronomy education projects around the country. Contact: Dennis Schatz, Pacific Science Ctr., 200 Second Ave. N, Seattle, WA 98109 (206-443-2886) [e-mail: schatz@pacsci.org]

8. Newsletters

Astronomical Society of the Pacific (ASP): Publishes free quarterly "Universe in the Classroom" newsletter for grades 3-12 teachers; co-sponsored by several other organizations. Requests must be sent on school stationery [*see address at end*]

Association of Astronomy Educators (AAE): Has a newsletter on teaching astronomy for its members [*see address at end*]

NASA: *Educational Horizons* Free newsletter brings news of NASA science and education activities to teachers. Contact: NASA Headquarters, Education Division, Code FE, Washington, DC 20546

9. Miscellaneous Projects

Astronomical Society of the Pacific (ASP): Universe Expo, a weekend "festival" of astronomy talks, seminars, and exhibits for the public, held in conjunction with the Society's annual meeting

Boston Museum of Science: *Touch the Stars* is a tactile

astronomy book in Braille for the visually impaired. Contact Noreen Grice, Museum of Science, Science Park, Boston, MA 02114 (617-589-0439)

Center for Astrophysics (CfA): Workshop on Research Techniques for Undergraduate Faculty at Small Observatories, offered to help instructors at 2- or 4-year teaching colleges who want to get students involved in hands-on observing. Contact Steve Leiker at the CfA address at end.

Coalition for Earth Science Education: An umbrella group of organizations interested in encouraging the teaching of earth science, which includes astronomy. Contact: Frank Ireton, AGU, 2000 Florida Ave., NW Washington, DC 20009 (202-462-6910)

Committee for the Scientific Investigation of Claims of the Paranormal (CSICOP): Group of scientists, educators, magicians, & skeptics that informs the public about the scientific perspective on such pseudosciences as astrology, psychic power, or the face on Mars. They publish the excellent *Skeptical Inquirer* magazine. Contact at: Box 703, Buffalo, NY 14226 (716-636-1425)

Earth and Sky Radio Series: a 2-minute daily radio series on astronomy and earth science. Highlights tape available. Contact at: P.O. Box 2203, Austin, TX 78703 (512-477-4441)

International Dark-Sky Association: Non-profit organization devoted to educating the public about the danger and waste of light pollution. Good information packets available. Contact: David Crawford, 3545 N. Stewart, Tucson, AZ 85716.

Mt. Wilson Observatory: Telescopes in Education Project allows classes to do remote observing with a 24-inch telescope using a computer and modem in their school. While the remote observing costs money for amateurs or colleges, NASA is funding the program free for schools on an experimental basis. Contact: TIE, Box 24, Mt. Wilson, CA 91023 (818-395-7579)

National Optical Astronomy Observatories: Has inaugurated a new program in education, with a number of projects that are just getting under way. Some are local to Arizona, others will be on the internet. For currents status, contact Suzanne Jacoby, NOAO, P.O. Box 26732, Tucson, AZ 85726.

National Undergraduate Research Observatory: A 31-inch telescope near Flagstaff, AZ and a consortium of colleges to operate it. Contact: Kathy Eastwood, Physics & Astronomy, N. Arizona Univ., P.O. Box 6010, Flagstaff, AZ 86011 (602-523-2661)

Smithsonian Astrophysical Observatory: Summer Intern Program for Undergraduates; especially targeted at women/minority students from small colleges. Contact: Kim Dow, CfA, 60 Garden St., Cambridge, MA 02138 (617-496-7586)

University of Arizona Astronomy Camps: Summer programs for youngsters and adults. Contact Don McCarthy, Steward Obs., U. of Arizona, Tucson, AZ 85721 (602-621-4079)

University of Texas McDonald Observatory: *StarDate* 2-minute daily radio program, broadcast on many stations, available on CD's or tape. A new Spanish-language version called *Universo* debuted in April 1995. Contact at RLM 15.308, U. of Texas, Austin, TX 78712 (512-471-5285)

ADDRESSES OF ORGANIZATIONS LISTED SEVERAL TIMES ABOVE:

American Association of Physics Teachers, 1 Physics Ellipse, College Park, MD 20740 (301-209-3300)

American Astronomical Society, Suite 400, 2000 Florida Ave., Washington, DC 20009 (202-328-2010)

Association of Astronomy Educators, c/o Katherine Becker, 5103 Burt St., Omaha, NE 68132

Astronomical Society of the Pacific, 390 Ashton Ave., San Francisco, CA 94112 (415-337-1100)

Center for Astrophysics, Education Dept, MS 71, 60 Garden St., Cambridge, MA 02138 (617-495-9798)

Lawrence Hall of Science, University of California, Berkeley, CA 94720

National Science Teachers' Association, 1840 Wilson Blvd., Arlington, VA 22201 (703-243-7100)

SELECTED RESOURCES FOR TEACHING ASTRONOMY

by Andrew Fraknoi
Foothill College &
The Astronomical Society of the Pacific

NOTE: This list is divided into three sections. The first offers books and articles that list or review written, audiovisual, and computer-based astronomy resources for K-12 teachers or astronomers who are visiting K-12 classrooms. The second contains compilations and examples of hands-on activities for teaching astronomy at this level. The third lists books and articles that describe or discuss the teaching of astronomy, student reasoning ability, and specific projects for improving astronomy education.

We have restricted the listings to materials that are widely available or published by educational or scientific organizations of long standing.

1. RESOURCES

A. Books on Teaching Resources

Friedman, A., et al. *Planetarium Educator's Workshop Guide, 2nd ed.* 1990, Lawrence Hall of Science, Astronomy Education Program, U. of California, Berkeley, CA 94720. Includes both resources and activities for teachers with access to a planetarium.

Gibson, B. *The Astronomer's Sourcebook.* 1992, Woodbine House, 5615 Fishers Lane, Rockville, MD 20852. A guide to astronomical equipment, publications, planetaria, organizations, events, etc.

Harrington, P. *Star Ware.* 1994, John Wiley. A guide written by an avid amateur for choosing, buying, and using telescopes and accessories.

Makower, J., ed. *The Air And Space Catalog: The Complete Sourcebook to Everything in the Universe.* 1989, Tilden/Random House. Lists hundreds of places to get information and teaching aids.

Saul, W. & Newman, A. *Science Fare: An Illustrated Guide and Catalog of Toys, Books, and Activities for Kids.* 1986, Harper & Row. Resources for school teachers in many areas.

Sneider, C., et al. *Planetarium Activities for Student Success.* 1990-4, Lawrence Hall of Science, University of California, Berkeley, CA 94720. Volume 3 of this very useful series of activity booklets is a resource guide to materials for astronomy teaching.

Regular book and resource reviews appear in *Sky & Telescope, Mercury, Astronomy,* and other astronomy magazines, as well as *The Science Teacher, The Physics Teacher,* and *Science Books & Films.*

B. Articles on Teaching Resources (except computer software)

Bruning, D. "Charting a Path Through the Night Sky" in *Astronomy,* Oct. 1993, p. 74. Comparative review of star atlases and guides.

Carter, C. "Sources for Space Science Projects" in *Science Books and Films,* Jan/Feb. 1990, p. 148.

Coyle, H. "Astronomy Texts from Abell to Zeilik" in *Sky & Telescope,* Nov. 1987, p. 487.

Fraknoi, A. "Science Fiction Stories with Good Astronomy" in *Mercury,* Jan/Feb. 1990, p. 26.

Fraknoi, A. "Scientific Responses to Pseudoscience Related to Astronomy" in *Mercury,* Sep/Oct. 1990, p. 144. Resources for debunking astrology, UFO's, ancient astronauts, etc.

Fraknoi, A. & Freitag, R. "Women in Astronomy: A Bibliography" in *Mercury,* Jan/Feb. 1992, p. 46. Part of a special issue on women in astronomy.

Kruglak, H. "Resource Letter: Laboratory Exercises for Elementary Astronomy" in *American Journal of Physics,* vol. 44, p. 828 (1976).

C. Articles on Computer-based Resources

Bruning, D. "Computers and Astronomy" in *Astronomy,* May 1994, p. 58. A thorough guide to astronomy software and how it can be used.

Goldman, S. "Browsing Computer Bulletin Boards" in *Sky & Telescope,* March 1994, p. 84.

Mosley, J. "The CD-ROM Comes of Age" in *Sky & Telescope,* July 1992, p. 77. Reviews of astronomical CD-ROM's by S&T's excellent software reviewer.

Mosley, J. "Exploring the Sky on Computer" in *Sky & Telescope*, June 1990, p. 650; Mar. 1992, p.318. Reviews of planetarium type programs; regularly updated in Mosley's column in the magazine.

Mosley, J. & Fraknoi, A. "Computer Software in Astronomy" in *Mercury*, May/June 1991, p. 87. Annotated listing, organized by types.

Sneider, C. & Berndt, H. "A Galaxy of Astronomy & Space Software" in *Technology & Learning*, May/June 1994, p. 20. Reviews of many software packages from a K-12 teaching perspective.

Trivette, D. "Stargazer's Delight: Six Programs for Astronomy Hobbyists" in *PC Magazine*, July 1993, p. 546. Good comparative reviews.

2. Activities

A. Selected Books of Classroom or Family Activities

Apfel, N. *Astronomy Projects for Young Scientists.* 1984, Arco.

Ball, N., et al., eds. *Project SPICA: A Teacher Resource to Enhance Astronomy Education.* 1994, Kendall-Hunt. A collection of 43 astronomy activities at the 6th - 12th grade level from a project at the Center for Astrophysics.

Bonnet, R. & Keen, G. *Space and Astronomy: 49 Science Fair Projects.* 1992, Tab Books.

Braus, J., ed. *Astronomy Adventures.* A special issue of Ranger Rick's NatureScope, from the National Wildlife Federation, 1412 Sixteenth St., NW, Washington, DC 20036. An 80-page guide to elementary school activities.

Coyle, H, et al. *Project STAR: The Universe in Your Hands.* 1993, Kendall Hunt. A high school astronomy supplement, full of sophisticated hands-on projects.

DeBruin, J. & Murad, D. *Look to the Sky: An All Purpose Interdisciplinary Guide to Astronomy.* 1988, Good Apple. Activities for grades 4-12 of varying quality and level.

Friedman, A., et al. *Planetarium Educator's Workshop Guide.* [See section 1 above.]

Gardner, R. *Projects in Space Science.* 1988, Messner.

Greenleaf, P. *Experiments in Space Science.* 1981, Arco.

Moeschl, R. *Exploring the Sky: Projects for Beginning Astronomers, 2nd ed.* 1993, Chicago Review Press. An eclectic collection of 60+ astronomy activities of varying quality.

Schatz, D. *Astronomy Activity Book.* 1991, Simon & Schuster. Basic observing activities for upper elementary and junior high level.

Schatz, D. *The Return of the Comet: An Activity Book for Skywatchers 9-14.* 1985, Pacific Science Ctr., 200 Second Ave., N., Seattle, WA 98109.

Schatz, D. & Cooper, D. *Astro Adventures: An Activity Based Astronomy Curriculum.* 1994, Pacific Science Center. Superb compilation of some of the very best K-12 astronomy activities; many of them are featured in *The Universe at Your Fingertips.*

Smith, P. *Project Earth Science: Astronomy.* 1992, National Science Teachers' Association. Collection of activities & resources for middle schools.

Sneider, C., et al. *Planetarium Activities for Student Success.* 1990-4, Lawrence Hall of Science, University of California, Berkeley, CA 94720. A wonderful series of 12 booklets on activities for portable planetaria, many of which can be used very well in the regular classroom.

Sneider, C. *Earth, Moon, and Stars: A Teacher's Guide for Grades 5-9.* Sutter, D., et al. *The Moons of Jupiter: A Teacher's Guide for Grades 4-9.* 1986, 1993, GEMS Program, Lawrence Hall of Science, U. of California, Berkeley, CA 94720. Excellent activity-oriented workbooks; part of a series of good *Great Explorations in Math & Science* guides in various fields.

Wooley, J. *Voyages through Space and Time: Projects for VOYAGER* [Software for the Macintosh]. 1992, Wadsworth. A series of high school or college level activites with this popular program that can show the sky at any time from any place on Earth.

B. Selected Articles on Classroom Activities

Chandler, D. "An Expanding Universe in the Classroom" in *Physics Teacher*, Feb. 1991, p. 103.

Dodson, S. "Simple Classroom Telescopes" in *Sky & Telescope*, Mar. 1991, p. 311. Simple designs you can build and how to use them in the classroom.

Harris, J. & Watson, F. "Some Astronomy Exercises for a High School Physics Course" in *Physics Teacher*, vol. 6, p. 394 (1968).

Kaufman, S. "Astronomy Spectra Experiments for Nonscience Students" in *Physics Teacher*, vol. 20, p. 170 (1982).

Lamark, D. "Set Your Class in Celestial Motion" in *Science Teacher*, Sep. 1989, p. 59. Having students enact and plot rotation, revolution, and retrograde motion.

Lamb, W. "Build-It-Yourself Astronomy" in *Science Teacher*, Mar. 1983, p. 16. On building a cross-staff and quadrant and using them.

Lebofsky, M. "A Skunk Is in the Sky or Is It a Plow?" in *Science Scope*, Mar. 1994, p. 26. On some activities for lower grades using sky mythology.

MacRobert, A. "Astronomy with a $5 Telescope" in *Sky & Telescope*, Apr. 1990, p. 384. On the Project STAR telescope kit and activities.

Murphy, R. & Kwasnoski, J. "Years, Days, and Solar Rays" in *Science Teacher*, Oct. 1983, p. 28. Measuring sunlight at different seasons.

Oates, C. & Hockey, T. "Twinkle Twinkle Little Star" in *Science Teacher*, Oct. 1993, p. 63. A scavenger hunt activity for astronomy information.

Riddle, B. "The Seasonal Sun: Rotating and Revolving through the Seasons" in *Science Teacher*, Oct. 1988, p. 29. Uses worksheets and computer software.

Sadler, P. "Projecting Spectra for Classroom Investigation" in *Physics Teacher*, Oct. 1991, p. 423.

Sarton, E. "Measuring the Moon's Orbit" in *Physics Teacher*, vol. 18, p. 504 (1980).

Zeilik, M. "An Expanding Universe Demonstration" in *American Journal of Physics*, vol. 50. p. 571 (1982).

C. Selected Articles on Sky Observing Activities

Bagnall, P. "Observing the Naked Eye Sky Glows" in *Astronomy*, June 1988, p. 46. On aurorae, zodiacal light, etc.

Bunge, R. "Nightscapes: Photos with Only Camera & Tripod" in *Astronomy*, June 1992, p. 72.

Burnham, R. "Observing the Sun" in *Astronomy*, Aug. 1984, p. 51.

Chaikin, A. "A Guided Tour of the Moon" in *Sky & Telescope*, Sep. 1984, p. 211.

Clark, G. "Starting an Astronomy Club" in *Astronomy*, Oct. 1981, p. 52.

Coco, M. "Staging a Moon Shot" in *Astronomy*, Aug. 1992, p. 62. Hints on photographing the moon for beginners.

Dyer, A. "Seeking the Best 35-mm Camera for Astrophotography" in *Astronomy*, Sep. 1993, p. 74; "Taking Pictures with Your Telescope" in *Astronomy*, Dec. 1992, p. 66.

Dyer, A. & Talcott, R. "Celestial Sights of the Future" in *Astronomy*, Aug. 1993, p. 64. Best sky events from 1994 to 2017.

Fortier, E. "An Observer's Guide to Sunspots" in *Astronomy*, May 1991, p. 62.

Harrington, P. "The Ten Best Double Stars" in *Astronomy*, July 1989, p. 78.

Hill, R. "Equipped for Safe Solar Viewing" in *Astronomy*, Feb. 1989, p. 66.

King-Hele, D. & Erberst, R. "Observing Artificial Satellites" in *Sky & Telescope*, May 1986, p. 457.

Levy, D. "Discovering Binocular Astronomy" in *Astronomy*, Dec. 1985, p. 44.

Lewis, S. "Eye on Io" in *Science Teacher*, May 1985, p. 42. Observing Jupiter's moons and figuring out their motion.

Ling, A. "An Eye on the Deep Sky: Objects Visible to the Naked Eye" in *Astronomy*, Jan. 1992, p. 68. 40 interesting objects you don't need a telescope to see.

Loudon, J. "Learning the Sky by Degrees" in *Astronomy*, Dec. 1986, p. 54. On angular measurement on the sky.

MacRobert, A. "Close-up of a Star" in *Sky & Telescope*, May 1985, p. 397. Hints for observing the Sun.

MacRobert, A. "An Observer's Guide to Saturn" in *Sky & Telescope*, Nov. 1993, p. 52.

MacRobert, A. "Photography Through a Telescope" in *Sky & Telescope*, Dec. 1986, p. 569.

Morel, P. "Drawing the Sun and the Moon" in *Sky & Telescope*, Apr. 1988, p. 404.

Olivarez, J. "Seeing the Most on Jupiter" in *Astronomy*, Mar. 1992, p. 85.

Porcellino, M. "Polar Aligning Your Telescope" in *Astronomy*, May 1992, p. 69.

Russo, R. "Astronomy By Day" in *Science Teacher*, Mar. 1982, p. 30; May 1983, p. 32. On measuring sunspots.

Sessions, L. "Observing Planets in the Daytime" in *Astronomy*, Jan. 1981, p. 40.

Sinnott, R. "Tracing Earth Satellites on Your Home Computer" in *Sky & Telescope*, May 1986, p. 501.

3. IDEAS AND PROGRAMS

A. Books on Astronomy Education

Berendzen, R., ed. *International Conference on Education In and History of Modern Astronomy.* 1972, New York Academy of Science. Published as vol. 198 of the *Annals of the NY Academy of Science*, with useful articles on education at a variety of levels.

Dubcek, L., et al. *Fantastic Voyages: Learning Science Through Science Fiction Films.* 1994, American Institute of Physics Press. Suggestions for using student interest in science fiction films for teaching concepts in physical science; high school or college level.

Hunt, J., ed. *Cosmos: An Educational Challenge.* 1986, European Space Agency. Proceedings of a European conference on education at various levels, held in Denmark in 1986.

Pasachoff, J. & Percy, J., eds. *The Teaching of Astronomy.* 1990, Cambridge U. Press. Proceedings of an international colloquium on astronomy education in the US and elsewhere; probably the best single source on astronomy education.

Pennypacker, C., ed. *Workshop on Hands-on Astronomy for Education.* 1992, World Publishing. Projects emphasizing computers and small telescopes.

B. Articles on Astronomy Education in General

Annett, C. "The Planetarium: A Sky for All Seasons" in *Mercury*, Nov/Dec. 1979, p. 136. An overview of planetaria.

Atkin, J. "Teaching Concepts of Modern Astronomy to Elementary-School Children" in *Science Education*, vol. 45, p. 54 (1961).

Baxter, J. "Children's Understanding of Familiar Astronomical Events" in *International Journal of Science Education*, vol. 11, p. 502 (1989).

Bishop, J. "U.S. Astronomy Education: Past, Present, and Future" in *Science Education*, vol. 61, p. 295 (1977, no. 3). Interesting review, with good historical accounts.

Bracher, K. "The Stars for All: The Centennial History of the Astronomical Society of the Pacific" in *Mercury*, Sep/Oct. 1989, the entire issue. Describes many of the educational activities of the Society over the years.

Cole, S. "Astronomy in our Schools" in *Astronomy*, Sep. 1988, p. 36. Overview of programs and projects.

Fraknoi, A. "Hearing the Music of the Spheres: Informal Astronomy Education" in Druger, M., ed. *Science For the Fun of It*. 1988, National Science Teachers Association. A review of astronomy learning outside the formal classroom.

Fraknoi, A. "Astronomy Education: The State of the Art" in *Mercury*, Jan/Feb. 1977, p.9. Report on a 1976 symposium on astronomy education.

Friedman, A. "Alternative Approaches to Planetarium Programs" in *Mercury*, Jan/Feb. 1973, p. 12. One of the first suggestions in print of the idea of participatory planetarium programs.

Haupt, G. "First Grade Concepts of the Moon" in *Science Education*, vol. 32, no. 4, p. 258 (Oct. 1948); vol. 34, no. 4, p. 224 (Oct. 1950).

Klein, A. "Children's Concepts of Earth and Sun: A Cross-Cultural Study" in *Science Education*, vol. 65, #1, p. 95 (1982).

Lightman, A. & Sadler, P. "The Earth is Round? Who Are You Kidding?" in *Science and Children*, Feb. 1988, p. 24. On studies of how children reason and learn.

Lightman, A. & Sadler, P. "Teacher Predictions Versus Actual Student Gains" in *Physics Teacher*, Mar. 1993. Results from Project STAR showed that teachers are often off the mark in predicting how much students would learn.

Lockwood, J. "Creating Gender-Friendly Astronomy Classrooms" in *Mercury*, Jan/Feb. 1994, p. 25. High school teacher discusses how to encourage girls in science.

Lockwood, J. "Latching on to the Whirlwind: The Changing Face of Astronomy Education" in *Mercury*, Mar/Apr. 1993, p. 56. On several innovative projects.

Norton, O. "Will Planetariums Become Extinct?" in *Sky & Telescope*, Dec. 1985, p. 534. On the history and current state of planetaria in the U.S.

Petersen, C. "There's No Place Like Dome" in *Sky & Telescope*, Sep. 1989, p. 255. On astronomy education at several planetaria and science centers.

Reed, G. "Is the Planetarium a More Effective Teaching Device than the Combination of the Classroom Chalkboard and Celestial Globe?" in *School Science and Mathematics*, vol. 70, no. 6, p. 487 (June 1970); vol. 72, no. 5, p. 368 (May 1972); vol. 73, no. 7, p. 553 (Oct. 1973).

Sagan, C. "Why We Need to Understand Science" in *Mercury*, Mar/Apr. 1993, p. 52. Eloquent overview of the need for public education in astronomy.

Schatz, D. & Lawson, A. "Effective Astronomy Teaching: Intellectual Development and Its Implications" in *Mercury*, Jul/Aug. 1976, p. 6. Clear introduction to how students reason and why educators at all levels need to understand more about the process.

Sneider, C. & Pulos, S. "Children's Cosmographies: Understanding the Earth's Shape" in *Science Education*, vol. 67, #2, p. 205 (1983). Useful for illuminating how youngsters think through problems in physical science in general.

Sneider, C., et al. "Understanding the Earth's Shape and Gravity" in *Learning*, vol. 14, no. 6, p. 43 (Feb. 1986).

Trimble, V. & Elson, R. "Astronomy as a National Asset" in *Sky & Telescope*, Nov. 1991, p. 485. Outlines some of the ways astronomy helps science education in the US.

Wall, C. "A Review of Research Related to Astronomy Education" in *School Science and Mathematics*, vol. 73, no. 8, p. 653 (Nov. 1973).

Yuckenberg, L. "Children's Understanding of Certain Concepts of Astronomy in First Grade" in *Science Education*, vol. 46, no. 2, p. 148 (Mar. 1962).

C. Articles on Specific Projects to Improve Astronomy Education

Bruning, D. "Astronomy in the Classroom: Building Astronomy's Future" in *Astronomy*, Sep. 1993, p. 40. Describes a number of K-12 projects in astronomy education.

Coyle, H. "The Universe in the Student's Mind" in *Mercury*, May/June 1994, p. 28. On Project STAR to improve astronomy teaching in high school through hands-on activities.

Coyle, H. "Education at CfA" in *Mercury*, Jul/Aug. 1994, p. 29. On programs at the Harvard-Smithsonian Center for Astrophysics in astronomy education.

Dahlmann, L. "Image Processing in Astronomy Education" in *Mercury*, Mar/Apr. 1994, p. 24. On work at the U. of Arizona helping students manipulate astronomical images.

Dierking, L. & Richter, J. "Project ASTRO: Astronomers and Teachers as Partners" in *Science Scope*, Special issue on Informal Science Education, Mar. 1995.

Greenberg, R. "Scanning the Images of Science" in *Science Teacher*, Nov. 1992, p. 14. On using image processing software in the classroom.

Hopper, E. & McCarthy, D. "Astronomy Camp" in *Mercury*, Nov/Dec. 1993, p. 25. On camps for youngsters and adults offered through the U. of Arizona.

Levy, D. "Project ARTIST" in *Sky & Telescope*, June 1994, p. 98. On the innovative curriculum and teacher training project at the U. of Arizona.

Morrison, M., et al. "Chabot's Century of Science Education" in *Sky & Telescope*, May 1983, p. 398. Story of a planetarium & observatory owned by an urban school system and run with the help of an active amateur club.

Mumford, G. "Science in the Imaginary Sky" in *Sky & Telescope*, Feb. 1992, p. 146. On a computer program that allows students to simulate telescope observations.

O'Meara, S. "The Worlds' Largest Solar System Scale Model" in *Sky & Telescope*, July 1993, p. 99. At the Lakeview Museum in Peoria, IL.

O'Meara, S. "Inside Boston's Hayden Planetarium" in *Sky & Telescope*, Mar. 1982, p. 293. How things work in a major public planetarium facility.

Pompea, S. & Blurton, C. "A Walk Through the Astronomy Village" in *Mercury*, Jan/Feb. 1995, p. 32. A multimedia package from NASA.

Richter, J. & Fraknoi, A. "Matches Made in the Heavens: Project ASTRO" in *Mercury*, Sep/Oct. 1994, p. 24. On a program at the ASP to foster partnerships between professional and amateur astronomers and 4th - 9th grade teachers.

D. Astronomy Education Through the Media

Byrd, D. "Astronomy on the Air" in *Mercury*, Jan/Feb. 1980, p. 7. On the *StarDate* radio program.

Goldman, S. "Prime Time Astronomers: *The Astronomers* TV Series" in *Sky & Telescope*, May 1991, p. 472.

Goldsmith, D. "Two Years in Hollywood: An Astronomer in Television Land" in *Mercury*, Mar/Apr. 1991, p. 34. An astronomer who wrote a public TV series recounts his experiences in the world of television.

O'Meara, S. "Jack Horkheimer: Star Hustler" in *Sky & Telescope*, May 1989, p. 544. On a planetarium director who does a series of short astronomy shows on public television.

Pierce, D. "Project Universe: Astronomy on Public TV" in *Mercury*, Sep/Oct. 1978, p. 118. An account of the making of an astronomy telecourse.

Reis, R. & Fraknoi, A. "A Syndicated Newspaper Column on Astronomy" in *Mercury*, Jan/Feb. 1977, p. 1.

Robinson, L. "How *Sky & Telescope* Came to Be" in *Sky & Telescope*, Nov. 1991, p.472. History of the main magazine that tied scientist, amateurs, and teachers together.

NOTE: I would like to acknowledge the contributions of Cary Sneider, Robert Dukes, Edna DeVore, and Katherine Becker in assembling this list.

INDEX